Materials Horizons: From Nature to Nanomaterials

Series Editor

Vijay Kumar Thakur, School of Aerospace, Transport and Manufacturing, Cranfield University, Cranfield, UK

Materials are an indispensable part of human civilization since the inception of life on earth. With the passage of time, innumerable new materials have been explored as well as developed and the search for new innovative materials continues briskly. Keeping in mind the immense perspectives of various classes of materials, this series aims at providing a comprehensive collection of works across the breadth of materials research at cutting-edge interface of materials science with physics, chemistry, biology and engineering.

This series covers a galaxy of materials ranging from natural materials to nanomaterials. Some of the topics include but not limited to: biological materials, biomimetic materials, ceramics, composites, coatings, functional materials, glasses, inorganic materials, inorganic-organic hybrids, metals, membranes, magnetic materials, manufacturing of materials, nanomaterials, organic materials and pigments to name a few. The series provides most timely and comprehensive information on advanced synthesis, processing, characterization, manufacturing and applications in a broad range of interdisciplinary fields in science, engineering and technology.

This series accepts both authored and edited works, including textbooks, monographs, reference works, and professional books. The books in this series will provide a deep insight into the state-of-art of Materials Horizons and serve students, academic, government and industrial scientists involved in all aspects of materials research.

More information about this series at http://www.springer.com/series/16122

Djalal Trache · Fouad Benaliouche · Ahmed Mekki
Editors

Materials Research and Applications

Select Papers from JCH8-2019

Springer

Editors
Djalal Trache
UER Procédés Energétiques
Ecole Militaire Polytechnique
Algiers, Algeria

Fouad Benaliouche
UER Physico-Chimie des Matériaux
Ecole Militaire Polytechnique
Algiers, Algeria

Ahmed Mekki
UER Physico-Chimie des Matériaux
Ecole Militaire Polytechnique
Algiers, Algeria

ISSN 2524-5384 ISSN 2524-5392 (electronic)
Materials Horizons: From Nature to Nanomaterials
ISBN 978-981-15-9222-5 ISBN 978-981-15-9223-2 (eBook)
https://doi.org/10.1007/978-981-15-9223-2

This Springer imprint is published by the registered company Springer Nature Singapore Pte Ltd.
The registered company address is: 152 Beach Road, #21-01/04 Gateway East, Singapore 189721, Singapore

Preface

This book encompasses a collection of selected papers presented at the 8th Chemistry Days (JCH8-2019), which was held on March 26–27, 2019, at the Ecole Militaire Polytechnique (EMP). This volume discusses the synthesis, processing, characterization, and applications of different types of materials. The book provides an overview of the recent advances in the areas of materials research and engineering and highlights a wide range in significant recent results in energetic materials, bio-based materials, ceramics, nanomaterials, among others, and their use for emerging applications. The volume reports on one comprehensive review article as well as several original research works from areas of materials science and technology with a focus on the preparation, properties, performance, and applications in various fields. The book begins by addressing different sections of materials and the assessment of some specific behaviors, where the first chapter provides a comprehensive review of one of the most exciting energetic materials, which has been developed in recent years. The second part of this book discusses the synthesis, preparation, processing, and application of some outstanding materials, such as nanocomposites, cellulose microcrystals, polymers, hydroxyapatite, to site a few. The following part of this volume will focus on the materials used for environmental uses. The last section will cover other emerging applications, including catalysis for methane conversion, the development of electrochemical impedance spectroscopy technique for the assessment of the electropolishing quality of steels, as well as the development of amine derivatives as corrosion inhibitors. The contents of this book will be relevant to researchers in academia and industry professionals working on the development of advanced materials and their applications.

The 8th Chemistry Days (JCH8-2019) was organized with the aim to provide a platform for academicians, faculty members, and young researchers from various public as well as academic and research institutes across the country and from abroad to meet together and interact with one another. This event allowed fruitful discussions with distinguished experienced scientists from various reputed national and international institutes and universities on the latest advances in this exciting area of research and add on new dimensions in various application areas. The broad list of interesting research fields of this conference comprises all areas of chemistry, chemical engineering, and materials such as polymers and composites, chemical processes, the environment and renewable energies, interfacial phenomena and reactivity, energetic materials and pyrotechnic systems, materials synthesis and characterization techniques, nanomaterials and novel materials.

The meeting format was built in order to favor detailed discussions among leading national and international researchers on a wide range of topics. Plenary lectures by outstanding speakers are meant to offer a critical survey of specific themes and pave the way to detailed contributions from the authors as well as general discussions within the audience. Serious effort is made to ensure that the results are readily available to a wide range of readers, and the level of technical accuracy at all stages, from paper presentations to written diffusion of results, is set at the highest professional standards. We applaud the effort of the conference participants and encourage their current and future personal and collective efforts in these important research fields of scientific endeavor.

It is worthy to note that the name of this event (JCH8) is expected to be changed to "Conference on Chemical and Chemical Engineering (3CE)" from the next edition, which will be held on 2021 at EMP.

It is our great pleasure to thank our scientific committee members for their substantial help during the reviewing process to raise the quality of the book. Without their efforts, the publication of this volume would have been impossible.

The editors acknowledge all the contributors to this book who have taken time to submit their manuscripts even with their busy schedules. We gratefully acknowledge the team of Springer Nature for their excellent editorial support. We truly hope that research scholars, scientists, academic and other professional workers will find this volume of this large collection of intriguing ideas, new data, and future developments, as a useful source to build on further studies.

With warm regards,

Algiers, Algeria

Prof. Djalal Trache
Prof. Fouad Benaliouche
Prof. Ahmed Mekki

Contents

About the Editors

Prof. Djalal Trache is an Associate Professor at Ecole Militaire Polytechnique (EMP), Algeria. He received his Engineer degree in Chemical Engineering, Magister degree in Applied Chemistry and D.Sc. in Chemistry at EMP. He has made several presentations at national and international conferences and published several research papers and book chapters. He is a reviewer of more than 50 international reputed journals. Prof. Trache has particular expertise in energetic materials, bio-based materials, polymer composites and their characterization. He also has interests in nanomaterials and their applications, phase equilibria and kinetics.

Prof. Fouad Benaliouche graduated from the Department of Mechanical and Energy Engineering, University of Batna (Algeria) in 1996. He received his postgraduate degree in Engineering Process from the Ecole Militaire Polytechnique (EMP), Algiers (Algeria). He pursued his Ph.D. degree (Applied Chemistry) in 2009 from the University of Sciences and Technology Houari Boumedienne (USTHB), Algiers (Algeria). After postdoctoral research at EMP, Fouad was promoted to an Associate Professor in 2014. Currently, he is the Head of the Porous Materials Laboratory in the EMP. Prof. Benaliouche is also interested in the adsorption phenomena related to gases, vapors and liquids. In addition, his current research interests are the porosity modulation and antibacterial activity studies for characterization of microporous zeolite/diatomite adsorbents.

Prof. Ahmed Mekki is an Associate Professor at the Ecole Militaire Polytechnique of Algiers in Algeria, where he obtained his Ph.D. and his habilitation in physical organic chemistry. He has written several scientific publications and book chapters. His main research interests are silane and aryl diazonium salts coupling agents, reactive and functional ultrathin polymer films via surface polymerization or "click" chemistry, carbon/polymers composites, molecularly imprinted polymer-based sensors, clay/polymer nano-composites and flexible films of conducting polymer-based gas sensors, thermoelectricity and energy conversion applications.

Chapter 1
TKX-50: A Highly Promising Secondary Explosive

Thomas M. Klapötke

Abstract TKX-50 is one of the most exciting new energetic materials which has been developed in recent years, since it shows great promise for future application as a secondary explosive. TKX-50 shows a combination of a low IS and FS with high thermal stability, density and detonation velocity, as well as a facile, inexpensive synthesis, and low toxicity, making it particularly attractive as a future high-performance explosive with increased safety. The interest in TKX-50 has been considerable since its first report in only 2012, and many varied aspects of its chemical and physical properties have been reported by groups from all over the world. The aim of this review is to provide a current overview of what has been achieved in the development of TKX-50.

Keywords TKX-50 · Secondary explosive · Energetic material · RDX · HMX · HATO · LASEM · EXPLO

1 Introduction

Energetic tetrazole compounds have been extensively investigated in recent years in attempts to find new nitrogen-rich secondary explosives (Gao and Shreeve 2011). Both neutral compounds such as H_2BTA (5,5′-Bis(1*H*-tetrazolyl)amine) (Klapötke et al. 2008) and salts containing many variations of the tetrazolate anion such as the CN_7^- azidotetrazolate anion have been reported (Klapötke and Stierstorfer 2009). The latter work has been extended to also include salts containing the azotetrazolate anion (Hammerl et al. 2002) as well as the 5,5′-bistetrazolate anion in which the –N=N– group connecting the two tetrazole rings is omitted, and the tetrazole rings are directly connected to one another instead (Fig. 1) (Singh et al. 2007). The salt TKX-50 is based on the latter, but with the additional *N*-oxide functional group. In

T. M. Klapötke (✉)
Department of Chemistry, Ludwig-Maximilians University Munich,
Butenandtstrasse 5-13 (Haus D), 81377 Munich, Germany
e-mail: tmk@cup.uni-muenchen.de

D. Trache et al. (eds.), *Materials Research and Applications*,
Materials Horizons: From Nature to Nanomaterials,
https://doi.org/10.1007/978-981-15-9223-2_1

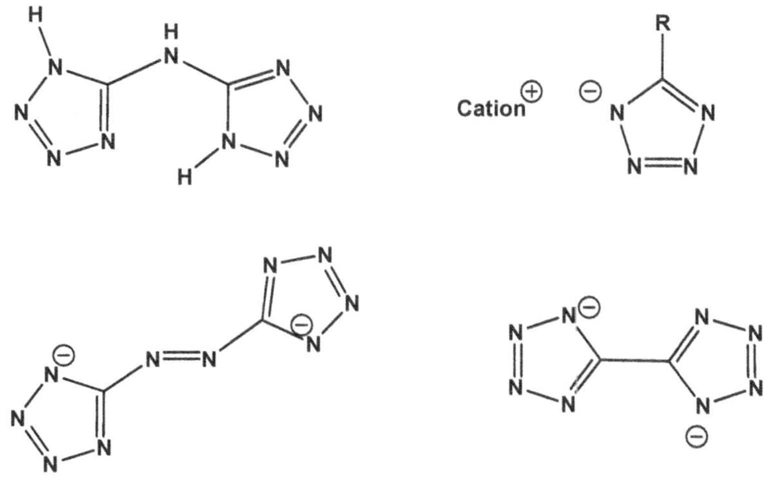

Fig. 1 Connectivities of tetrazole-based energetic materials: the neutral molecule H_2BTA, salts containing variations of the tetrazolate anion (e.g., R = –H, $-NH_2$, $-NHNH_2$, –CN, $-N_3$, $-C(NO_2)_3$), the bis (azotetrazolate) anion, and the bistetrazolate anion

each of the four examples given in Fig. 1, one or more rings are present which incorporate several N–N linkages. This is a common feature of the compounds which are nowadays investigated as possible new candidates as secondary explosives. This is because the tetrazole ring has the ability of linking several N atoms to one another, but due to delocalization of electrons within the ring, there are no –N–N– single bonds such as in hydrazine (N_2H_4)—which have low bond energy and are easily broken (Szwarc 1949)—and pseudoaromatic systems result. Non-cyclic compounds containing only catenated N-chains have so far not found promise as energetic materials for applications due to their instability. Nitrogen-rich energetic materials based on cages rather than on rings or chains have also been investigated and two hugely interesting compounds have been reported, namely Hexanitrohexaazaisowurtzitane (CL-20, also abbreviated as HNIW) (Nair et al. 2005) and Octanitrocubane (Eaton et al. 2000). Although CL-20 has been investigated to an enormous extent over the past years (Nair et al. 2005), it possesses the disadvantages of necessitating a many-step, complicated and expensive synthesis, and also shows a very high sensitivity to impact (4 J, ε-CL-20), which classifies it as sensitive under the UN classification for the transport of dangerous goods (UN classifications for IS: <4 J = very sensitive, 4–35 J = sensitive) (Klapötke 2017). For large-scale production, the considerable cost of CL-20 production is highly problematic, especially since established secondary explosives such as RDX are relatively cheap to produce. Another aspect which is important for secondary explosives is whether any phase transitions occur in relevant temperatures ranges. This problem can also be illustrated using CL-20 as an example, since several polymorphs of CL-20 are known (β-, γ-, ε-, and the high-pressure polymorph ζ-, as well as the hydrate CL-20 · $^1/_2H_2O$ which is often incorrectly referred to as being the α- polymorph of CL-20 in the literature); however, it is only the

ε-polymorph which has found relevance with respect to application. Unfortunately, however, it is the γ-polymorph—and not the ε-polymorph—which is the thermodynamically most stable one, and phase transition of the ε- to γ-phase occurs in the temperature range of approximately 157–176 °C (Foltz et al. 1994a). Other reports in the literature state that both the β- and ε-phases slowly convert to the γ-phase at elevated temperatures (Simpson et al. 1997), meaning that elevated temperatures have to be avoided in order to preserve phase purity of the ε-CL-20. Furthermore, solvents and the method of crystallization have a considerable effect on causing phase changes (Foltz et al. 1994b; Sun et al. 2012). Ideally, in contrast to CL-20, a secondary explosive would only show one polymorph, which would eliminate the considerable issues which can arise if phase purity isn't present.

The second cage compound mentioned above, Octanitrocubane, is quite remarkable in terms of its structure and energetic properties and on paper would seem to be a clear stand-out candidate as a future energetic material. However, its promise on paper has so far not been able to transfer to real-life promise because only very small quantities have so far been reported to have been prepared (<10 g) and because of its high sensitivity (Eaton et al. 2000). The cage strain and eight $-NO_2$ groups which make its energetic [VoD = 9800 m/s at $\rho = 2.00$ g cm^{-3} (Keshavarz and Haz 2009); VoD = 10,100 m/s at $\rho = 2.00$ g cm^{-3} (Cooper 1996)] and oxidizing properties so outstanding are also the reason it is so difficult to prepare on large quantities in a facile synthesis (current synthetic route is >10 steps). Therefore, nitrogen-rich ring systems have proven to be the most promising approach for synthesizing high-nitrogen compounds which possess the necessary stability and energetic properties. Five-membered rings containing three (triazoles) or four nitrogen atoms (tetrazoles) have received the most attention, since neutral pentazole compounds of the type R-N_5 have so far not proven to be stable enough to be isolated unless larger –R groups are employed as stabilizers (Wallis and Dunitz 1983), and salts containing the pentazolate, N_5-anion have only very recently been reported for the first time (Xu et al. 2017).

Recently published studies have established that often the energetic properties of a specific tetrazole can be improved by converting it into the corresponding *N*-oxide. This has been shown to generally result in higher densities and stabilities, as well as lower sensitivities and an improved oxygen balance for the *N*-oxide analogue of a given tetrazole compound (Klapötke and Stierstorfer 2014; He et al. 2016; Klapötke 2007). However, it is not by any means that any nitrogen-rich *N*-oxide compound will be stable—the recently prepared TTTO (Fig. 2) (Klenov et al. 2016) has been reported to possess a calculated enthalpy of formation ($\Delta H_f°$) of 1080 kJ/mol (Klenov et al. 2015) and calculated detonation velocity of 9710 at a density of 1.98 g cm^{-3} (Zhai et al. 2019); however, it exhibits a relatively low decomposition temperature of only 183–186 °C (Klenov et al. 2016). Furthermore, so far only small quantities have been prepared and extensive characterization of the energetic properties and compatibility of TTTO remain to be investigated. Therefore, it is a very difficult task to design a tetrazole *N*-oxide which will incorporate as many of the positive properties of *N*-oxide compounds such as TTTO, but avoid the problems of low thermal stability and difficult synthetic routes. However, as the remainder of this chapter will show,

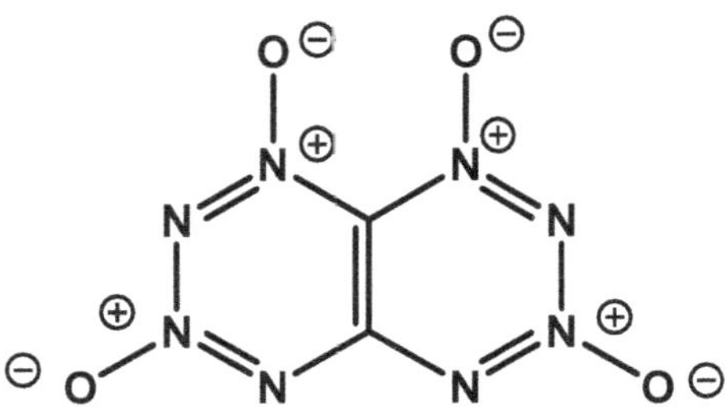

Fig. 2 Connectivity of the bicyclic tetrazine *N*-oxide [1,2,3,4]tetrazino[5,6-e][1,2,3,4]tetrazine 1,3,6,8-tetraoxide (TTTO) (Klenov et al. 2016)

the synthesis of TKX-50—a salt containing a bistetrazolate anion in its *N*-oxide form—has resulted in a most promising secondary explosive candidate with realistic potential for use in military applications.

2 Synthesis and Characterization of TKX-50

2.1 Synthesis and Improved Synthetic Routes

The synthesis of TKX-50 was reported in 2012 by two different routes (Fischer et al. 2012), both of which involved the synthesis of the neutral 5,5′-bistetrazole-1,1′ diole (1,1′-BTO) molecule, followed by its double deprotonation using hydroxylamine (Fig. 3) (Fischer et al. 2012). The two routes differed only in the preparative method used for the 1,1′-BTO precursor molecule (Fig. 3). However, both synthetic methods showed considerable problematic issues which would require a solution if TKX-50 was to be investigated on a larger scale, let alone on an industrial scale. Synthetic route A produced only poor yields of the 1,1′-BTO molecule, and the oxidation method used by this research group for the preparation of many other tetrazole *N*-oxide compounds unfortunately resulted in the 2,2′-isomer of BTO as the major product and not the desired 1,1- isomer (Fischer et al. 2012). Route B for the preparation of the neutral 1,1′-BTO molecule was reported in 2001 by *Tselinskii* (Tselinskii et al. 2001). However, this route requires isolation of diazidoglyoxime— which on cyclization forms 1,1′-BTO—which is both highly friction and impact sensitive, meaning that even small-scale reactions are hazardous and that the route is not suitable for larger-scale production (Fig. 4). The 1,1′-BTO molecule is the precursor for the TKX-50 salt.

A crucial advance was made by *Klapötke and co-workers,* by reducing the synthetic procedure for 1,1′-BTO to a one-pot reaction from dichloroglyoxime, and crucially negating the necessity of isolating the highly sensitive diazidoglyoxime (Fig. 5) (Fischer et al. 2012). Instead, a strategy was utilized in which the formation and subsequent cyclization of the hazardous diazidoglyoxime occurred in solution. Following the successful synthesis of 1,1′-BTO using the improved method illustrated in Fig. 5, the target compound TKX-50 can then be easily prepared

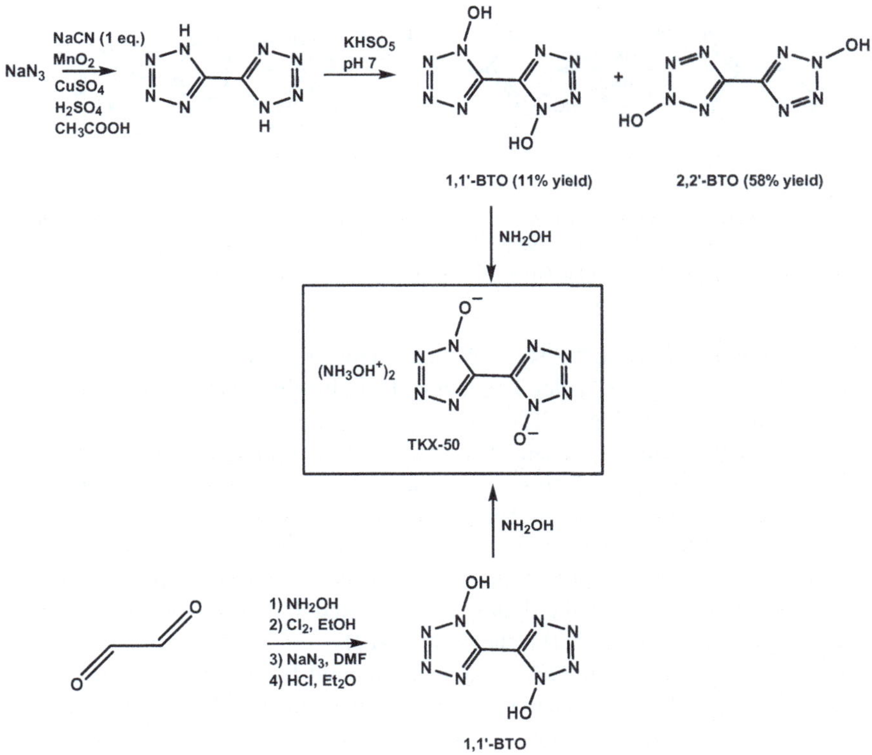

Fig. 3 Two initial synthetic routes for TKX-50—the routes differ only in the route used to form 1,1′-BTO. In both cases (routes A and B), the 1,1′-BTO which is formed is doubly deprotonated by hydroxylamine. In route A, 5,5′-bistetrazole is prepared and subsequently oxidized to form the corresponding neutral (protonated *N*-oxide molecule) 1,1′-BTO using aqueous potassium peroxymonosulfate ($KHSO_5$). In route B, glyoxal was transformed to diazidoglyoxime which was subsequently cyclized to form 1,1′-BTO

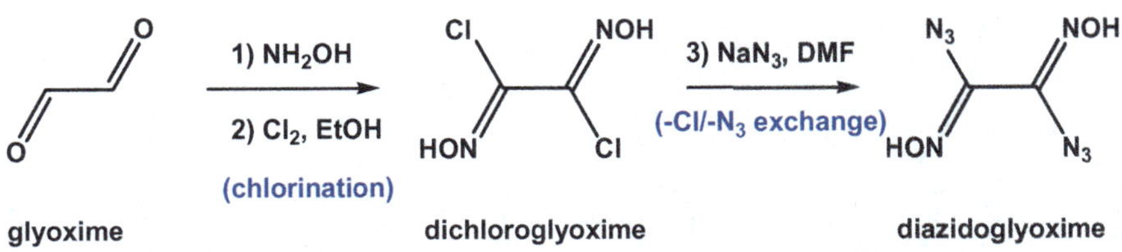

Fig. 4 Formation of 1,1′-BTO according to route B involves chlorination of glyoxime to form dichloroglyoxime, which then undergoes –Cl/–N_3 exchange using NaN_3 in DMF to form the highly hazardous diazidoglyoxime. The 1,1′-BTO molecule is then formed as a result of cyclization of diazidoglyoxime in acidic conditions

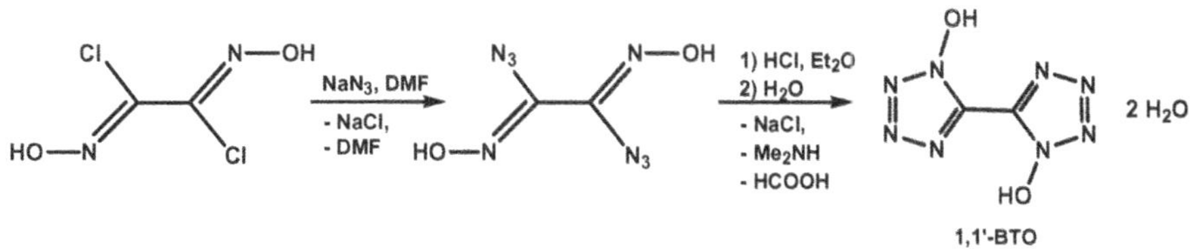

Fig. 5 New, improved one-pot synthesis of 1,1′-BTO starting from dichloroglyoxime in which the generation of and subsequent further reaction of the hazardous diazidoglyoxime occurs in solution, thereby avoiding the isolation of hazardous diazidoglyoxime (Fischer et al. 2012)

by the deprotonation of 1,1-BTO (5,5′-bistetrazole-1,1′ diole) by dimethylamine (Fischer et al. 2012). The latter is generated in situ by cleaving DMF in acidic conditions according to Fig. 6. The dimethylamine deprotonates 1,1′-BTO once to form the dimethylammonium salt of the monoanion, namely dimethylammonium 5,5′-bistetrazole-1,1′-diolate. This compound is then isolated and recrystallized in order to ensure high purity before being re-dissolved in boiling water. In the subsequent reaction of this salt with hydroxylammonium chloride, cation exchange occurs in addition to a second deprotonation which results in the formation of the bis-hydroxylammonium salt of the doubly deprotonated 1,1′-BTO—namely TKX-50 which crystallizes out (Fig. 6) (Fischer et al. 2012).

TKX-50 was obtained in 82% yield using method B for the formation of 1,1′-BTO, which was then deprotonated by the addition of aqueous hydroxylamine to an aqueous 1,1′-BTO solution to form TKX-50 (Fig. 3) (Fischer et al. 2012). Using the

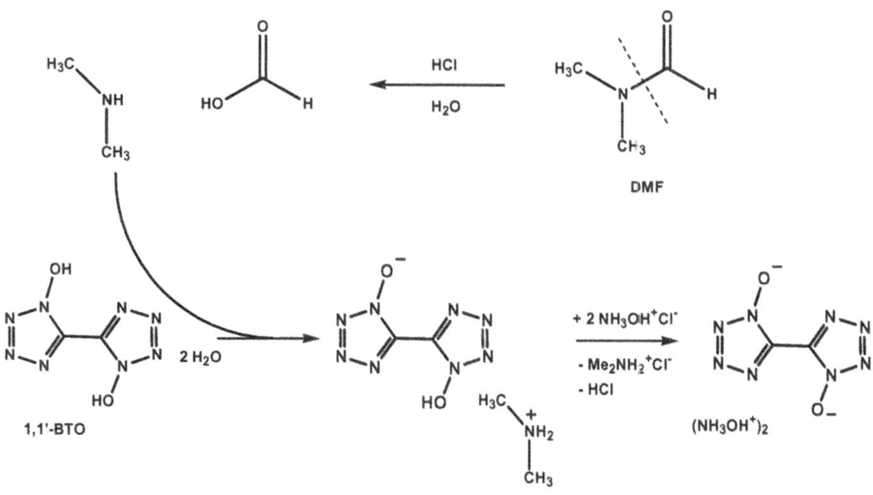

Fig. 6 Synthetic route for TKX-50 starting from DMF which undergoes cleavage under acidic conditions to form dimethylamine, which subsequently deprotonates 1,1′-BTO to form the dimethylammonium salt of the monoanion. This salt undergoes cation exchange and a second deprotonation to form the bis-hydroxylammonium salt of the dianion—TKX-50 (Fischer et al. 2012)

safer multi-step one-pot reaction outlined in Fig. 6, employing DMF, TKX-50 was obtained in 74.6% yield, and the yield could further be increased using the so-called NMP method, in which the disodium salt of the 1,1′ bistetrazolediolate anion was isolated, re-dissolved in a minimum of boiling water and reacted with a concentrated aqueous hydroxylamine solution to form TKX-50 in 85.1% yield (Fischer et al. 2012). At room temperature, TKX-50 is a white solid which is not hygroscopic and also not light sensitive. It is not only stable at room temperature, but shows a decomposition temperature of 221 °C (Fischer et al. 2012), well above the generally accepted requirement of >200 °C for secondary explosives. It should be pointed out however, that despite being a salt with an anion based on a bistetrazolate, TKX-50 has a nitrogen content of a modest 59.3% which is considerably lower than those of many other nitrogen-rich compounds (e.g., HN_3 97.7% N, hydrazinium azide 93.3% N, ammonium azide 93.3% N); however, despite this, it shows excellent detonation performance. The synthesis of TKX-50 in which the explosive diazidoglyoxime (DAG) is present as an intermediate was also modified very recently by a Korean research group that used the strategy of replacing DAG as an intermediate by ditetrahydropyranyl diazidoglyoxime (THP-DAG), since the latter is much less sensitive to external mechanical stimuli (Lee and Kwon 2019). The basis for this approach was the introduction of protecting groups to dichloroglyoxime (DCG) prior to azidization which thereby avoids the sensitive DAG intermediate (Fig. 4) (Lee and Kwon 2019). This modification is highly advantageous, since on larger-scale production of TKX-50 it would be important to eliminate highly sensitive intermediate compounds for safety reasons. A comparison of the IS, FS and ESD of diazidoglyoxime (DAG) and the new intermediate THP-DAG was made by the authors: IS (BAM) = 10.95 J THP-DAG, 1.5 J DAG; FS (BAM) = 352.8 N THP-DAG, <5 N DAG; ESD (OZM) = 50 mJ THP-DAG, 7 mJ DAG (Lee and Kwon 2019).

THP-DAG was prepared by suspending dichloroglyoxime in dichloromethane, followed by the addition of pyridinium *p*-toluenesulfonate and 3,4-dihydro-2*H*-pyran. After dilution of the solution, washing with $NaHCO_3$ and NaCl solutions followed by filtration from a suspension in hexane, THP-DCG was prepared (Lee and Kwon 2019). The THP-DCG prepared by this method was then dissolved in DMF and reacted with sodium azide at 0 °C followed by heating to 100 °C. After cooling the reaction mixture and pouring into water, the precipitate was filtered and dried resulting in the less sensitive THP-DAG intermediate in 79% yield (Fig. 7) (Lee and Kwon 2019). The further reaction of THP-DAG to form TKX-50 was achieved by bubbling HCl gas into an ether suspension which resulted in the deprotection and cyclization of THP-DAG to form 1,1′-BTO (Fig. 8) (Lee and Kwon 2019). The 1,1′-BTO which was formed was then further reacted with an aqueous hydroxylamine solution as in the original synthesis to form TKX-50 (Fig. 8) (Fischer et al. 2012; Lee and Kwon 2019). Very recently, the use of DFT methods has been reported in the literature to investigate using a computational approach the transition states, reactant complexes and product complexes, in order to propose the calculated reaction pathway for the synthesis of TKX-50 (Li et al. 2019a). The solubility of TKX-50 in various solvents has been measured using the gravimetric method at temperatures between 293.15 and 323.15 K, and the solubility of TKX-50 was found to

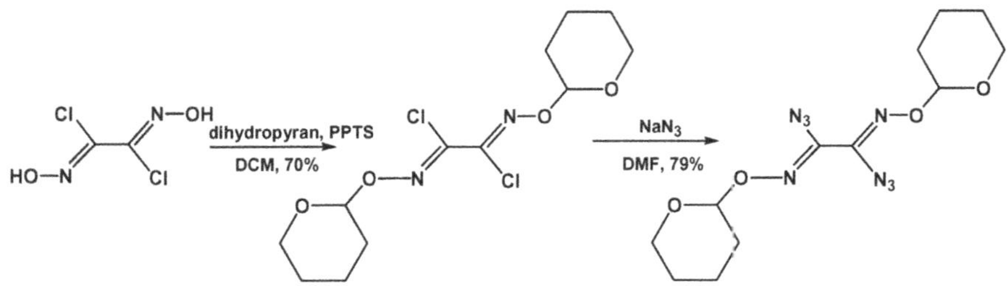

Fig. 7 Synthetic route for the formation of THP-DAG. By introducing the protecting groups to dichloroglyoxime prior to azidization, it is possible to synthesize TKX-50 using THP-DAG instead of the highly sensitive diazidoglyoxime (DAG) intermediate) (Lee and Kwon 2019)

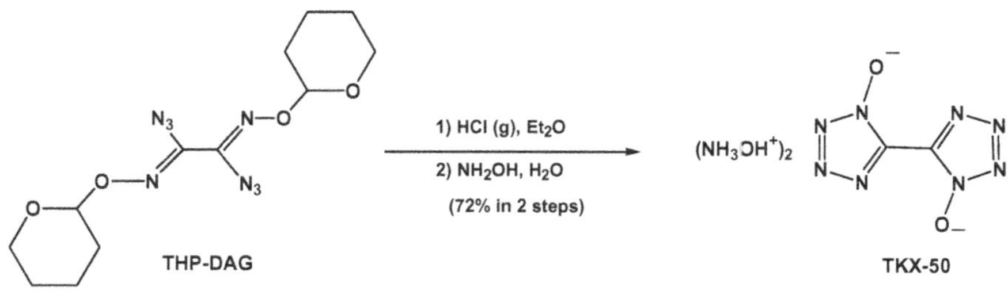

Fig. 8 TKX-50 can be synthesized using the much less sensitive intermediate THP-DAG by deprotection and cyclization to form 1,1′-BTO, which in a further reaction is doubly deprotonated by aqueous hydroxylamine solution for form TKX-50 (Lee and Kwon 2019)

increase in the following order: ethyl acetate < ethanol < toluene < deionized water < DMF < formic acid < DMSO (Zhang et al. 2016). In order to improve the purity of synthesized samples of TKX-50 to the highest level, the solubility of TKX-50 in different solvents was of importance—even although the purity of TKX-50 samples is routinely >99%. It was found that the solubility of TKX-50 in the solvents used increased in all cases with increasing temperature (Zhang et al. 2016). The highest solubility was shown by far in DMSO, whereas the strongest dependency of the solubility on temperature was shown with DMF (Zhang et al. 2016). The measured mole fraction solubilities of TKX-50 in the above-mentioned solvents can be found in more detail in the literature; however, Fig. 9 shows the temperature dependence of the mole fraction solubilities of TKX-50 in deionized water, DMF, formic acid and DMSO as a guide (Zhang et al. 2016). Furthermore, very recently the stability of TKX-50 in a variety of solvents has been investigated at different temperatures (Jia et al. 2019). It was found that TKX-50 is stable in the following solvents for 14 days at 25 °C or for 48 h at 50 °C: DMSO, deionized water, ethyl acetate, acetonitrile, methanol, ethanol, petroleum ether, and hexane (Jia et al. 2019). However, TKX-50 was found to react with DMF to form different products depending on the temperature (Jia et al. 2019). For example, at a molar ratio of DMF/TKX-50 lower than 9:1 at 25 °C DMA · BTO was obtained, whereas at higher molar ratios of DMF/TKX-50 the compound 2DMA · BTO was obtained (Jia et al. 2019). TKX-50 could also

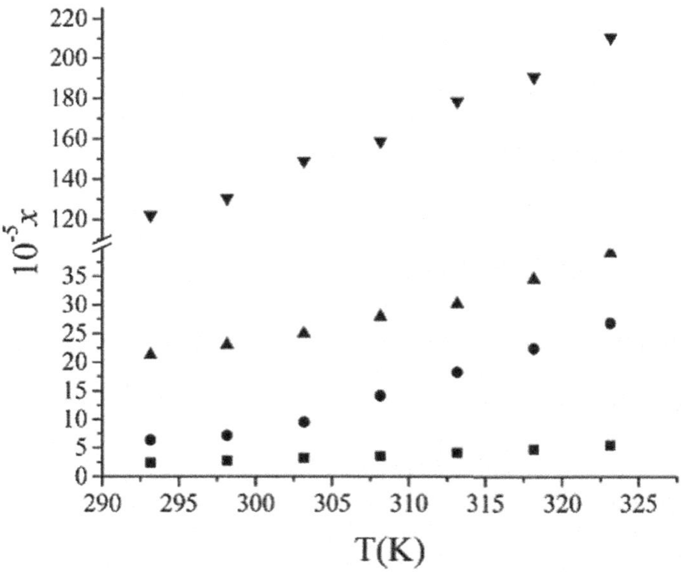

Fig. 9 Temperature dependence of the mole fraction solubilities of TKX-50 in deionized water, DMF, formic acid, and DMSO (Zhang et al. 2016) Reproduced with kind permission from the publisher from Zhang et al. (2016)

not be recrystallized form hot DMF, and instead 2DMA · BTO was formed and ABTOX on increasing the temperature and time (Jia et al. 2019). Figure 10 shows the interconversion paths of TKX-50 and its products on attempted recrystallization

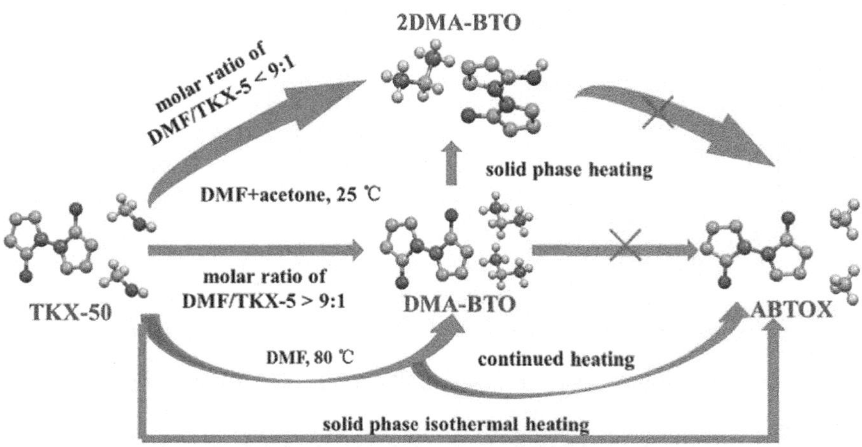

Fig. 10 Interconversion pathways of TKX-50 and its products on attempted recrystallization form DMF at different molar ratios and temperatures (Jia et al. 2019) Reproduced with kind permission from the publisher from Jia et al. (2019)

from DMF at different molar ratios and temperature (Jia et al. 2019). TX-50 was not only unstable in DMF at 25 and 50 °C, but also in DMF/acetone at 25 °C and in DMF/DMSO at 25 °C (Jia et al. 2019). This led the authors to conclude that the best solvents for the recrystallization of TKX-50 out of the solvents studied were deionized water and DMSO (Jia et al. 2019).

2.2 *Spectroscopic Characterization*

In addition to establishing the energetic properties of TKX-50, standard spectroscopic characterization was performed. In solution, ^{1}H NMR spectroscopy in d_6-DMSO at 25 °C showed one signal at 9.66 ppm (singlet) corresponding to the NH_3OH^+ atoms of the hydroxylammonium cations (Fischer et al. 2012). The $^{13}C\{^{1}H\}$ NMR spectrum of TKX-50 dissolved in d_6-DMSO solution shows only one signal as expected for the C atom of each of the tetrazole rings at 135.5 ppm (Fischer et al. 2012). The Raman spectrum of a single crystal of TKX-50 at ambient pressure was recently reported (Fig. 11) (Dreger et al. 2015) and the vibrations categorized into three main groups: $\leq$200 cm^{-1} correspond to collective vibrations of anions and cations and are either lattice modes or are mixed with low-frequency molecular vibrations (Dreger et al. 2015). Various distortions of the anion ring as well as NH_3 rotations and OH wagging vibrations of the hydroxylammonium cations are observed at intermediate wavenumbers of 200–1650 cm^{-1}, with the modes >1650 cm^{-1} being attributed to

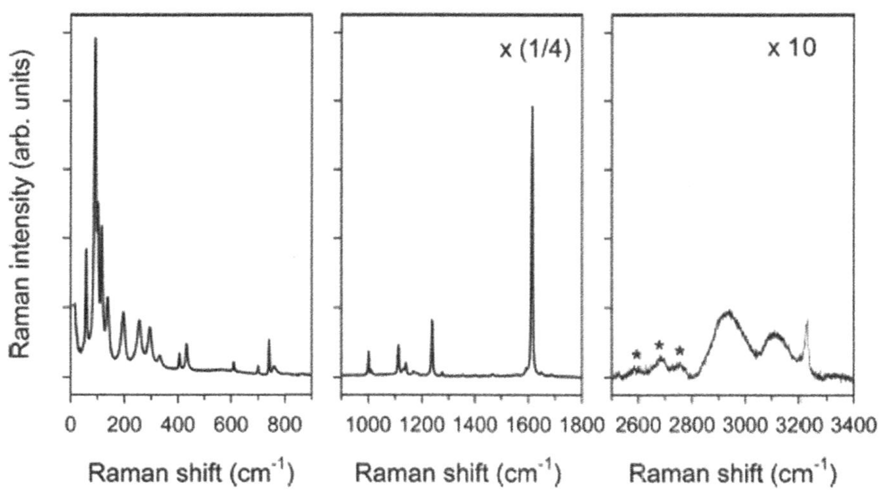

Fig. 11 Raman spectrum of a single crystal of TKX-50 under ambient conditions separated into three areas (in order to make it easier to see the peaks other than the peak at 1615 cm^{-1} which is four times stronger than any other peak in the spectrum). The * indicate peaks which had very low intensity and correspond to OH and NH_2 stretching vibrations (Dreger et al. 2015). Reproduced with kind permission from the publisher from Dreger et al. (2015)

NH and OH rocking as well as NH_2 and OH stretching vibrations (Dreger et al. 2015). The full assignment of the Raman spectrum is given in Table 1. The Raman spectrum of a single crystal of TKX-50 under pressure (2.1–48.1 atm.) shows some generalizations: A blue shift of peaks (except for two modes) occurs with increasing pressure, the sensitivity of the modes to increasing pressure in the range 4–10 GPa decreases, some peaks not being observed at pressure between 10 and 13 GPa whereas new peaks are observed in the pressure range 32–36 GPa and coupling of various modes on increasing pressure (Dreger et al. 2015). However, importantly, even pressures up to 50 GPa did not cause irreversible changes and these pressure-induced changes in the Raman spectra were completely reversible on reducing the pressure back down ambient pressure (Dreger et al. 2015). This was proposed to support the low-sensitivity TKX-50 shows, and also suggest high chemical stability—typically observed by single crystals of IHEs (Dreger et al. 2015). Based on the pressure dependencies of the Raman shifts, the intermolecular interactions in TKX-50 are possibly stronger than those observed in crystals of typical energetic materials such as RDX (Dreger et al. 2015). Changes in the Raman spectrum in the pressure range 4–10 GPa, however, have not been attributed to a phase transition and are suggested instead to correspond to a structural adjustment instead, and spectral change in the pressure range 10–13 GPa are also not concurrent with a phase transition, whereas the changes observed at very high pressures of 32–36 GPa may be related to a phase transition although the evidence is not conclusive (Dreger et al. 2015). Therefore, using Raman spectroscopy in a diamond anvil cell, it was concluded that TKX-50 doesn't undergo a phase transition from pressures between ambient and 32 GPa (Dreger et al. 2015). That the single-crystal structure of TKX-50 under ambient conditions is maintained even at greatly elevated pressures has been attributed to the extensive hydrogen bonding network observed in the crystal structure of TKX-50 at ambient conditions which is also maintained at high pressures (Dreger et al. 2015). Further computational work from this study on the compressibility of TKX-50 is included in Sect. 2.3. An interesting new description of a shear-induced phase transition in TKX-50 was recently reported (Lu et al. 2017b). The authors used a combined theoretical and experimental approach to examine the structural response of TKX-50 crystals under shear. An investigation of the response of an energetic material to shear is not trivial and also not unimportant, since shear is always experienced by an energetic material due to external thermal-mechanical loading on the energetic material (Lu et al. 2017b). Moreover, since the resistance that an energetic material to the mechanical stimulation induced shear/slip is one of the main causes of hotspot production and growth, the lower the steric hindrance as a result of crystal packing during shear/slip is, the lower the mechanical sensitivity tends to be (Lu et al. 2017b). Therefore, an analysis of the response of an energetic crystal toward shear is of importance also with respect to the mechanical sensitivity of the explosive. The above-mentioned work of *Dreger* et al. in which Raman spectroscopy was used to show that at high pressures of between 32 and 36 GPa a phase transition may occur; however, no structure was proposed or determined (Dreger et al. 2015). Furthermore, computational work on the plastic and shear behavior of TKX-50 using large-scale MD simulations found no evidence for a phase transition (non-reactive force field) (An et al. 2014). However,

Table 1 Assignment of the experimentally observed peaks in the Raman spectrum of a single crystal of TKX-50 under ambient conditions

Wavenumber (cm^{-1})	Assignment		Wavenumber (cm^{-1})	Assignment	
	Anion	Cation		Anion	Cation
58	Rock (in-plane rotation)	Rotation	1114	Ring distortion (N=N stretching, N–N and N–N′ stretching, in-plane)	
92	Twist	Rotation	1131	Ring distortion (N–N stretching, N–N′ stretching opposite, in-plane)	
103	Wag	Rotation	1142		
116	Wag	Translation	1169		NH rock, OH rock
122	Twist	Rock	1183	Ring distortion (N=N–N′ symmetric stretching, in-plane)	
138		Translation	1239	Ring distortion (N=N–N′ symmetric stretching, in-plane)	
196	Twist	Rotation	1277	In-plane C–C rock, complex ring distortion	
220		Rock	1468	In-plane C–C rock, ring distortion (C–C–N′ scissor, in-plane)	

(continued)

Table 1 (continued)

Wavenumber (cm^{-1})	Assignment		Wavenumber (cm^{-1})	Assignment	
	Anion	Cation		Anion	Cation
256		NH_3 rotation	1486	In-plane C–C rock, ring distortion (C–C–N′ scissor, in-plane)	NH rock, OH rock
295	Ring rock, N–O rock	NH_3 rotation	1591		NH rock, OH rock
332	C–C wag, ring wag	NH_3 rotation	1593	C–C stretch	NH rock, OH rock
406	C–C stretch, ring in-plane shifting		1615	C–C stretch	NH rock, OH rock
433	C–C wag, N–O, ring wag		1650		NH_2 scissor, NH rock, OH rock
610	Ring in-plane rock		2931		NH symmetric stretch, OH symmetric stretch
701	C–C wag, ring distortion (half-chair, out-of-plane)		3096	Combination (1486–1615)	
742	C–C wag, ring distortion (envelope, out-of-plane)		3170		NH_2 symmetric stretch
761	Ring distortion (ring stretching)	O–H wag	3216		NH_2 asymmetric stretch
1002	C–C rock in-plane, ring distortion (C–N′ (N′ atom bonded with O), N=N–N scissor, in-plane)		3228	Overtone (2 × 1615)	

(continued)

Table 1 (continued)

Wavenumber (cm^{-1})	Assignment		Wavenumber (cm^{-1})	Assignment	
	Anion	Cation		Anion	Cation
1013		N–O stretch and breath			

Additional calculated vibrations which were not observed experimentally have been omitted (Dreger et al. 2015)

during recent work investigating the shear behavior of TKX-50, it was possible to find a new phase of TKX-50, the shear-induced γ-TKX-50 (Lu et al. 2017b).

For the shear calculations, each slip system had a series of shear trains imposed upon it in order to simulate TKX-50 against shear loading (Lu et al. 2017b). In order to ascertain the preferred slip system of TKX-50 toward shear, a closer look at the hydrogen bonding in the TKX-50 crystal system at room temperature was undertaken using a computational approach. The authors stated that out of the even hydrogen bonds in the TKX-50 unit cell (there are 8 hydrogen bonds around one NH_3OH^+ cation and twelve around one $C_2N_8O_2{}^{2-}$ anion) which contribute to supporting the TKX-50 crystal, five relatively strong hydrogen bonds are distributed within the layers along the (010) plane (interlayered) whereas only two relatively weak hydrogen bonds are present between the interlayers (Lu et al. 2017b). Therefore, it can be expected that compression along the [100] orientation (*b* axis) should occur more readily than along the [100] (*a* axis) and [001] (*c* axis) orientations (Lu et al. 2017b). The hydrogen bond arrangement in the TKX-50 crystal is also responsible for the layered arrangement of the cations and anions within the crystal, and this facilitates the smooth sliding along the (100) plane (Lu et al. 2017b). This sliding is important, since through this process, the stimuli from external sources can be better dispersed, which in turn prevents the formation of hotspots which are known to cause detonation and thereby lowers the mechanical sensitivity of TKX-50 (Lu et al. 2017b). DFT calculations were then performed for shearing along this plane with different orientations, e.g., [100], [101], etc. (Fig. 12) and the dependencies of the shear strain along various orientations was determined (Fig. 12) (An et al. 2015). The energy dependence on the shear strain that TKX-50 is subjected to along various slip systems is plotted in Fig. 12. It can be concluded from Fig. 12 that the shear barriers are as follows: 18.6 kcal/mol for (010)/[101], 28.04 kcal/mol for (010)/[@@@10$\bar{1}$], 28.31 kcal/mol for (010)/[001] and 38.40 kcal/mol for (010)/[100] slip orientations, respectively (Fig. 12) (Lu et al. 2017b). Since the shear of the (010) plane along other orientations shows lower barriers relative to the (100)/[001] slip system (except for the (010)/[100] which shows a slightly higher barrier), this supports the suggestion that the slip occ\bar{1} urs preferentially along the (010) plane (Fig. 12) (Lu et al. 2017b). The lowest energy barrier of 18.6 kcal/mol is calculated for slip along (010)/[101] (Fig. 12) which is in agreement with the above-discussed analysis of the hydrogen bond arrangement (Lu et al. 2017b). Importantly, since the slip barriers are energetically significantly lower than the energy barrier for decomposition of the

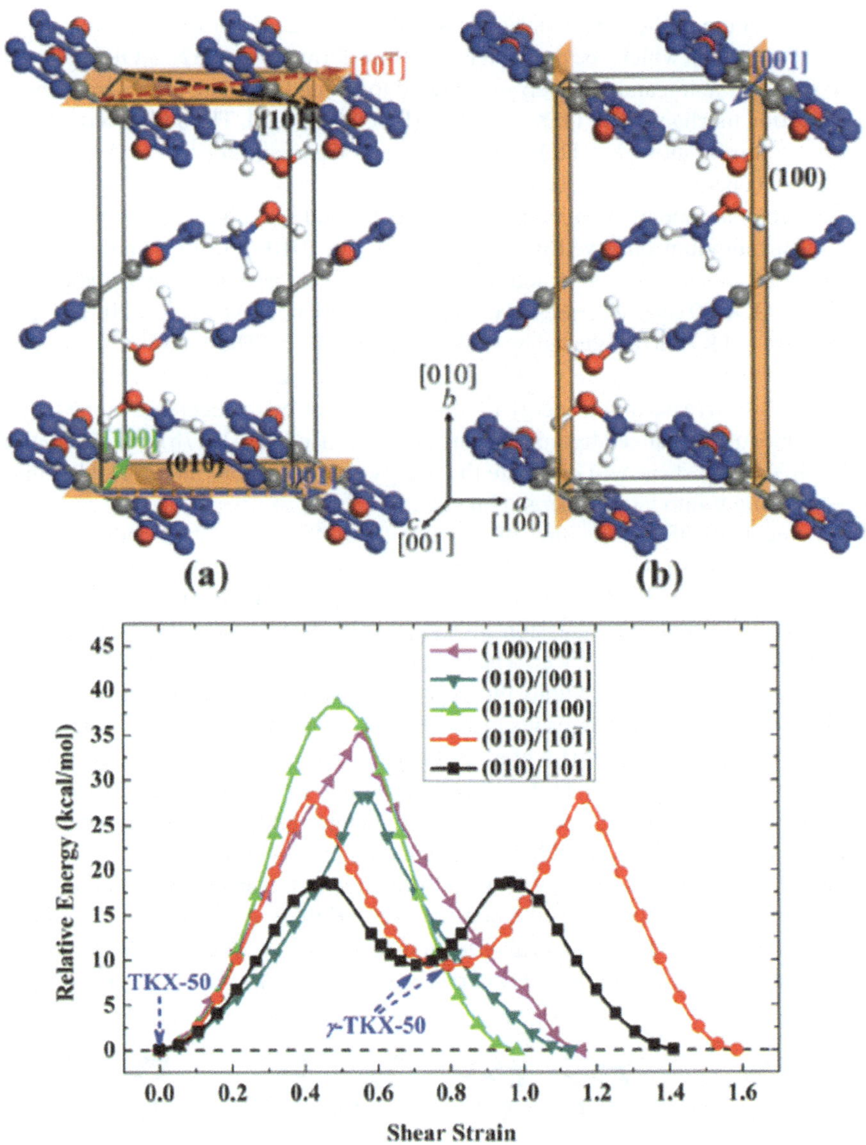

Fig. 12 Slip planes and orientations of shear-loaded TKX-50 (left) (Lu et al. 2017a, b) and the dependency of the energy on the shear strain along various orientations (right) (Lu et al. 2017a, b). Reproduced with kind permission from the publisher from (Lu et al. 2017a, b)

$C_2N_8O_2^{2-}$ anion in TKX-50 which corresponds to 45.1 kcal/mol or 37.2 kcal/mol for the $C_2HN_8O_2^-$ (in which one proton has been transferred from the NH_3OH^+ cation to the $C_2N_8O_2^{2-}$ anion and is widely considered to be the first stage in the decomposition reaction mechanism of TKX-50) anion (Lu et al. 2017b). The authors concluded that these results show that TKX-50 can undergo shear-slip in order to disperse the mechanical energy it is exposed to which in turn prevents or reduces hotspot formation which consequently results in a lower mechanical sensitivity (Lu et al. 2017b). The local minimum observed in the (010)/[101] or (010)/[10$\bar{1}$] orientation corresponds each to a metastable structure which the authors assign to a new metastable phase: γ-TKX-50 (the structures of both minim are identical) (Lu et al. 2017b). This metastable γ-TKX-50 is induced by shear (Lu et al. 2017b). The energy of γ-TKX-50 is reported to be approximately 9.4 kcal/mol higher in energy than TKX-50 (Lu et al. 2017b). This increase in the energy is compensated for by the decrease in the lattice energy due to the changed hydrogen bond situation (Lu et al. 2017b). The formation of γ-TKX-50 by shear will dissipate the energy resulting from the external stimuli by converting it into lattice energy (Lu et al. 2017b).

A comparison of the structures of "normal" TKX-50 and the shear-induced phase γ-TKX-50 is shown in Fig. 13. It has been proposed by the authors that the possible new phase observed by *Dreger* et al. at high pressure corresponds to γ-TKX-50, since non-hydrostatic compression always causes shear, and γ-TKX-50 resulted from exposing TKX-50 to shear. The effects of shear on the hydrogen bonding in TKX-50 are several. For example, the hydrogen bonding is weakened after shear by ~63 kJ/mol (unit cell) which results in a reduction in the lattice energy in comparison with TKX-50 prior to shear (Lu et al. 2017b). However, some—but not all—of the hydrogen bonds observed in TKX-50 are present in γ-TKX-50 and interestingly, the order of strength of the hydrogen bonds also changes after shear, for example, the strongest hydrogen bond in TKX-50 is of the type O–H····O, whereas in γ-TKX-50 it is a hydrogen bond of the type N–H····O which is the strongest (Lu et al. 2017b). Overall, it was concluded that the contributions of the H····O and H····N contacts decreases in γ-TKX-50 which is asserted to cause result in weaker hydrogen bonds between the cations and anion and between cations, while the $^{\delta+}$H····H$^{\delta+}$ electrostatic

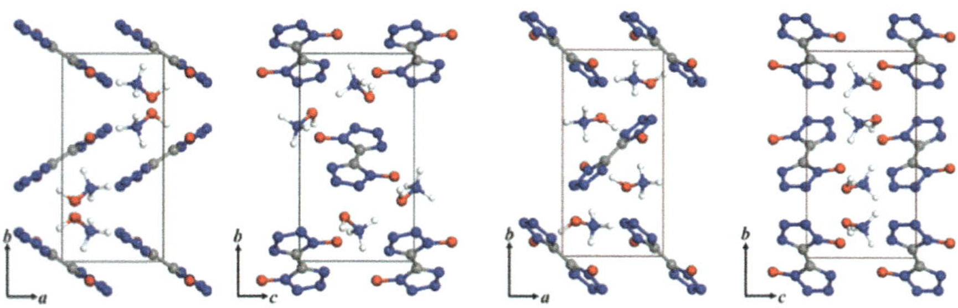

Fig. 13 Comparison of the calculated structures of "normal" TKX-50 (left) and the shear-induced γ-TKX-50 (right) (Lu et al. 2017a, b). Reproduced with kind permission from the publisher from Lu et al. (2017)

repulsion between the cations is increased. This is important because it means that it is more difficult for proton transfer to occur from the NH_3OH^+ cation to the $C_2N_8O_2^{2-}$ anion (Lu et al. 2017b) which has been discussed extensively in the literature as being the first step in the decomposition reaction mechanism of TKX-50. Overall, the effect of shear on TKX-50 can be summarized as follows: Shear along the appropriate orientation results in the new metastable phase of TKX-50, namely γ-TKX-50 which is 9.4 kcal/mol higher in energy relative to TKX-50. However, shear is suggested to dynamically stabilize TKX-50 as γ-TKX-50, since the energy barrier for cation–anion proton transfer is higher in γ-TKX-50 due to shear (Lu et al. 2017b). Finally, Table 2 gives an overview of the calculated lattice parameters and properties of γ-TKX-50 in comparison with the calculated lattice parameters of TKX-50 prior to shear ("normal" TKX-50) (Lu et al. 2017b).

As is discussed at considerable length in this section, several investigations have established that at elevated temperature at ambient pressure, TKX-50 undergoes first-stage decomposition via hydrogen transfer from the NH_3OH^+ cation to the anion. Therefore, consequently, strong hydrogen bonds are considered to be a limiting factor on the thermal stability of TKX-50 since strong interionic hydrogen bonds of a particular type should promote hydrogen transfer between the cation and anion. However, on the other hand, it was also shown that under ambient temperature and elevated pressure, TKX-50 maintains the monoclinic structure observed under ambient conditions but undergoes highly anisotropic compressibility. The maintaining of the ambient conditions structure was attributed to the strong interionic hydrogen bonding being

Table 2 Comparison of the calculated optimized lattice parameters for the shear-induced metstable phase γ-TKX-50 with that calculated for "normal" TKX-50

Crystals	TKX-50		γ-TKX-50	
	PBE+D2	OptB86-vdw	PBE+D2	OptB86-vdw
Crystal system	Monoclinic	Monoclinic	Monoclinic	Monoclinic
Space group	$P2_1/c$	$P2_1/c$	$P2_1/a$	$P2_1/a$
a (Å)	5.5589	5.5262	5.7008	5.6215
b (Å)	11.3567	11.3830	11.7409	11.7507
c (Å)	6.3025	6.4228	6.3025	6.3313
α (°)	90	90	90	90
β (°)	96.293	96.089	91.808	91.407
γ (°)	90	90	90	90
V_0 (Å^3)	401.76	401.74	421.63	418.09
ρ (g cm^{-3})	1.952	1.952	1.860	1.876
B_0	24.9	25.7	28.1	
B_0'	6.9	6.5	5.5	
C_v (J K^{-1} mol^{-1})		251.5		253.0

B_0 = bulk modulus, B_0' is the first derivative of the bulk modulus, C_v = the standard molar specific heat capacity at 298.15 K (Lu et al. 2017a, b)

present which were strengthened under pressure. So at high temperatures, strong interionic hydrogen bonds are proposed to facilitate hydrogen transfer (cation → anion) and thereby consequently the first-stage decomposition reaction of TKX-50; however, at high pressure, the overall structure of TKX-50 remains intact due to strong hydrogen bonding! This clearly raises the question of the influence of both elevated temperature (273–760 K) *and* pressure (8 GPa) at the same time on a crystal of TKX-50 since the two opposing effects of the interionic hydrogen bonds will then compete with each other (Dreger et al. 2017a). This investigation using Raman spectroscopy resulted in the establishment of the pressure–temperature phase diagram of TKX-50 (Dreger et al. 2017a). Heating the TKX-50 crystal (β = 1 K/min) to 296 K, followed by 533 K and finally 610 K at 0.8 GPa pressure revealed characteristic features, on increasing the temperature, namely: decrease in the intensity of the Raman peaks, new peak patterns in the ranges corresponding to the lattice and internal modes, and visible changes in the sample morphology and appearance from optic images (Dreger et al. 2017a), which the authors stated were indication of two new forms intermediate 1 and intermediate 2 (Dreger et al. 2017a). However, the Raman spectra of intermediate 1 and intermediate 2 were found to be significantly different from that of TKX-50 and also the Raman spectrum of intermediate 1 was found to be different from that of intermediate 2. The major differences were found to correspond to the lattice vibrations as well as the NH/OH stretching vibrations which is indicative of TKX-50, intermediate 1 and intermediate 2 possessing a different hydrogen bond network (Dreger et al. 2017a). At higher pressure, similar observations were made; however, the transition temperature to intermediates 1 and 2 was found to increase as the pressure was increased (Dreger et al. 2017a). Importantly, these high pressure-high temperature (HP-HT) forms of TKX-50 were recovered as polycrystalline samples at ambient conditions although in very small quantities (Dreger et al. 2017a). The authors were then able to establish that intermediate 1 corresponded to the TKX-50 decomposition product ABTOX which has been described in several other reports (Dreger et al. 2017a). They also concluded that intermediate 2 was potentially most likely to be either diammonium 5,5′-bistetrazolate or dihydroxylammonium 5,5′-bistetrazolate (Dreger et al. 2017a). The transformation of TKX-50 during isobaric heating is shown in Fig. 14 as proposed by the authors (Dreger et al. 2017a). Based on these findings using Raman spectroscopy, as well as a close consideration of the decomposition reactions of TKX-50, the authors constructed the following pressure–temperature phase diagram (Fig. 15) (Dreger et al. 2017a). A further report appeared in the literature with respect to a heat-induced solid–solid phase transformation of TKX-50 (Lu et al. 2017a, b). The authors reported a solid–solid phase transition to a new metastable phase which they called Meta-TKX-50 at 180 °C. The basis for this conclusion was a combination of Raman spectroscopy and TGA-DSC measurements in combination with computational work (Lu et al. 2017a, b). The authors asserted that rotation of the NH_3OH^+ cations with a small volume expansion of 3% results in the formation of Meta-TKX-50 which is evidenced by a lowering of the symmetry of the space group from $P2_1/c$ for "normal" TKX-50 to P-1 for Meta-TKX-50 (Lu et al. 2017a, b). However, almost concurrently, a different publication appeared in which the authors briefly commented that the Raman spectra

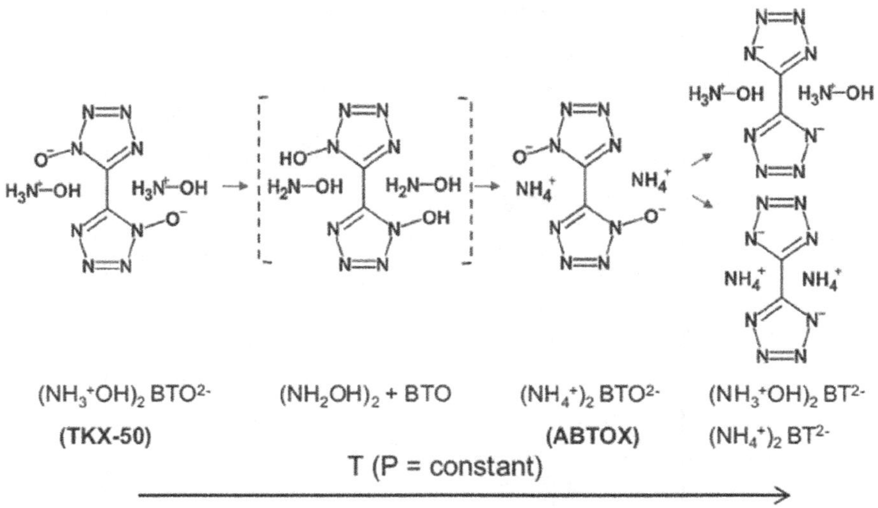

Fig. 14 Proposed transformation of TKX-50 during isobaric heating. Compounds in brackets indicate compounds which were not observed in the experiments that were undertaken (Dreger et al. 2017a, b). Reproduced with kind permission from the publisher from Dreger et al. (2017a, b)

Meta-TKX-50 "resemble to some degree the spectra of intermediate 1" which the authors were able to show corresponded to the primary first-stage decomposition product of TKX-50, namely ABTOX (Badgujar and Talawar 2017). Therefore, the question as to whether Meta-TKX-50 is actually a new phase of TKX-50 or ABTOX is unclear at this stage and is currently undergoing further investigation.

Using Raman spectroscopy at different elevated temperatures, the authors concluded that the thermal behavior of TKX-50 could be classified into three groups: (i) no change in the Raman spectra up to 170 °C, (ii) new peaks becoming visible at 180 °C and increase in intensity up to 90 °C after which temperature the new peaks decrease in intensity, while the old peaks remain meaning that at 200 °C the Raman spectrum is similar to that obtained at 180 °C, and (iii) the peaks in the Raman spectra decrease in intensity and disappear at temperatures of around 210 °C (Lu et al. 2017a, b). The authors state that the observation of (i) in their work is in agreement with previous reports on the thermal stability of TKX-50 which stated that no gas was evolved on heating TKX-50 for 5 h at 170 °C. They also found that the decomposition of TKX-50 begins at ~210 °C ($\beta = 5$ °C/min) (Lu et al. 2017a, b). The authors concluded that the middle stage (ii) corresponds to the formation of new phase or heat-induced phase transition (Lu et al. 2017a, b). Using TGA-DSC curves, the authors detected a very small endothermic peak at 170–185 °C (Fig. 16) which they attributed to the phase transition, in addition to the two major exothermic peaks (220–245 °C and 245–270 °C) (Lu et al. 2017a, b). They point out that the TGA-DSC step at 170–185 °C is in the same temperature range as the new intermediate peaks which appear in the Raman spectrum at 180 °C (Lu et al. 2017a, b). The authors point out the absence of mass loss in the TGA in this region which they conclude

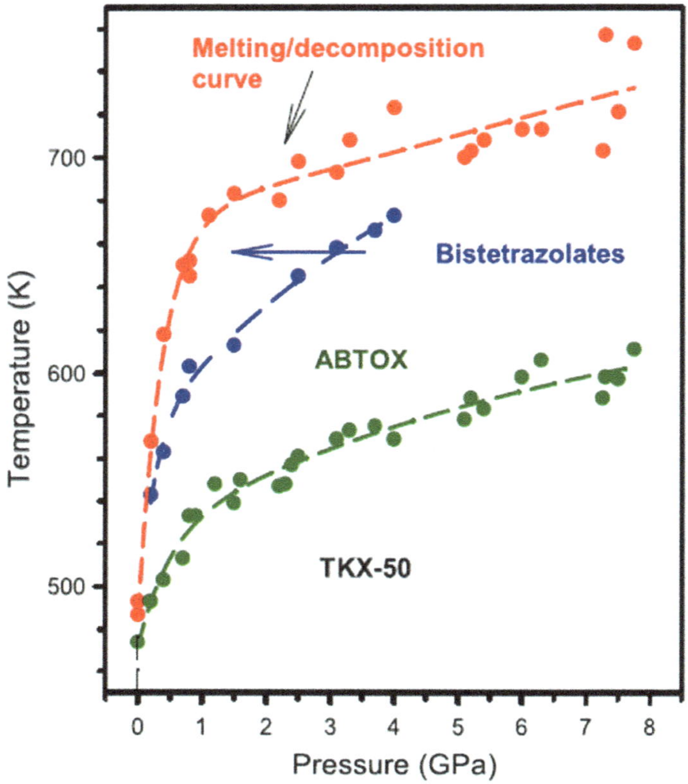

Fig. 15 Phase diagram of TKX-50. The colors indicate transition temperatures between different species: green = TKX-50 and intermediate 1 (ABTOX); blue = intermediate 1 (ABTOX) and intermediate 2 (diammonium 5,5′-bistetrazolate or dihydroxylammonium 5,5′-bistetrazolate); red = intermediate 2 (diammonium 5,5′-bistetrazolate or dihydroxylammonium 5,5′-bistetrazolate) and decomposed sample (Dreger et al. 2017a, b). Reproduced with kind permission from the publisher from Dreger et al. (2017a, b)

suggests a second-order phase transition without latent heat (Lu et al. 2017a, b) and that a new phase of TKX-50, namely Meta-TKX-50, and a second-order structural phase transition of TKX-50 occur at ~180 °C (Lu et al. 2017a, b). Furthermore, the authors state that the decomposition of a crystal of TKX-50 beginning at ~205 °C is evidenced by the slow mass loss observed in the TGA curves and is in good agreement with previous reports which reported obvious gas loss from TKX-50 at ~200 °C which the authors of the current study claim is consistent with the decomposition of the new peaks in the Raman spectrum assigned to Meta-TKX-50 (Lu et al. 2017a, b). They conclude, therefore, that this provides evidence for the decomposition of TKX-50 initiating from Meta-TKX-50 and that the phase transition from TKX-50 to Meta-TKX-50 is an intermediate step to the decomposition of TKX-50 (Lu et al. 2017a, b). The morphology of a single crystal of TKX-50 was also observed to change above 170 °C by which the crystal becomes opaque and the homogeneous

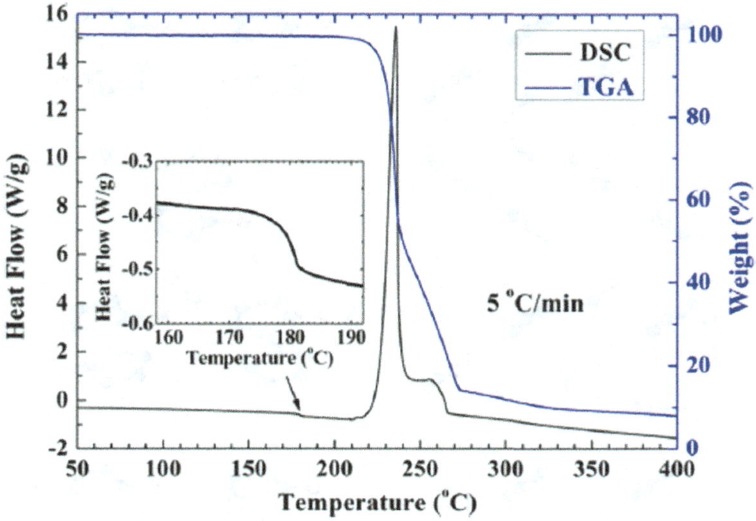

Fig. 16 TGA-DSC curves of TKX-50 in the temperature range of 23–400 °C showing the two exothermic decomposition peaks as well as the very small endothermic step at 170–185 °C which the authors attribute to a new TKX-50 phase, Meta-TKX-50 (Lu et al. 2017a, b). Reproduced with kind permission from the publisher from Lu et al. (2017a, b)

nucleation of a new phase at the edges of the crystal which then spreads across the crystal was reported (Lu et al. 2017a, b). A change in the crystal was again subsequently observed at 200 °C in which erosion of the crystal edges was observed to begin, culminating with the production of bubbles of gas above 210 °C (Lu et al. 2017a, b).

Using computational approaches, the authors found that rotation of the NH_3OH^+ cations clearly occurs before hydrogen transfer. The latter corresponds to the first-stage decomposition of TKX-50 (Lu et al. 2017a, b). The crystal structure of Meta-TKX-50 was calculated and lattice dynamics calculations were performed which predicted the *P*-1 space group for Meta-TKX-50 (Lu et al. 2017a, b). The calculated unit cells of TKX-50 and of Meta-TKX-50 are shown in Fig. 17 for comparison in which the orientation of the $C_2N_8O_2^{2-}$ anions remain, but the NH_3OH^+ cations are rotated in Meta-TKX-50 with respect to those in TKX-50 (Lu et al. 2017a, b). The calculated Raman spectra for Meta-TKX-50 are also reported to be in good agreement with the experimentally observed additional peaks at 180 °C which were assigned to TKX-50 (Lu et al. 2017a, b).

The authors continued to assess the effect the Meta-TKX-50 phase has on the thermal stability of TKX-50 (Lu et al. 2017a, b). They concluded that the phase transition to Meta-TKX-50 promotes the decomposition of TKX-50 since it facilitates the hydrogen transfer from the cations to anion (Lu et al. 2017a, b) and lowers the energy barrier of hydrogen transfer (Lu et al. 2017a, b). The thermal stability of TKX-50 was initially established using DSC at a heating rate of 5 °C/min which

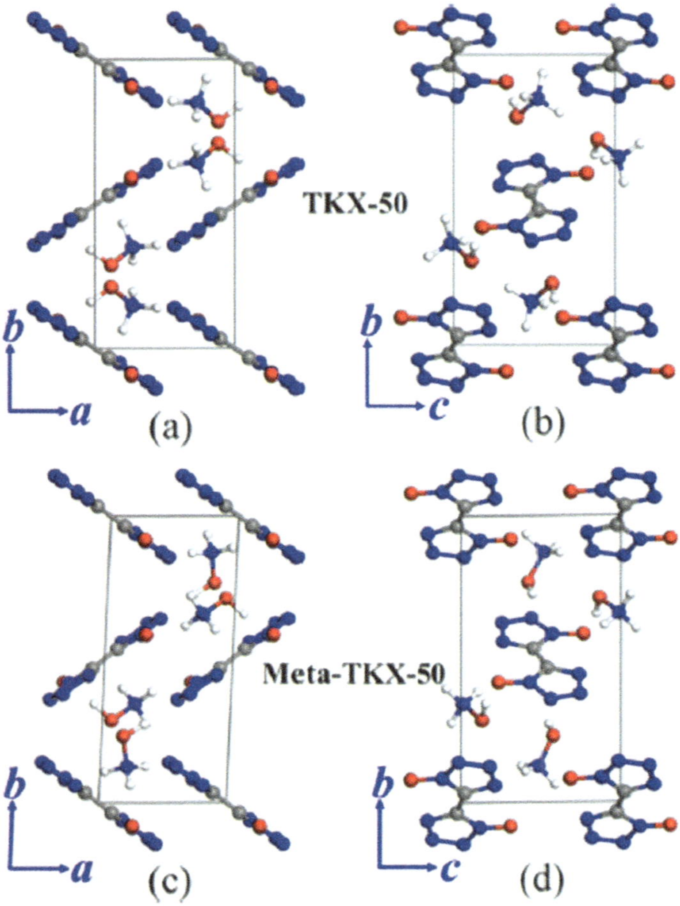

Fig. 17 Calculated crystal structures of TKX-50 (above) and of Meta-TKX-50 (below) showing the orientation of the $C_2N_8O_2{}^{2-}$ anions remains unchanged, however the NH_3OH^+ cations undergo rotation (Lu et al. 2017a, b). Reproduced with kind permission from the publisher from Lu et al. (2017a, b)

showed only a decomposition exotherm at 221 °C and no endotherm indicative of either melting or a phase transition at temperatures between room temperature and 221 °C under the conditions used was reported (Fischer et al. 2012). In fact, low-temperature single-crystal X-ray crystallography showed that TKX-50 was present in the same phase at 100, 173 and 298 K (Fischer et al. 2012). These results are somewhat comparable with DSC measurements reported in (Huang et al. 2015a, b) at a heating rate of 10 °C/min under a dynamic nitrogen atmosphere (50 mL/min flow rate, 0.1 MPa pressure) which showed only the presence of one exotherm at 245.72 °C (followed by a second at 278.53 °C) and no endotherm corresponding to melting or a phase change at lower temperatures (Huang et al. 2015a, b). However, despite only a 5 °C/min difference in the heating rate, the exotherm peak maximum differed by 25 °C (Huang et al. 2015a, b). The widely used techniques for establishing the decomposition temperature of a newly synthesized energetic material are DSC and TGA. Even although *Muravyev* et al. obtained results from DSC spectra which agree excellently with those from Ref. (Fischer et al. 2012) using the Kissinger method, they have questioned the validity of using the Arrhenius parameters obtained from DSC and TGA to describe the complex two-stage decomposition of TKX-50 and propose the ARC (accelerating rate calorimetry) technique as more reliable (Muravyev et al. 2017). A comparison of the Arrhenius parameters for the primary stage of TKX decomposition estimated by different groups for TKX-50 is given in Table 3. However, the decomposition temperature of TKX-50 obtained by *Muravyev* et al. is in good agreement with those reported by other groups also using DSC/TG and shows that the onset of decomposition (DSC, $\beta = 10$ K/min) occurs at 238 °C with an exothermic peak maximum at 243 °C which agrees well with the exothermic peak observed at 221 °C in the DSC results ($\beta = 5$ °C/min) reported in the original synthetic report (Fischer et al. 2012). The DSC/TG curves of TKX-50 are shown in Fig. 18 (Muravyev et al. 2017).

Table 3 Comparison of the exothermic processes and Arrhenius parameters for TKX-50 decomposition obtained by various groups using DSC/TG techniques

Method	T_{dec} (K)	E_a (kJ/mol)	$\log(A/s^{-1})$	References
DSC/Kissinger (1 mg sample, β = 5–20 K/min)		237.6	23.9	Huang et al. (2015a, b)
DSC/Kissinger (1 mg sample, β = 1–20 K/min)		143.2	12.3	Fischer et al. (2013)
DSC, non-isothermal		156.6	13.7	Sinditskii et al. (2015)
Isothermal manometric measurements at 443–473 K, autocatalytic kinetic scheme 1st stage 2nd stage		166.9 190.0	13.8 18.1	Sinditskii et al. (2015)
DSC (1 mg sample, 10 K/min)	511 (onset), 516 (exotherm peak max), 546 (second exotherm peak)			Muravyev et al. (2017)
TG (1 mg sample, β = 10 K/min)	510 (55% mass loss)			
DSC (0.7 mg sample, β = 2–20 K/min)		143.6	12.6	Wang et al. (2017)

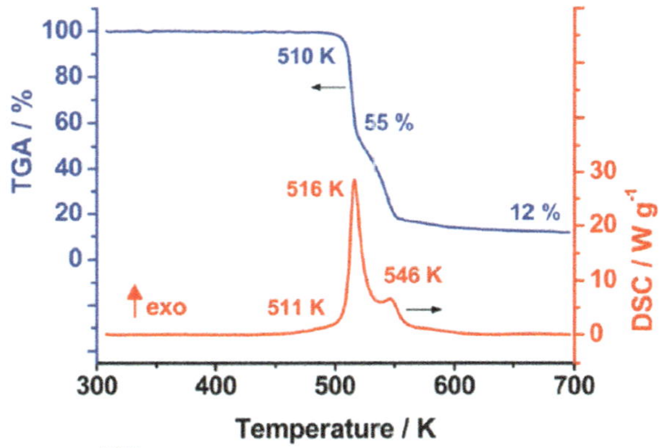

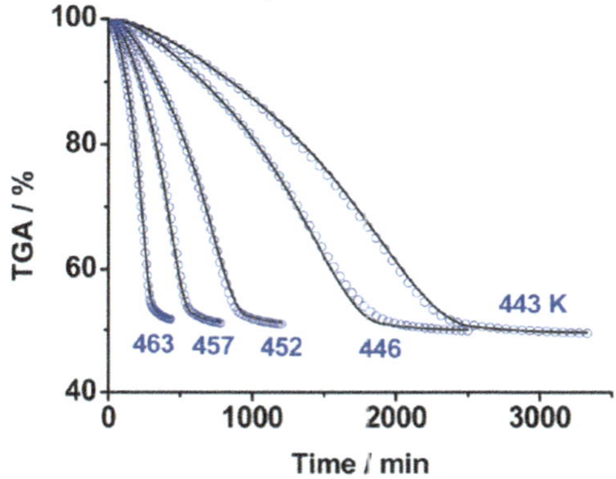

Fig. 18 DSC and TG curves of TKX-50 (1 mg sample, $\beta = 10$ K/min) (left) (Muravyev et al. 2017) and isothermal TGA mass loss curves over the temperature range 443–463 K for TKX-50 decomposition under Ar flow (0.1 MPa) (right, circles are values obtained from experimental data whereas the solid lines are from formal kinetic modeling) (Muravyev et al. 2017). Reproduced with kind permission from the publisher from Muravyev et al. (2017)

Also in the DSC/TG reported by *Muravyev* et al., crucially there was no endothermic peak observed indicative of a phase transition below the decomposition temperature of TKX-50 (Fig. 18) (Muravyev et al. 2017). Furthermore, isothermal TGA mass loss curves for the decomposition of TKX-50 performed under a flow or Argon (0.1 MPa) in the temperature range 443–463 K showed that the average value for the total mass loss is 49% independent of the temperature (Fig. 18) (Muravyev et al. 2017). In addition, *Muravyev* et al. concluded that the average heat effect of decomposition for TKX-50 (Al pans covered with pierced lids) is 2320 ± 190 J/g which is higher than the values obtained for Octogen (β-HMX, 1600 J/g), Hexogen (RDX, 2100 J/g) and slightly lower than that of CL-20 (2600 J/g) (Muravyev et al. 2017).

The decomposition temperature of TKX-50 obtained using no-isothermal DSC shows that the decomposition temperature moves to higher temperature on increasing the heating rate (Fig. 19) (Muravyev et al. 2017). The exotherm reported at 245.72 °C using a heating rate of 10 °C/min (Huang et al. 2015a, b) is in agreement with the exotherm observed at approximately 523 K using heating rate of 10 K/min in Fig. 19 (Muravyev et al. 2017), and the value of 221 °C which was observed using a heating rate of 5 °C/min in the original report on TKX-50 (Fischer et al. 2012) is however

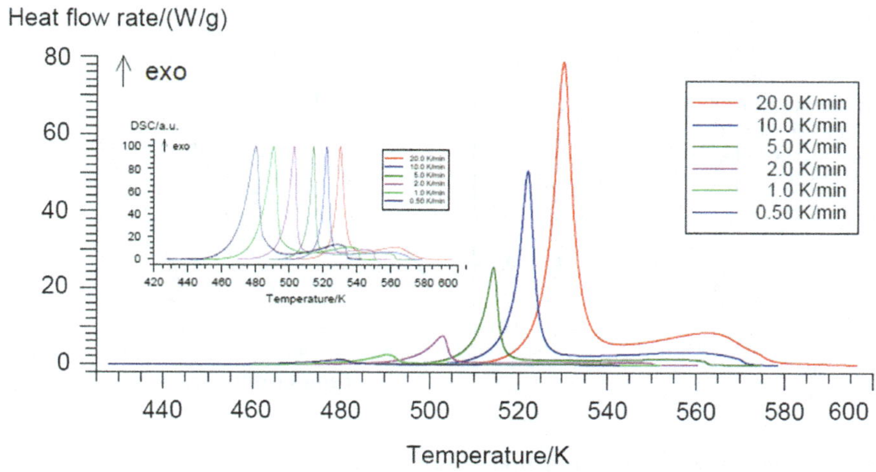

Fig. 19 Raw non-isothermal DSC curves showing the two-stage decomposition of TKX at different heating rates (normalized DSC curves shown in inset) (Muravyev et al. 2017). Reproduced with kind permission from the publisher from Muravyev et al. (2017)

lower than the value of approximately 241 °C which is shown in Fig. 19 (Fischer et al. 2012).

The ARC technique was used to investigate the thermal stability of TKX-50 by determining the dependency of self-heating rates on the temperature (Muravyev et al. 2017). The onset self-heating temperature T_0 of TKX-50 was determined to be 454 K (Muravyev et al. 2017), and TKX-50 was found to undergo self-heating under the adiabatic conditions used until thermal explosion occurred (Muravyev et al. 2017). The dependency of self-heating rates on the temperature were also determined for ADN, CL-20, RDX, HMX, Fox-7 and TNT for comparison (Muravyev et al. 2017). The self-heating onset for ADN at ~370 K showed it to be the thermally least stable, with the most thermally stable group being HMX, Fox-7 and TNT (onset temperatures at ~490 K) with CL-20, RDX and TKX-50 showing thermal stability in the middle range with onset temperatures of ~450 to 465 K (Fig. 20) (Muravyev et al. 2017). These results indicate that using this method, the onset temperature of ~450 K for TKX-50 indicates a similar thermal stability for TKX-50 as that of CL-20, but slightly lower than that of RDX and ~45 K lower than that of HMX (Fig. 20) (Muravyev et al. 2017), in contrast to the original findings in DSC data (Fischer et al. 2012). However, again it must be pointed out, that usually in the literature, for energetic materials the decomposition temperature is usually based on DSC/TGA data, and therefore, in comparing the onset temperature for TKX-50 from ARC experiments with the thermal stability of other energetic materials, the ARC onset temperatures for the other energetic materials must also then be established and used for comparison.

Very recent work reporting on the compatibility of TNT/TKX-50 with respect to possible use as melt-castable mixtures was also undertaken, and the compatibility

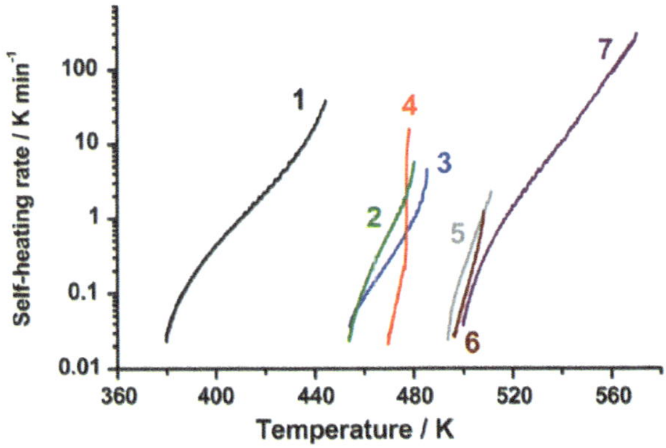

Fig. 20 Results of ARC experiments showing the self-heating rate dependencies on temperature for several energetic materials, including TKX-50: 1 = ADN, 2 = TKX-50, 3 = CL-20, 4 = RDX, 5 = HMX, 6 = Fox-7, 7 = TNT (Muravyev et al. 2017). Reproduced with kind permission from the publisher from Muravyev et al. (2017)

data are summarized in a later section. However, before a comparison could be made of the compatibility the thermal properties of each of the individual pure components (i.e., TNT and TKX-50) has to first be established (Li et al. 2019b). The DSC of TKX-50 (β = 10 °C/min) showed two exothermic peaks at 251 °C and 308 °C which correspond to the first and second stage of decomposition respectively (Li et al. 2019b). Of particular importance is the exothermic peak corresponding to the first stage of decomposition which is shown in Fig. 21 to shift to lower temperature on decreasing the heating rate, β in good agreement with the results shown in Fig. 19 (Li et al. 2019b). Reports on the thermal stability of TKX-50 continue—particularly using DSC/TGA techniques (Talawar et al. 2018). A recent report stated that from the TG profile of TKX50, it could be concluded that the decomposition of TKX-50 is a two-step process with the major weight loss of 73.50% occurring in the second stage of decomposition in the temperature range of 195–330 °C, whereas only a minor weight loss of 13.58% occurred in the first stage of decomposition in the temperature range 130–195 °C at a heating rate of 10 °C/min (Fig. 22) (Talawar et al.

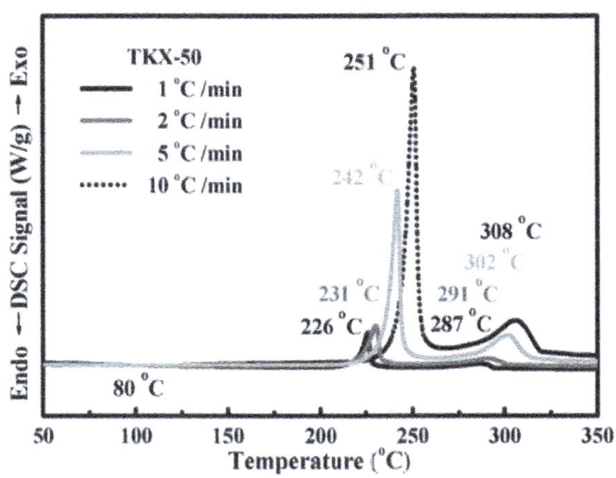

Fig. 21 DSC curves showing the two-stage decomposition of TKX at different heating rates (Li et al. 2019b). Reproduced with kind permission from the publisher from Li et al. (2019b)

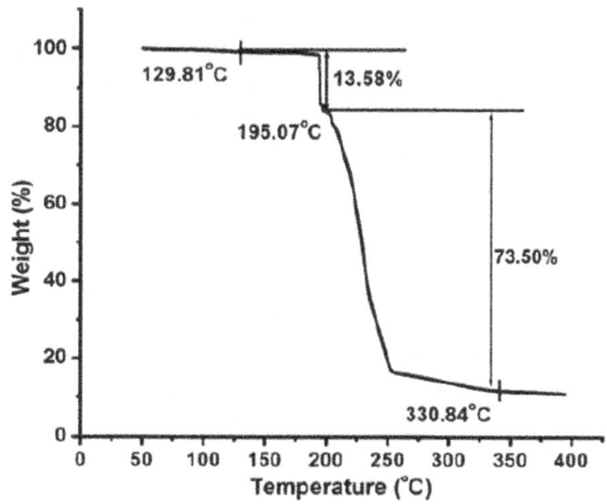

Fig. 22 TGA curve (weight vs. temperature) of TKX-50 ($\beta = 10$ °C/min) showing the first stage of decomposition and minor mass loss of 13.58% in the temperature range 130–195 °C and the second stage of decomposition with major weight loss of 73.50% in the temperature range 195–330 °C Talawar et al. (2018). Reproduced with kind permission from the publisher from Talawar et al. (2018)

2018). The DSC curve of TKX-50 was reported by the same authors to show an exothermic decomposition at 247.8 °C, without prior melting (Talawar et al. 2018).

The non-isothermal decomposition of TKX-50 was investigated using TG-DTG and TG-IR-MS, and the main gases decomposition products were found to be N_2, N_2O, NH_3 and H_2O, with traces of HCNO and HCN being detected in microscale quantities (Niu et al. 2016). Again, the TG/DTG curves indicated the first mass loss of 66.10% occurs at 225.10 °C and ended at 247.83 °C, with a peak maximum in the DTG curve at 235.50 °C ($\beta = 10$ °C/min) (Niu et al. 2016). The second stage decomposition was reported under the same conditions at 247.83 °C ending at 272.66 °C with a mass loss of 26.97% and peak maximum in the DTG curve at 256.33 °C (Niu et al. 2016). The DSC curves agreed with these findings showing two exothermic processes occurring at 238.0 °C and 288.3 °C ($\beta = 10$ °C/min) with the weak peak maximum at 288.3 °C being attributed by the authors to the decomposition of residual TKX-50 (Niu et al. 2016). The authors calculated the activation energy to decomposition for TKX-50 using a number of different methods which gave the following results: $E_a = 142.2$ kJ/mol (Kissinger), 143.07 kJ/mol (Ozawa), 135.98 kJ/mol (Maccallum-Tanner, $\beta = 2$ °C/min), 136.5 kJ/mol (Šatava-Šesták, $\beta = 2$ °C/min), 135.49 kJ/mol (Agrawal, $\beta = 2$ °C/min) (Niu et al. 2016).

In order to further assess the thermal stability of TKX-50, the SADT simulation of 25 kg of TKX-50 was performed in which the TKX-50 was packaged in a polyethylene can which gave a value of 129.01 °C for TKX-50 and a separate calculation of the TMR_{ad} of TKX-50 gave a value of 142.12 °C, which the authors concluded indicated that TKX-50 shows good thermal properties (Niu et al. 2016).

Using in situ X-ray diffraction, it has been shown that below 175 °C, the main process which TKX-50 undergoes on heating is anisotropic thermal expansion which consequently results in a change in the intermolecular forces and expansion, but also in the flipping of some of the molecules (Jia et al. 2017).

In the anisotropic thermal expansion (which was investigated using in situ non-isothermal X-ray diffraction), a shorter cation–anion distance results as a consequence of the negative thermal expansion along the *a* axis of the unit cell, meaning that proton transfer from the cation to the anion is induced (Jia et al. 2017). In addition, the β angle was also observed to decrease (−0.27% from 30 to 175 °C). In contrast, the *b* axis increased by 0.96%, the *c* axis by 0.50% and the unit cell volume, *V*, by 1.42%, which corresponded to positive thermal expansion (Jia et al. 2017).

On increasing the temperature further, isothermal in situ X-ray diffraction was used to determine that the first-stage decomposition occurs producing the decomposition product ABTOX, which has been identified in several reports as a first-stage decomposition product of TKX-50 (Jia et al. 2017). The second-stage decomposition was observed to occur at approximately 280 °C using TG-DSC (Jia et al. 2017). The extent of conversion of TKX-50 to the intermediate product during the stage 1 decomposition is given in Fig. 23, as determined by in situ isothermal X-ray diffraction after grinding the sample (Jia et al. 2017).

Finally, hot stage microscopy was used to investigate the effect of thermal expansion followed by first-stage decomposition on the crystal (Jia et al. 2017). Morphological changes were observed using this technique, which indicated that between 30 and 175 °C the morphology was unaffected, whereas at 198 °C the crystal began to slowly change in appearance with cracks appearing and liquid and bubbles being produced (Jia et al. 2017). This was attributed to the partial decomposition of the crystal which was largely complete after 150 min indicated by the crystal turning from transparent (single crystal) to opaque (polycrystalline) and that ABTOX had been formed during this primary decomposition stage of TKX-50 (Jia et al. 2017). Based on the results of their investigations, the authors proposed a thermal decomposition mechanism which agreed well with those published previously and is illustrated in Fig. 24 (Jia et al. 2017). The authors point out that NH_2OH is highly unstable at room temperature and the NH_2OH which is formed during decomposition continues

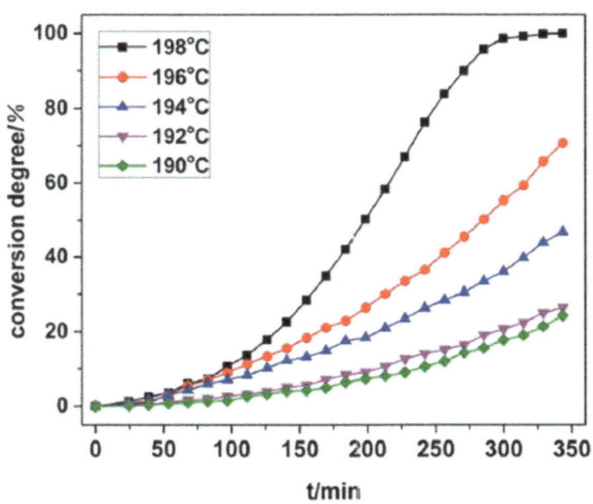

Fig. 23 Degree of conversion of TKX-50 to the intermediate decomposition product (stage 1 decomposition) at different isothermal temperatures (Jia et al. 2017). Reproduced with kind permission from the publisher from Jia et al. (2017)

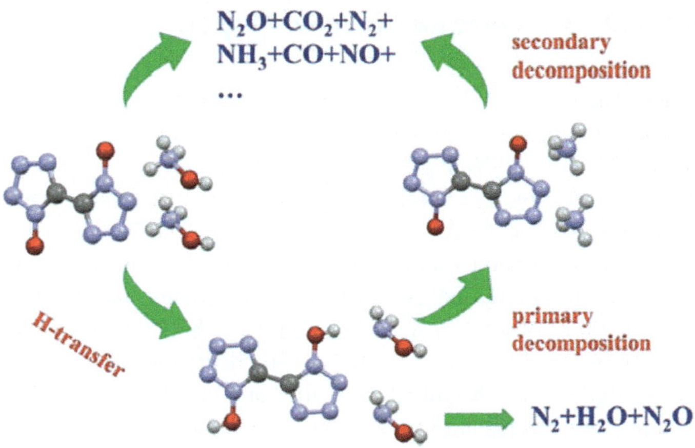

Fig. 24 Proposed thermal decomposition mechanism of TKX-50 (Jia et al. 2017). Reproduced with kind permission from the publisher from Jia et al. (2017)

to decompose forming NH_3, H_2O, and other decomposition products at elevated temperature (Jia et al. 2017). Using FT-IR spectroscopy, the authors detected the formation of large amounts of H_2O at 198 °C (Jia et al. 2017).

Overall, the results of these investigations showed that between 30 and 175 °C only anisotropic thermal expansion occurs in which negative thermal expansion is shown by the *a* axis and β angle, whereas the *b* axis, *c* axis, and unit cell volume show positive thermal expansion, which leads to a change in the intermolecular forces which are present and the flipping of some molecules (Jia et al. 2017). Due to NTE along the *a* axis, the interionic distance is decreased which results in a more facile hydrogen transfer being possible which consequently facilitates the first-stage decomposition of TKX-50 in which proton transfer has been proposed by several reports in the literature as the first stage of the thermal decomposition of TKX-50 (Jia et al. 2017). The formation of ABTOX as the primary decomposition product in the first-stage decomposition of TKX-50 was also confirmed, with secondary decomposition not occurring before 270 °C (Jia et al. 2017). The enthalpy of fusion has been calculated empirically recently to be 38.45 kJ/mol using the following equation; however, no melting point has been reported for TKX-50, and it is proposed to decompose in the solid state (Keshavarz et al. 2018):

$$\Delta H_{\text{fus}} = 0.542a + 1.490b + 2.044c + 1.252d + 1.839e + 9.848\Delta H_{\text{Inc,fus}} - 675\Delta H_{\text{Dec,fus}}$$

2.3 Solid-State Structure and Related Properties

The solid-state structure of TKX-50 was determined using low temperature (100 K, 173 K) and room temperature (298 K) single-crystal X-ray diffraction (crystallographic data at 298 K: $C_2H_8N_{10}O_4$, Fwt. = 236.18, monoclinic, $P2_1/c$, crystal size = 0.10 × 0.25 × 0.40 mm, a = 5.4408(6) Å, b = 6.5612(9) Å, c = 6.5612(9) Å, β = 95.071(11)°, V = 417.86(9) Å^3, $\rho_{calc.}$ = 1.877 g cm^{-3}) (Fischer et al. 2012). At all three temperatures, the same monoclinic $P2_1/c$ phase was observed (Fischer et al. 2012). The cell volume increased as expected on increasing the temperature (V = 408.97(10) Å^3 at 100 K, 409.58(7) Å^3 at 173 K and 417.86(9) Å^3 at 298 K; Z = 2 for all three temperatures), while the calculated density decreased from 1.918 g cm^{-3} at 100 K to 1.915 g cm^{-3} at 173 K and 1.877 g cm^{-3} at 298 K (Fischer et al. 2012). The unit cell contains two anions and four cations. The value for the molecular volume (V) is also highly useful for calculation of the lattice energy of an energetic salt and consequently the standard enthalpy of formation, $\Delta H°_f$ which is discussed in more detail in Sect. 3.2. If the low temperature densities obtained from X-ray diffraction are corrected to give the room temperature value—and consequently, crucially the TMD—using the equation given below and compared with the density value from X-ray diffraction at room temperature (Table 4), then although the conversion gives an approximate value, conversion of the value obtained using single-crystal X-ray crystallography at 100 K is too low and that at 170 K is only marginally too high in comparison with the single-crystal X-ray diffraction value at 298 K. The similarity of the converted low-temperature single-crystal values with the 298 K value adds support to the TMD of TKX-50 being 1.877 g cm^{-3} which is high. The density is not a trivial value with respect to energetic materials, since the detonation velocity is proportional to the density, and therefore, the density has a more marked influence on the detonation performance of an explosive than does the heat of formation.

$$d_{298\,K} = \frac{d_T}{1 + \alpha_V(298 - To)} \quad \text{where } \alpha_V = 1.5\,x\,10^{-4}K^{-1}$$

The solid-state structure of TKX-50 was determined using single-crystal X-ray diffraction at low temperature (100 K) and is shown in Fig. 25 (Fischer et al.

Table 4 Comparison of the room temperature density of TKX-50 obtained using single-crystal X-ray diffraction (TMD) and by converting the densities obtained using low-temperature X-ray diffraction to the room temperature (298 K) value using the equation above

Value	Density (g cm^{-3})
100 K, single crystal X-ray diffraction	1.918
170 K, single crystal X-ray diffraction	1.915
298 K, single crystal X-ray diffraction (TMD)	1.877
100 K value converted to value at 298 K using eqn. given above	1.863
170 K value converted to value at 298 K using eqn. given above	1.879

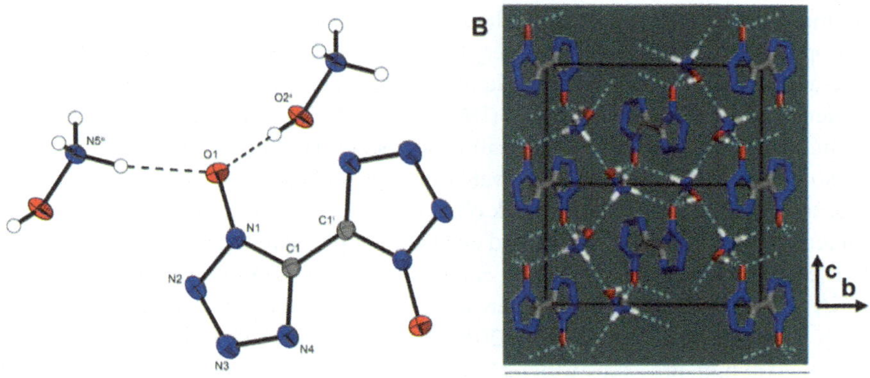

Fig. 25 Solid-state molecular structure of TKX-50 as determined using single-crystal X-ray diffraction at 100 K. The formation of interionic hydrogen bonds between the NH_3OH^+ cations (donors) and O atom of the N-oxide functionality of the $C_2N_8O_2{}^{2-}$ dianion are shown (left) (Fischer et al. 2012) and a view of the extensive network of 3-D hydrogen bonding in the unit cell (right) (Dreger et al. 2015). Reproduced with kind permission from the publisher from Dreger et al. (2015)

2012). Despite the presence of strong hydrogen bonds between the cations and anion, NH_3OH^+ cations and $C_2N_8O_2{}^{2-}$ anions are clearly present. The extensive 3-D hydrogen bond network which results is crucial in the stabilization of TKX-50. The 5,5′-bis(tetrazolate-1 *N*-oxide) anion shows the two tetrazole rings to be linked via a C–C bond and a *trans* orientation of the two *N*-oxide groups. Each O atom of the *N*-oxide group forms two principal hydrogen bonds and acts in both cases as the acceptor: One involves the H atom of the –OH group of the hydroxylammonium cation, whereas the other involves the H atom of the $-NH_3$ group of a second hydroxylammonium cation (Fig. 25) (Fischer et al. 2012). In total, however, there are eight different types of hydrogen bonds observed in the solid-state structure of TKX-50 at 298 K: three N–H···N (weak), three N–H···O, one O–H···N, and one O–H···O (strong) type [25, 31]. The N–N bond lengths range from 1.3128(17) to 1.3513(18) Å, which are considerably shorter than the N–N single bond length in hydrazine (1.451 Å) and markedly shorter than the N≡N bond length in N_2(1.095 Å), indicating a bond order of >1 but <3 for all three ring N–N bonds and the presence of delocalization in the ring. The C–C bond length of 1.445(3) Å is shorter than the C–C single bond in ethane (d(C–C) (g) = 1.54 Å). Interestingly, the density of the neutral 1,1′-BTO · $2H_2O$ is only 1.811 gcm^{-3} and thereby distinctly lower than that of TKX-50, despite its extensive hydrogen bond network (Fischer et al. 2012).

In work by *Dreger, Gupta, and co-workers*, the large difference in the hydrogen bond strengths of the N–H···N (weak) and O–H···O (strong) instigated investigations into the effect high-pressure compression has on TKX-50 single crystals, since they proposed that this large energy difference could result in anisotropy in the intermolecular interactions in TKX-50 (Dreger et al. 2015). To undertake this work, single crystals of TKX-50 were exposed to high pressures generated using a diamond anvil cell (DAC) and Raman spectroscopy was performed on the samples under high pressure and the results compared with results from DFT calculations. Using Raman spectroscopy and exposing TKX-50 to very high pressures in a diamond anvil cell, it was concluded that TKX-50 doesn't undergo a phase transition from pressures between ambient and 32 GPa (see Sect. 2.2) (Dreger et al. 2015). It was also elucidated that the single-crystal structure of TKX-50 which is observed under ambient conditions is also observed even at greatly elevated pressures. This has been attributed to the extensive hydrogen bonding network observed in the crystal structure of TKX-50 at ambient conditions being maintained even at high pressures (Dreger et al. 2015).

The effect of pressure on the unit cell parameters of TKX-50 was also calculated (GGA (PBE-D) method) for pressures up to 30 GPa, and no change in the space group was observed (Dreger et al. 2015), in agreement with Raman spectroscopy performed at high pressures which also suggested the absence of a phase change up to pressures of 32 GPa (Dreger et al. 2015). The calculated pressure-induced changes in the unit cell parameters of TKX-50 show that the compressibility of the TKX-50 single crystal to be highly anisotropic with extremely low compressibility along the *a* axis ($a/a_0 = 97\%$, $b/b_0 = 83\%$, $c/c_0 = 88\%$) (Dreger et al. 2015), which has been attributed to the shorter intermolecular distances along the *a* axis suggesting stronger interactions along the *a* axis and the strongest hydrogen bonds between cations and anions having components along the *a* axis (Dreger et al. 2015). Finally, calculation of the unit cell volume at ambient and high pressure (30 GPa) shows that only a 70% reduction in size occurs which is less than that usually observed for other secondary explosives, but in a scenario in which temperature has an influential role in the decomposition, then since the temperature increase under shock compression is related to the volume change, and the lower the compressibility of a single crystal the lower the temperature that will result, then the calculated low compressibility of TKX-50 would result in a smaller increase in the temperature on shock compression (in comparison with other secondary explosives), meaning that TKX-50 should initiate at higher stress—which is in good agreement with its marked low impact sensitivity (Dreger et al. 2015). The compressibility, κ_T (under isothermal conditions) is the relative volume change in response to pressure and can be obtained from molecular dynamics simulations. Using this approach, a value of $3.863 \times 10^{-1}1$ m^3/J at room temperature was predicted for TKX-50 using molecular dynamics (An et al. 2015), and a value of 3.167×10^{-11} m^3/J at 0 K was predicted using quantum mechanical methods (An et al. 2015).

The crystal structure was also calculated using the PBE-ulg approach ($a = 5.116$ Å, $b = 12.260$ Å, $c = 6.775$ Å, $\beta = 95.9°$) (An et al. 2015) and were shown to be in good agreement with those determined using single-crystal X-ray diffraction at 298 K ($a = 5.4408(6)$ Å, $b = 6.5612(9)$ Å, $c = 6.5612(9)$ Å, $\beta = 95.071(11)°$) (Fischer et al. 2012). A comparison of various cell parameters for TKX-50 from single-crystal X-ray diffraction at 100, 173, and 298 K as well as from computational work is given in Table 5.

Furthermore, the temperature dependency of the unit cell parameters of TKX-50 was calculated using the FF approach in the temperature range 100–400 K (Fig. 26) and resulted in a calculated coefficient of thermal expansion, α_p of 64.8×10^{-6} K^{-1} which is approximately three times less than that of RDX (An et al. 2015),

Table 5 Comparison of the experimentally determined (single-crystal X–ray diffraction) and calculated unit cell parameters for TKX-50 under different conditions (space group in all cases is $P2_1/c$ except where indicated otherwise)

Method	*a* (Å)	*b* (Å)	*c* (Å)	α (°)	β (°)	γ (°)	*V* (Å^3)	$\rho_{calc.}$ (g cm^{-3})
Single crystal X-ray diffraction at 100 K Fischer et al. (2012)	5.4872(8)	11.5472(15)	6.4833(9)	90	95.402(12)	90	408.97(10)	1.918
Single crystal X-ray diffraction at 173 K Fischer et al. (2012)	5.4260(5)	11.6597(12)	6.5013(7)	90	95.256(9)	90	409.58(7)	1.915
Single crystal X-ray diffraction at 298 K Fischer et al. (2012)	5.4408(6)	11.7514(13)	6.5612(9)	90	95.071(11)	90	417.86(9)	1.877
Calcd., PBE-ulg (An et al. 2015)	5.116	12.260	6.775	90	95.9	90		
Calcd., FF at room temperature (An et al. 2015)	5.25(2)	12.11(4)	6.60(1)	90	93.5	90	418(2)	1.875(8)
Calcd., PBE+D2 (VASP, PAW) Lu et al. (2017a, b, Meng et al. (2016)	5.5589	11.3567	6.4025	90	96.293	90	401.76	1.952
Calcd., OptB86-vdw (VASP, PAW) Lu et al. (2017a, b)	5.5262	11.3830	6.4228	90	96.089	90	401.74	1.952
Single crystal X-ray diffraction at 20 K Tidey et al. (2017)	5.49040(10)	11.4912(3)	6.45560(10)	90	95.5710(16)	90	405.368(14)	1.935
Calcd., DFT D2 (Lu et al. 2017a, b)	5.5589	11.3567	6.4025	90	96.293	90	401.76	
Meta-TKX-50								
Calcd., DFT-D2, Triclinic, *P*-1 (Lu et al. 2017a, b)	5.2696	11.9230	6.5964	89.366	91.859	87.4993	413.79	
Calcd., OptB86-vdw, Triclinic, *P*-1 Lu et al. (2017a, b)	5.2809	11.8956	6.6040	89.360	91.727	87.429	414.22	

The calculated unit cell parameters for Meta-TKX-50 are also included

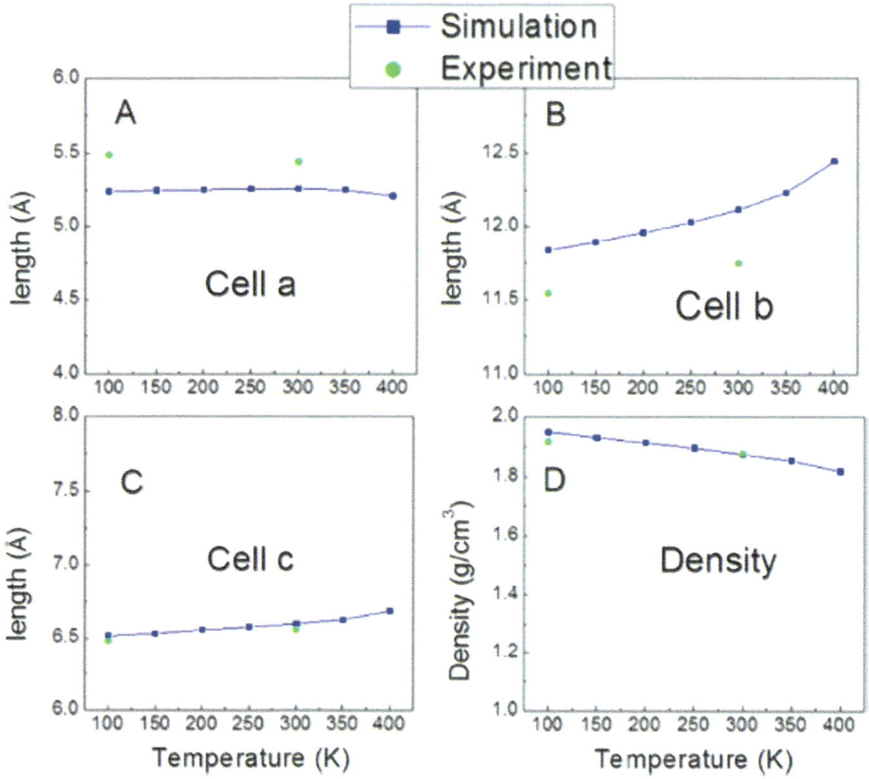

Fig. 26 Calculated temperature dependence of the unit cell parameters (lengths *a*, *b*, and *c* as well as the density) of TKX-50 in the temperature range 100–400 K simulated using molecular dynamics (An et al. 2015). Reproduced with kind permission from the publisher from An et al. (2015)

due to the fact that TKX-50 is an ionic compound which results in less expansion than is observed for a neutral molecule such as RDX (An et al. 2015). It can be concluded from Fig. 26 that as the temperature increases, the value of length *a* remains relatively unchanged, and only a small increase in length *c* occurs, whereas length *b* increases clearly considerably more with increasing temperature, demonstrating the anisotropic expansion behavior of TKX-50 (An et al. 2015).

The *trans* orientation of the two *N*-oxide groups which was observed in the single-crystal X-ray diffraction study (Fig. 25) (Fischer et al. 2012) was also mirrored in computational work for the neutral $C_2H_2N_8O_2$ molecule. Calculation of the N–C–C–N torsion barrier using both force field and quantum mechanical methods for the neutral diol molecule also predicted the *trans* arrangement to be lower in energy than the *cis* arrangement of the N–OH groups (Fig. 27) (An et al. 2015). An interesting experimental charge density study of TKX-50 was reported in which the intra- and intermolecular bonding was scrutinized in more detail using charge density data obtained using high-resolution X-ray diffraction data obtained at the very low

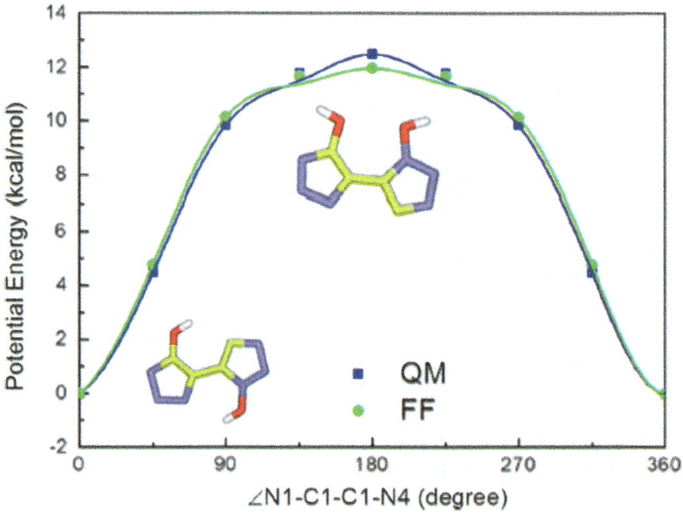

Fig. 27 Computational investigations of the N–C–C–N torsional barrier in the neutral diol molecule (protonated TKX-50 anion) shows that the *trans* arrangement of the *N*-oxide groups observed in the crystal structure of TKX-50 is mirrored by a *trans* arrangement of the N–OH groups in the diol molecule. Both FF and QM approaches were used (An et al. 2015). Reproduced with kind permission from the publisher from An et al. (2015)

temperature of 20 K, which was then compared with computational results (Tidey et al. 2017). Scrutiny of the hydrogen bonding which is present in TKX-50 is important, since strong hydrogen bonding motifs are generally found in energetic materials which have high densities, and it was asserted that if strong hydrogen bonding motifs could be established in energetic materials, that this would decrease the sensitivity of energetic materials to external stimuli (Tidey et al. 2017). Since TKX-50 shows less sensitivity toward friction and impact than other energetic materials such as RDX or CL-20, it is clearly of interest to then examine how the hydrogen bonding in TKX-50 has an effect on its sensitivity. TKX-50 is in this sense also interesting, since the layered motifs formed by arrangements of stronger and weaker hydrogen bonds should enable sliding to dissipate the mechanical energy from external stimuli thereby enhancing the mechanical stability of TKX-50; however, hydrogen bonding in TKX-50 is also expected to decrease its thermal stability, since the first stage of thermal decomposition of TKX-50 involved proton transfer from the cation to the anion! Another important feature of TKX-50 with respect to its high mechanical stability is the face-to-face $\pi \cdots \pi$ stacking, which decreases the formation of hotspots (Tidey et al. 2017).

The solid-state structure of TKX-50 at 20 K was recently determined using single-crystal X-ray diffraction (Tidey et al. 2017). Furthermore, the authors analyzed TKX-50 based on its experimentally determined charge density distribution from X-ray diffraction at 20 K, in order to provide a more detailed description of the structure (Tidey et al. 2017). The unit cell parameters of TKX-50 from this study at 20 K (monoclinic, space group $P2_1/c$) are included in Table 5 (Tidey et al. 2017). A topological analysis of the electron density of TKX-50 shows that the shortest and strongest interionic hydrogen bond is of the O–H····H type, and that the crystal packing is heavily determined by the cation–cation and cation–anion hydrogen bonds, in addition to face-to-face π····π between anions (Tidey et al. 2017). There are two bonding interactions between face-to-face interacting tetrazole rings which results in a total of 11.0 kJ/mol (Tidey et al. 2017). There are in addition two inequivalent interlayer interactions between $C_2N_8O_2^{2-}$ anions which form a zig-zag arrangement which results in interactions between a dianion and its two nearest neighbors in each adjacent layer (Tidey et al. 2017). The three strongest hydrogen bond interactions involve the hydrogen atoms of the –OH group of the hydroxylammonium cation and the oxygen of the anion *N*-oxide group (Tidey et al. 2017). The fourth strongest hydrogen bond forms the strongest interlayer interaction and occurs between the oxygen atom of one hydroylammonium cation with the hydrogen of an $-NH_3^+$ group of a second hydroxylammonium cation (Tidey et al. 2017). The hydrogen bonds involving the N atom of the *N*-oxide group described in previous reports were not found in this work (Tidey et al. 2017). An analysis of the static deformation density maps for the TKX-50 cations and anions (Fig. 28) (Tidey et al. 2017) gives topological $N_{(tetrazole)}$–$N_{(tetrazole)}$ bond values of 1.39, 1.37, and 1.40 which are slightly higher than the value for $N_{(tetrazole)}$–$C_{(tetrazole)}$ ($\eta_{topo} = 1.12, 1.22$), and higher than the η_{topo} value of 1.12 for the C–C linking the two tetrazole groups, which indicates a non-negligible degree of conjugation for this bond which is consistent with the planarity of the N–C–C–N moiety (0.3°) (Tidey et al. 2017). The $N_{(tetrazole)}$–O of the *N*-oxide group however shows significant multiple bond character ($\eta_{topo} = 1.42$), although the oxygen doesn't participate in classical π-delocalization in the system and a σ-hole is located at the oxygen as expected (single bonded sp^3 center) from deformation density (Tidey et al. 2017). Computational work has shown that the >1 bond order for the N–O bond has been shown in the gas phase to result from several LP → σ* hyperconjugation interactions between the O atom of the *N*-oxide group and the tetrazole ring (Tidey et al. 2017). Furthermore, BCP has indicated that the N–O bond of the *N*-oxide group is strongly polar in contrast to the bond within the tetrazole rings (Tidey et al. 2017).

An analysis of the atomic charges of TKX-50 was performed by integrating the total electron density over the atomic basins as defined by the zero-flux surfaces (Tidey et al. 2017). The total charges for the cation were +0.807 e and for the dianion −1.610 e (Tidey et al. 2017). The N atoms of the tetrazole rings all showed minor negative charges (~−0.1 e), except the N4 atom (−0.637 e) (N–C–N–O connectivity) which is similar to the high negative charge of the oxygen atom of the *N*-oxide group (−0.661 e), whereas the C atom of the tetrazole rings showed a high positive charge (+0.870) which was concluded to indicate a significant level of charge partitioning occurring within the tetrazole ring (Tidey et al. 2017). Despite the significant partitioning of the atomic charges, in the electrostatic potential surface, this large charge partitioning is not reflected in highly positive or negative regions and instead a relatively featureless EPS is obtained for TKX-50 (Fig. 29) (Tidey et al. 2017).

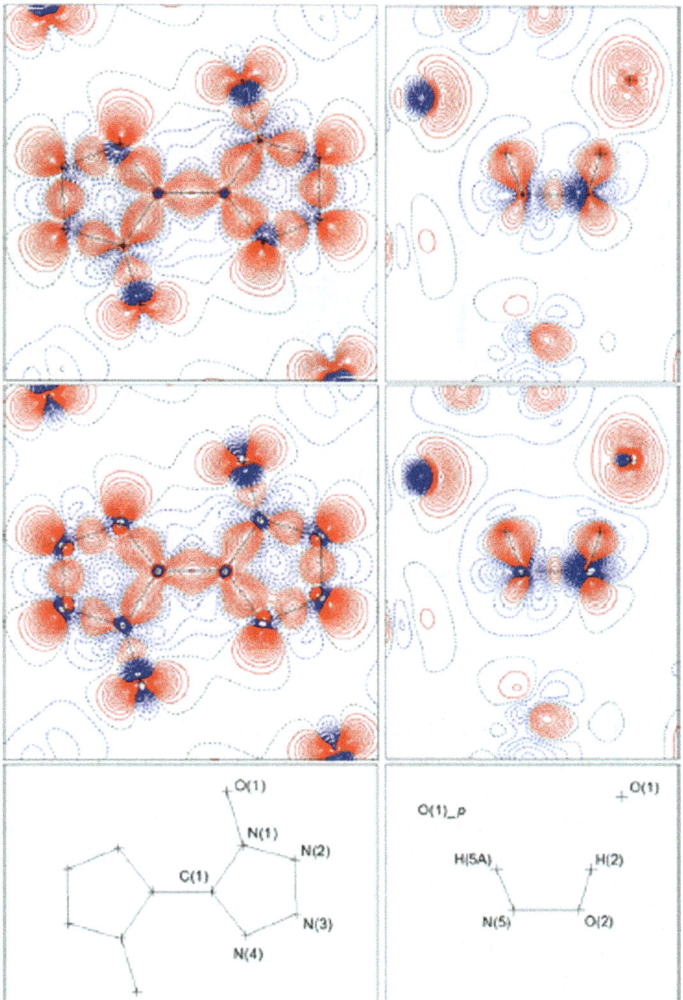

Fig. 28 Static deformation density maps of the anions (right) and cations (left) which make up TKX-50: MPR_{anh} (top), MPR_{theor} (middle) and atomic numbering scheme (bottom). Contours are given at 0.05 e $Å^{-3}$ intervals. Positive deformation density is indicated in red, whereas negative is indicated in blue and the zero contour in black (dashed) (Tidey et al. 2017). Reproduced with kind permission from the publisher from Tidey et al. (2017)

The structural response of TKX-50 crystals to high pressure was also investigated using synchroton single-crystal X-ray diffraction measurements at pressures of approximately 10 GPa (Dreger et al. 2017a, b). Interestingly, it could be shown that not only does TKX-50 show highly anisotropic compression, but lower volume compressibility than other common energetic materials show (Dreger et al. 2017a, b). Additionally, negative linear compression was observed along the *a* axis on exposing

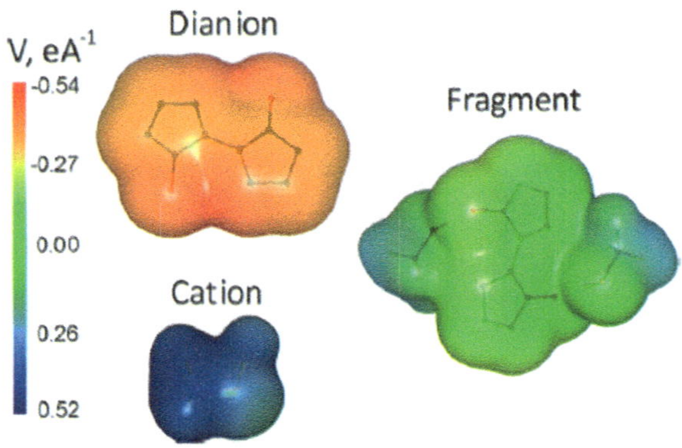

Fig. 29 Color-map images of MPR_{anh} of the electrostatic potential drawn on the 0.001 au isosurface for the $C_2N_8O_2^{2-}$ dianion NH_3OH^+ cation and complete molecular fragment of TKX-50 (Tidey et al. 2017). Reproduced with kind permission from the publisher from Tidey et al. (2017)

a single crystal of TKX-50 to 8 GPa pressure and is a consequence of the strong interionic hydrogen bonds (Dreger et al. 2017a, b). The strong and highly anisotropic hydrogen bonds are concluded to control the structural stability and also possibly the shock sensitivity of TKX-50 (Dreger et al. 2017a, b). These experimental observations were in good agreement the authors concluded with their previous predictions based on DFT calculations. In more detail, the pressure dependence of the unit cell parameters is shown in Fig. 30, from which it can be concluded that while the monoclinic $P2_1/c$ space group remains unchanged up to 10.3 GPa, the lattice parameters demonstrate highly anisotropic compressibility (Dreger et al. 2017a, b). For example, at 10 GPa, the *a* axis showed an increase of 0.2%, whereas the *b* and *c* axes showed a decrease of 12% and 8%, respectively (Dreger et al. 2017a, b). Interestingly, however, the situation is slightly more complicated, since up to 3 GPa the a axis increase by 0.8%; however, above 3 GPa up to 10 GPa, a contraction is then observed meaning that the *a* axis exhibits anomalous compressibility, i.e., negative up to 3 GPa and positive above 4 GPa (Dreger et al. 2017a, b). This behavior has been attributed by the authors to be a result of the different pressure response of intermolecular and intramolecular interactions within the TKX-50 crystal (Dreger et al. 2017a, b). Specifically, the authors state that increasing the pressure causes the distance between cations to significantly decrease due to the weak interlayer interactions. However, this in turn, causes a compression of two strong hydrogen bonds linking two anions with two cations) which in turn pushes the $C_2N_8O_2^{2-}$ anions back by the rigid hydrogen bonds resulting in an increase in the length of the *a* axis (Dreger et al. 2017a, b).

Using synchroton single-crystal X-ray diffraction, the authors described in detail the hydrogen bonding present in TKX-50 (Dreger et al. 2017a, b). The extensive hydrogen bonding in TKX-50 is hugely important in understanding the energetic properties of TKX-50. The authors state that eight hydrogen bonds are present of four different types: N–H····N, N–H····O, O–H····O, and O–H····N (Fig. 31) (Dreger et al. 2017a, b). The strongest hydrogen bonds are of the type O–H····O, N–H····O, whereas the two weakest are of the type O–H····N and N–H····N (Dreger et al. 2017a, b).

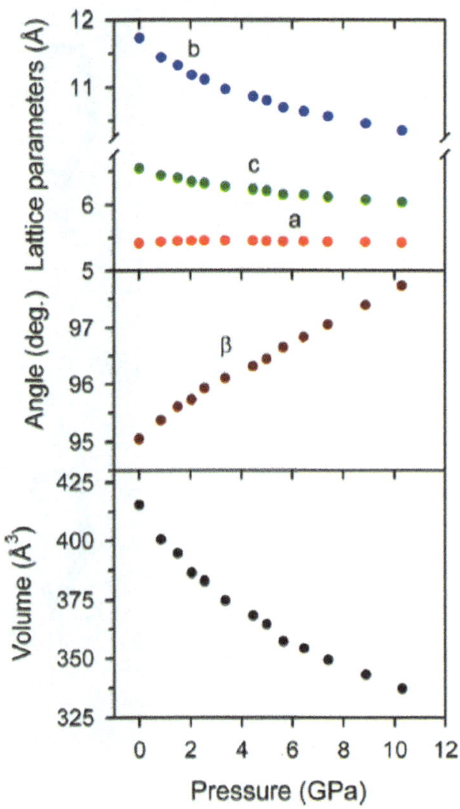

Fig. 30 Pressure-induced changes in the unit cell parameters of TKX-50 (Dreger et al. 2017a, b). Reproduced with kind permission from the publisher from Dreger et al. (2017a, b)

Under the influence of elevated pressure, the hydrogen bond motifs remain, however the hydrogen bonds all decrease in length and the decrease in length is larger for the strong hydrogen bonds than for the short hydrogen bonds (Dreger et al. 2017a, b). The angles, however, are less sensitive to pressure changes, and most angles remain essentially unchanged (Dreger et al. 2017a, b). All of the hydrogen bonds are expected to increase in strength on increasing the pressure except two: N····H–O (bond 2 in Fig. 31) which will show relatively little change and O····H–N (bond 4 in Fig. 31) which will decrease in strength on increasing the pressure (Dreger et al. 2017a, b). Even at higher pressures, the O–····H–N (bond 4 in Fig. 31) and O····H-O (bond 1 in Fig. 31) bonds remain the strongest (Dreger et al. 2017a, b). It is also bonds 1 and 4 (Fig. 31) that are discussed above as being responsible for the low and anomalous compressibility of TKX-50 along the *a* axis (Dreger et al. 2017a, b).

A further analysis of the hydrogen bonding and its effect on the stability of TKX-50 was undertaken in order to compare the two competing effects: (i) strong hydrogen bonds within layers and weak hydrogen bonds between layers resulting in low impact sensitivity and (ii) lower thermal stability due to strong hydrogen bonds facilitating

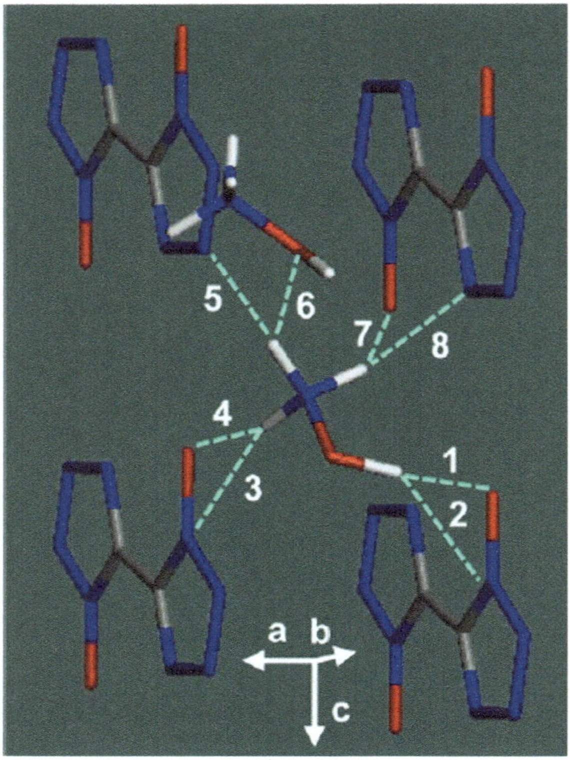

Fig. 31 There are eight hydrogen bonds in TKX-50 which can be categorized into four types: N–H····N, N–H····O, O–H····O, and O–H····N (Dreger et al. 2017a, b). Reproduced with kind permission from the publisher from Dreger et al. (2017a, b)

hydrogen transfer. In more detail, (i) the strong hydrogen bonds result in a layer-like arrangement of the cations and anion in the (010) plane whereas the adjacent layers are only connected by the two weakest types of hydrogen bonds present which makes it easier for external kinetic energy to be converted into sliding of the layers which has the consequence of TKX-50 exhibiting low impact sensitivity and (ii) the extensive and often strong hydrogen bonds which are present in TKX-50 facilitate hydrogen transfer between the cation and the anion which has been predicted by many different research groups using various methods to be the initial step in the first-stage decomposition of TKX-50 (Meng et al. 2016). In their interpretation of the hydrogen bonding present in TKX-50 using the QTAIM method to analyze the electron density obtained from ab initio calculations, they conclude that seven interionic hydrogen bonds are present of the following types: three N–H····O, three N–H····N, and one O–H····O hydrogen bond (Meng et al. 2016). Furthermore, the planar $C_2N_8O_2^{2-}$ anions are found to be face-to-face π-stacked along the [100] direction (Meng et al. 2016) and v-shaped π-stacked along the [010] direction. The hydroxylammonium cations are located between these anions (Meng et al. 2016). They asserted that hydrogen bonding was observed not only between cations and anions, but also between neighboring cations (Meng et al. 2016). Furthermore, one anion is surrounded by eight cations in which twelve hydrogen bonds are involved (Meng et al. 2016). They continue that the cations act both as hydrogen bond donors and acceptors and that all of the H atoms in the hydroxylammonium cations participate in the formation of hydrogen bonds (Meng et al. 2016). Importantly, the authors stated that the dissociation energies of the hydrogen bonds in TKX-50 with other common CHNO explosives shows that the interionic hydrogen bonds in TKX-50 are stronger and a more extensive interionic hydrogen bond network is present (Meng et al. 2016). The hydrogen bonding should enhance the stability since these bonds must be broken for reaction of the explosive to occur (Meng et al. 2016). The hydrogen bonds between planes are, however, much weaker, and only two hydrogen bonds connect the planes with dissociation energies of 8.5 and 8.4 kJ/mol (Meng et al. 2016). This means that the stronger intralayer hydrogen bonds should be able to remain intact when the interlayer hydrogen bonds are unable to sustain their integrity during, for example, interlayer sliding (Meng et al. 2016). Using computational methods, the authors were able to show that the sliding barrier of TKX-50 is higher than that of TATB, but lower than that of β-HMX, which is in agreement with the trend observed in these three explosives with respect to impact sensitivity (Meng et al. 2016). Attempts to improve the morphology and surface properties of TKX-50 crystals using carboxymethylcellulose sodium (CMC) which is an environmentally friendly (biocompatible, biodegradable, and non-toxic), anionic polyelectrolyte using the antisolvent method have very recently been reported (Dong et al. 2019). CMC is a cellulose derivative in which β-linked glucopyranose residues and carboxymethyl substituents are present and has found use as a binder for energetic materials (Dong et al. 2019). The presence of CMC can result in a phase transfer occurring on the interface of the aqueous CMC solution and an organic phase in which the CMC coprecipitates with the crystal which can result in a controlling of the morphology of the crystal and surface modification, including coating (Dong et al. 2019). SEM was used to show that TKX-50 granules with low aspect ratios could be prepared in the presence of CMC which was a significant change in the morphology compared to the rod-like crystals of raw TKX-50 and a change in the crystal size and distribution was also reported (Dong et al. 2019). The low aspect ratio of the formed crystals was attributed to promoted growth of the (011) and (110) planes in the presence of CMC (Dong et al. 2019). Furthermore, the authors used molecular dynamics simulations to explore morphological change that was observed (Dong et al. 2019). It was found experimentally, that the highest coating efficiency was obtained using a 4% CMC content (Dong et al. 2019). The use of non-isothermal kinetic studies showed that the thermal stability was also improved in comparison with the raw TKX-50 (Dong et al. 2019). For example, DSC curves of raw TKX-50 and various CMC/TKX-50 samples gave the following results ($\beta = 2$ °C/min): $T_{dec} = 223.69$ °C, $E_a = 163.98$ kJ/mol (Kissinger) $E_a = 164.02$ kJ/mol (Ozawa) raw TKX-50; $T_{dec} = 215.30$ °C, $E_a = 173.36$ kJ/mol (Kissinger) $E_a = 172.76$ kJ/mol (Ozawa) TKX-50/CMC (2%); $T_{dec} = 218.65$ °C, $E_a = 195.42$ kJ/mol (Kissinger) $E_a = 193.79$ kJ/mol (Ozawa) TKX-50/CMC (3%);$T_{dec} = 215.21$ °C, $E_a = 188.68$ kJ/mol (Kissinger) $E_a = 187.32$ kJ/mol (Ozawa) TKX-50/CMC (4%); $T_{dec} = 217.86$ °C, $E_a = 183.36$ kJ/mol (Kissinger) $E_a = 182.32$ kJ/mol (Ozawa) TKX-50/CMC (5%) (Dong et al. 2019). Not only has molecular dynamics been used extensively to predict the thermal decomposition mechanism of TKX-50, it has also been used to investigate the influence of the solvent on the morphology of TKX-50 crystals and therefore the interaction of TKX-50 with various solvents (Chen et al. 2019). Since crystallization occurs in two main stages, namely the nucleation of crystals and then the subsequent growth of the crystals and in solution, these two stages depends not only on the arrangement of the molecules inside the crystal lattice, but also on the external environment, obviously the solvent which is used is a highly important and influential factor in determining the crystallization process. The morphology of the resulting crystal is determined by the relative growth rates of the different crystal faces. Using computational approaches, it could be shown that the dominant growth surfaces with respect to the morphology for TKX-50 in vacuum were the (020), (100), (011), (110), (11-1), and (12-1), and the crystal morphology was that of a sheet shape with the six dominant faces listed above and an aspect ratio of 1.98 (Chen et al. 2019). Furthermore, it could be predicted using these investigations that ethylene glycol should be the solvent which promotes spheroidization of TKX-50 crystals (Chen et al. 2019). As is discussed in this review, crystal shape is important for energetic materials, since it can have a pronounced effect on the sensitivities to external stimuli such as IS and FS that an energetic material shows. Usually, the aim is to obtain energetic materials with a regular morphology are more likely to be less sensitive (Chen et al. 2019). Therefore, being able to obtain the desired crystal morphology for an energetic material is important. The site of growth of a crystal is the crystal surface and the solution with which it is in contact. The classical model to describe crystal growth proposes that solvent molecules diffuse from the solution to the growth interface and attach there to the surface of the crystal, meaning that the solutes and the solvent are both in competition to attach to this site and are therefore consequently in competition with respect to crystal growth. In order for the crystal to grow, solute molecules must attach to this growth interface; however in order to do so, first any solvent molecules attached to this crystal surface have to be removed. How this process proceeds will affect the growth of the crystal and therefore also the final morphology of the crystal (Chen et al. 2019). Therefore, the interaction between solvent molecules and the crystal surface at the growth interface is important for crystal morphology and is expressed by the interaction energy, E_{int} (Chen et al. 2019). The larger the value of E_{int}, the stronger the solvent interacts with the crystal surface (Chen et al. 2019) and the crystal growth rate will be slower on that crystal surface than on a different surface of the same crystal with a lower E_{int} value (Chen et al. 2019). The lower the E_{int} value, then the weaker the interaction is between a specific solvent and a specific crystal face, and therefore, it will be easier to remove solvent molecules from that crystal face and replace them with solute molecules, and therefore, crystal growth on that surface will be faster (Chen et al. 2019). For example, for TKX-50, it was calculated that with ethylene glycol solvent, the modified attachment energies E^*_{att} has a maximum value for the (11-1) face and a minimum value for the (100) face which are different than the E_{att} values predicted for the vacuum (Chen et al. 2019). The overall order of the absolute values of the modified attachment energies on TKX-50 habit faces was calculated to be: (11-1) > (011) > (020) > (12-1) > (110) > (100) (slowest) (Chen et al. 2019). For other polar solvents, such as H_2O, DMSO, DMF, EtOH, and toluene, it was found

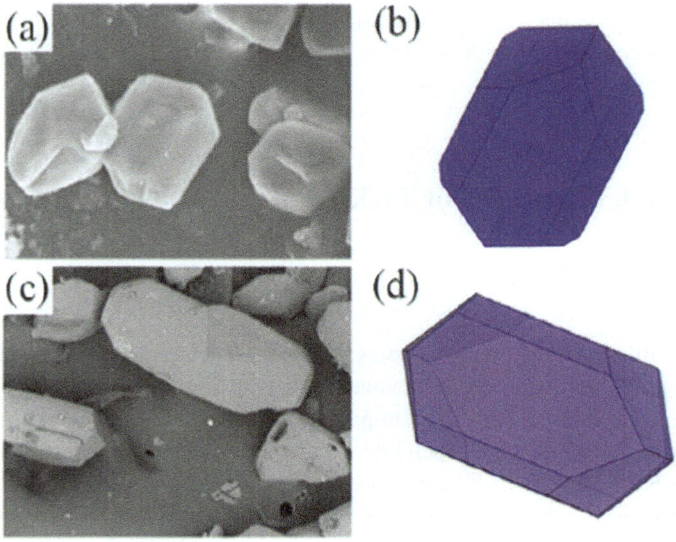

Fig. 32 SEM photographs of TKX-50 crystals from **a** ethylene glycol solvent and **c** DMF showing the different morphologies and the predicted morphologies from **b** ethylene glycol solvent and **d** DMF (Chen et al. 2019). Reproduced with kind permission from the publisher from Chen et al. (2019)

that these solvents interacted readily with the crystal surfaces of TKX-50 (Chen et al. 2019). The predicted morphologies of the TKX-50 crystals were fusiform from water and DMSO and flakes from EtOH and toluene (Chen et al. 2019). All of the crystals showed a different morphology than in the vacuum (Chen et al. 2019). An example of two different morphologies for TKX-50 crystals obtained from ethylene glycol solvent in one case and from DMF solvent on the other are shown in Fig. 32 (Chen et al. 2019). The authors concluded from their work that in comparison with crystallization in vacuum, the area of the (020) and (011) faces of TKX-50 crystals decreases in all solvents whereas the area of the (110) and (12-1) faces increases (Chen et al. 2019). As the polarity of the solvent decreases, the area of the (100) face increases and that of the (110) face decreases (Chen et al. 2019). The (11-1) faces were observed to disappear (apart from in EtOH solvent), and the (12-1) faces occupy a large area (Pang et al. 2018). Furthermore, the crystal morphology of TKX-50 is dominated by five faces in ethylene glycol and DMF solvents, whereas in water it is only dominated by two faces (Chen et al. 2019). Consideration of the aspect ratio, *R* shows that ethylene glycol has the value of R which is minimally closer to 1 than the other solvents which were considered: $R = 3.33$ in H_2O, 3.12 in DMSO, 3.05 in ethylene glycol, 3.27 in DMF, 3.13 in ethanol, and 3.20 in toluene (Chen et al. 2019). (the closer the aspect ratio is to 1, the more spherical the predicted morphology tends to be) (Chen et al. 2019). The aspect ratio is important, because it has been shown that usually, a spherical crystal morphology of a given explosive results in a lower sensitivity in comparison with other morphology types such as flakes or needles for the same explosive.

3 Energetic Properties of TKX-50

3.1 IS, FS, and ESD

One of the first eye-catching properties of TKX-50 which made it stand out as a possible highly interesting energetic material was its low impact and friction sensitivity. Using the BAM method, the impact sensitivity of TKX-50 (2 kg mass) was determined to be 20 J which is much lower than the values for the common secondary explosives RDX (7.5 J), β-HMX (7 J), and ε-CL-20 (4 J) (Fischer et al. 2012). In contrast to RDX, β-HMX, and ε-CL-20 which require the use of desensitizers to enable their use in practical applications, the low impact sensitivity of TKX-50 suggests that this may not be necessary for TKX-50 (Fischer et al. 2012). The friction sensitivity is another important factor in assessing the safe handling of an energetic material. A comparison of the friction sensitivity of TKX-50 (120 N, BAM) with those of RDX (120 N), β-HMX (112 N), and ε-CL-20 (48 N) shows that the sensitivity of TKX-50 is lower than those of β-HMX and ε-CL-20 and comparable with that of RDX (Fischer et al. 2012). Finally, an often neglected but equally important sensitivity is the electrostatic sensitivity, ESD. Since the human body can generate up to 25 mJ of static electricity, it is important that the ESD of an energetic material should be as low as possible to ensure safe handling. Some primary explosives such as lead azide or silver fulminate are so sensitive toward electrostatics that they can easily be set off by the electrostatic energy generated by the human body. Although the ESD of TKX-50 determined to be 0.100 J is a lower value than those of RDX (0.20 J), β-HMX (0.20 J), and ε-CL-20 (0.13 J) meaning it is more sensitive to electrostatics, it is still crucially higher than the 25 mJ which the human body can generate (Fischer et al. 2012). It is this combination of high performance but low sensitivity—and therefore, consequently, increased safety—which makes TKX-50 an exciting new emerging energetic material. It has very recently been shown that the IS of TKX-50 can be lowered by pre-treating the TKX-50 sample with the ball-milling process for 90 min, which has the effect of lowering the particle size distribution of the TKX-50 particles (Xing et al. 2019). After undergoing the ball-milling process for 90 min, further milling does not significantly change the particle size further (Fig. 33) (Xing et al. 2019) and the highest volume fraction percentage of around 150 *um* (Xing et al. 2019). The IS was reduced from 56 J in the raw sample to 50 J in the sample exposed to 90 min ball milling (Xing et al. 2019). The apparatus used to determine the IS has an important effect on the values obtained and was not specified in this paper; however, the value of 56 J for raw TKX-50 was markedly higher than the 20 J value

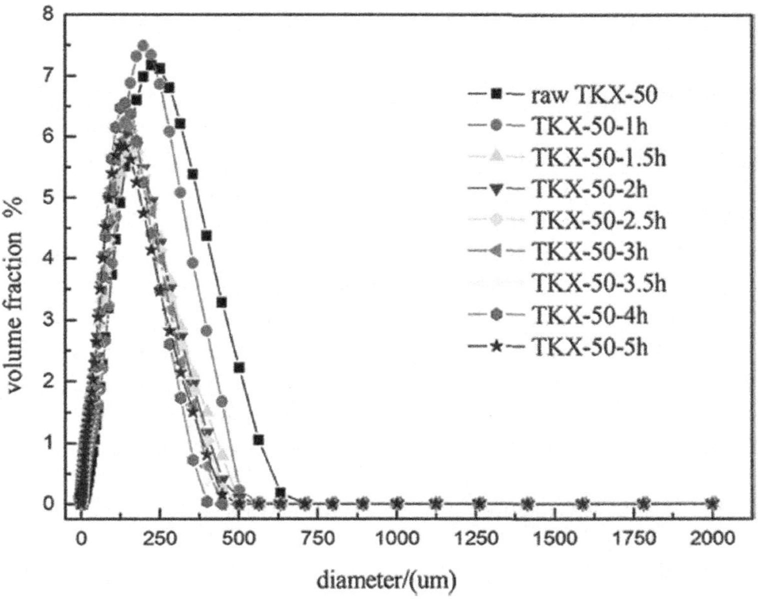

Fig. 33 Effect of ball milling on the particle size distribution of TKX-50 particles (Xing et al. 2019). Reproduced with kind permission from the publisher from Xing et al. (2019)

(BAM apparatus, 1 of 6) reported in the original report on TKX-50 (Fischer et al. 2012). Furthermore, the predicted h_{50} value for TKX-50 using a 2.5 kg dropping mass using the empirical approach of *Keshavarz* was 71.25 cm (Zhao et al. 2019). The effect of ball milling a sample of TKX-50 for 90 min on the friction sensitivity of TKX-50 is similar to that on the impact sensitivity, since the friction sensitivity decreases from 50 N for the raw TKX-50 sample to 40 N for the ball-milled sample (Xing et al. 2019). However, the method used to determine the FS was not states and varies considerably from that reported in the original investigation of TKX-50 in which a FS (using the BAM method, 1 of 6) of 120 N was found (Fischer et al. 2012). Based on computational work concerned with calculation of the change in the unit cell volume at high pressure (30 GPa) compared to ambient pressure, it was concluded that only a 70% reduction in size occurs (Dreger et al. 2015). It was further proposed that in a scenario in which temperature has an influential role in the decomposition, then since the temperature increase under shock compression is related to the volume change, and the lower the compressibility of a single crystal the lower the temperature that will result, then the calculated low compressibility of TKX-50 would result in a smaller increase in the temperature on shock compression (in comparison with other secondary explosives), result in the conclusion that TKX-50 should initiate at higher stress—which is in good agreement with its marked low impact sensitivity (Dreger et al. 2015).

3.2 Heats of Formation, Combustion, and Explosion

The standard enthalpy of formation, $\Delta H°_f$ for an energetic compound is hugely important, since it has a direct influence on the energetic performance of an energetic material. While the enthalpy of formation of a compound has less influence on the detonation velocity (VoD) than the density does, it still has an important influence. Computational results for $\Delta H°_f$ values are known to notoriously depend on the level of theory applied and can vary by crucially large amounts. One approach which has been proposed as a method for calculating the $\Delta H°_f$ values involves calculation of the lattice energy using volume-based thermodynamics and subsequent application of this value in combination with the calculated gas-phase enthalpy of formation using the CBS-4 M method to obtain the $\Delta H°_f$ value (Glasser et al. 2014). Using this approach, a lattice energy of −1506 kJ/mol was calculated for TKX-50 (Glasser et al. 2014), which, in conjunction with the CBS-4 M calculated gas-phase enthalpy of formation, resulted in a $\Delta H°_f$ value of 447 kJ/mol for TKX-50 (Glasser et al. 2014). This calculated value was in good agreement with the value of 439 kJ/mol determined experimentally using bomb combustion calorimetry, and higher than the empirically calculated value for the condensed phase heat of formation of 396.26 kJ/mol (Keshavarz et al. 2018). Using isodesmic reactions, a calculated heat of formation value of 367 0.1 kJ/mol was also reported (Zhao et al. 2019). Others suggest that the heat of formation of TKX-50 should be lower for liquid TKX-50 for which they propose a surprisingly low value of 149.5 kJ/mol (Sinditskii et al. 2015).

3.3 Detonation Velocity and Pressure, Explosion Temperature, and Volume of Gaseous Detonation Products

Recently, the detonation velocity (VoD) of TKX-50 was estimated experimentally using the LASEM technique. The Laser-induced Air Shock from Energetic Materials (LASEM) method has one big advantage over traditional experimental methods to determine the VoD and that is that it requires only milligram quantities of the explosive for the tests (Gottfried and Witkowski 2017). This is especially important, since promising new secondary explosives are only produced on very small scales until the sensitivities and properties have been determined. However, it is also important to establish early on in the investigation of a new energetic candidate material the VoD, since materials which show only low VoDs are highly unlikely to be of further interest for scaling-up for applications. The LASEM method only estimates the maximum attainable VoD of a test substance, since the test substance is not detonated by the laser. However, the laser air-shock velocity that is generated can be measured, and for energetic materials which have VoDs that have been determined using large-scale tests, the LASEM estimated VoD could be correlated to these values to obtain the LASEM estimated VoD values (Gottfried and Witkowski 2017). A comparison of

Table 6 Comparison of the detonation performance of TKX-50 determined using various methods

Method	ρ (g cm^{-3})	$-\Delta_{ex}U°$ [kJ kg^{-1}]	T_{ex} [K]	$P_{C\text{-}J}$ [kbar]	VoD [m s^{-1}]	V_0 [L kg^{-1}]
Empirical (Keshavarz et al. 2018)	1.877		3724	418.95	9650	
EXPLO5.05 (Fischer et al. 2012)	1.877	6024	3954	424	9698	846
EXPLO5 6.01 (Gottfried and Witkowski 2017)	1.877	5892	3623	40.0	9767	913
LASEM (Gottfried and Witkowski 2017)	TMD				9560	
CHEETAH 8.0 (Gottfried and Witkowski 2017)	1.877	5862	2845	42.4	9735	925
Large-scale detonation tests (Gottfried and Witkowski 2017)	TMD				9432	

the LASEM estimated VOD values for several secondary explosives in comparison with those calculated using EXPLO5 v6.01 showed an average difference of 3.1%, and an average difference of 3.9% with CHEETAH 8.0 predicted values (Gottfried and Witkowski 2017). The estimated VoD from LASEM for TKX-50 was 9.56 ± 0.28 km/s at TMD, whereas the calculated values were 9.767 m/s at a density of 1.877 g cm^{-3}using EXPLO5 6.01 and 9.735 km/s at a density of 1.877 g cm^{-3}using CHEETAH 8.0. The measured VoD using large-scale detonation tests for TKX-50 was 9.432 km/s at TMD (Gottfried and Witkowski 2017). These results are contained in the summary in Table 6. Using an empirical approach, the detonation velocity of TKX-50 (and of other energetic compounds of the type formula $C_aH_bN_cO_d$) was calculated using the relationship D = 1.453 I_{sp} ρ_0 + 1.98, to be 9650 m/s at a density of 1.877 g cm^{-3} (Keshavarz et al. 2018), which is similar to the experimentally determined value of 9698 m/s (Fischer et al. 2012), but higher than 9190 m/s reported in (Sinditskii et al. 2015). Using a similar approach, the detonation pressure was also calculated using the following empirical method $P = 44.4\ I_{sp}\ \rho_0^{2-2}1$ to be 418.95 kbar (Keshavarz et al. 2018), in good agreement with the previously calculated EXPLO value of 424 kbar (Fischer et al. 2012). Finally, again using an empirical method ($T_{det} = 5136 - 190.1a - 56.4b + 115.9c + 148.4d - 466.0(d/a)$ $-700.8(b/d) - 282.9n_{NHx}$), the detonation temperature was predicted to be 3724 K (KESHAVARZ ET AL. 2018). Table 6 gives an overview of the detonation performance of TKX-50 determined using various methods. If the VoD of TKX-50 (9.56 ± 0.28 km/s) estimated by the LASEM technique is compared with that of RDX

Table 7 Theoretically predicted values of TKX-50 and RDX as potential ingredients for gun propellant applications Talawar et al. (2018)

Parameter	TKX-50	RDX
Density (g/cc)	1.88	1.82
Flame *T* (K)	3770	4090
Force constant (J/K)	1471.73	1412.59
Internal energy (cal/g)	1138.79	1230.20
Explosion pressure (MPa)	365.53	346.88
No. of moles (moles/g)	0.04657	0.04051
Covolume (cc/g)	0.97381	0.9278
C_v (cal/gmol/K)	0.34964	0.3363
C_p (cal/gmol/K)	0.44293	0.4189
Gamma	1.2668	1.2453
Mean MW of gases (g/mol)	21.297	24.073
Calorific value (cal/g)	1243.43	1349.74
Gas volume (cc/g)	1051.74	930.49
Oxygen balance	−27.1	−21.61

(8.85 ± 0.19 km/s) using the same technique (Gottfried and Witkowski 2017), then a significantly higher VoD is estimated for TKX-50 than for RDX. In fact, the value for TKX-50 is the same as that estimated for CL-20 (9.56 ± 0.24 km/s) using the same technique (Gottfried and Witkowski 2017), and much higher than the value for TNT (6.99 ± 0.23 km/s) and HNS (7.20 ± 0.21 km/s) (Gottfried and Witkowski 2017). One of the most important features of ε-CL-20 is its very high VoD, and therefore, it is particularly promising that TKX-50 shows an estimated VoD the same as that of ε-CL-20, but without the high impact sensitivity of ε-CL-20, complicated synthesis and many polymorphs that CL-20 that are major drawbacks of CL-20. Whereas the EXPLO5 6.01 calculated VoD of TKX-50 (9767 m/s) is predicted to be slightly higher than that of CL-20 (9673 m/s), the CHEETAH v8.0 calculations predict the opposite, namely that the VoD of TKX-50 (9735 m/s) is slightly lower than that of CL-20 (9833 m/s), with the differences in both cases being approximately only 100 m/s (Gottfried and Witkowski 2017).

Using the method of Kamlet and Jacobs, the VoD was calculated as 9250 m/s at a calculated density of 1.93 g cm^{-3}, with a detonation pressure, *P* of 29.53 GPa and a calculated heat of detonation, *Q* of 11,351.17 cal/g (Zhao et al. 2019). The specific impulse, I_{sp}, of TKX-50 was calculated using the EXPLO5.05 computer code to be 261 s assuming isobaric (60 bar) conditions (Fischer et al. 2012), whereas using the empirical approach suggested by *Keshavarz and co-workers*, a value of −2.56 N · s/g was estimated (Keshavarz et al. 2018). Recently, computational work (MMFGDP code) on TKX-50 as a potential ingredient for gun propellant application were also reported and compared with those of RDX (Table 7) (Talawar et al. 2018). The computational results for TKX-50-based solid rocket propellant formulations (22% energetic binder, 20% oxidizer, 18% Al, 40% TKX-50) indicated that the density impulse for TKX-50 was higher than that for HMX, but lower than that of CL-20

Table 8 Comparison of the theoretically predicted (NASA SP-273 code) performance of TKX-50 and other explosive-based solid rocket propellant formulations (22% energetic binder, 20% oxidizer, 18% Al, 40% TKX-50), where [a] = density and [b] = impulse density (Talawar et al. 2018)

Compound	ΔH_f (cal g^{-1})	ρ (g cm^{-3})[a]	T_f (K)	I_{sp} (s)	T_f (MW)	C^* (m/s)	ID[b]
RDX	−180	1.7922	3658	271	140.5	1620	486
HMX	−190	1.8245	3655	270.9	140.4	1620	494
CL-20	−130	1.8766	3829	270.2	138.6	1618	507
TKX-50	−30	1.8139	3536	275	143.8	1679	499
FOX-7	−300	1.8192	3522	264.2	134.1	1580	481
NTO	−312	1.8359	3381	255.2	124.7	1561	469
TATB	−273	1.8369	2933	248.8	116.4	1485	457

(Talawar et al. 2018). Prediction of the parameters of rocket propellant formulations using the NASA SP-273 code showed that the TKX-50 containing formulation shows a superior specific impulse value compared to the formulations containing explosives other than TKX-50 (Talawar et al. 2018). The authors concluded based on this work that the theoretically predicted data indicates the potential for application of TKX-50 in rocket and gun propellants (Table 8) (Talawar et al. 2018).

3.4 *Explosive Power and UNDEX*

The explosive power of TKX-50 has been predicted using the following empirical approach (Keshavarz et al. 2018):

$$EP_{\text{Trauzl}(\%TNT)} = 373.2 - 6525(a/Mw) - 5059(c/Mw) + 21.74(\Delta_f H^\theta(c)/Mw)$$

in which EP_{Trauzl} is the explosive power from the Trauzl test, MW is the molecular weight of TKX-50, and *a* and c are obtained from the empirical formula $C_aH_bN_cO_d$ and a value for TKX-50 of 140% that of TNT was predicted, which is higher than the value of 129% TNT which was predicted using the same method for HMX (Keshavarz et al. 2018). The explosive performance of TKX-50 was also assessed with a small-scale reactivity test (SSRT), and its performance using this test compared with those of ε-CL-20 and RDX (Fischer et al. 2012). Measurement of the size of the dent caused by the explosion of the energetic material in the aluminum block, gives an estimation of the explosive performance of the energetic material. From this investigation, it was concluded that due to the larger size of the dent caused by TKX-50 than by either ε-CL-20 or RDX, that the explosive performance of TKX-50 is higher than those of ε-CL-20 and RDX (Fischer et al. 2012). One advantage of the SSRT is that only a small quantity of explosive is required in comparison with the Koenen test. To further assess the explosive performance of TKX-50, a Koenen test was performed. Although the Koenen test is of importance for assessing the transport

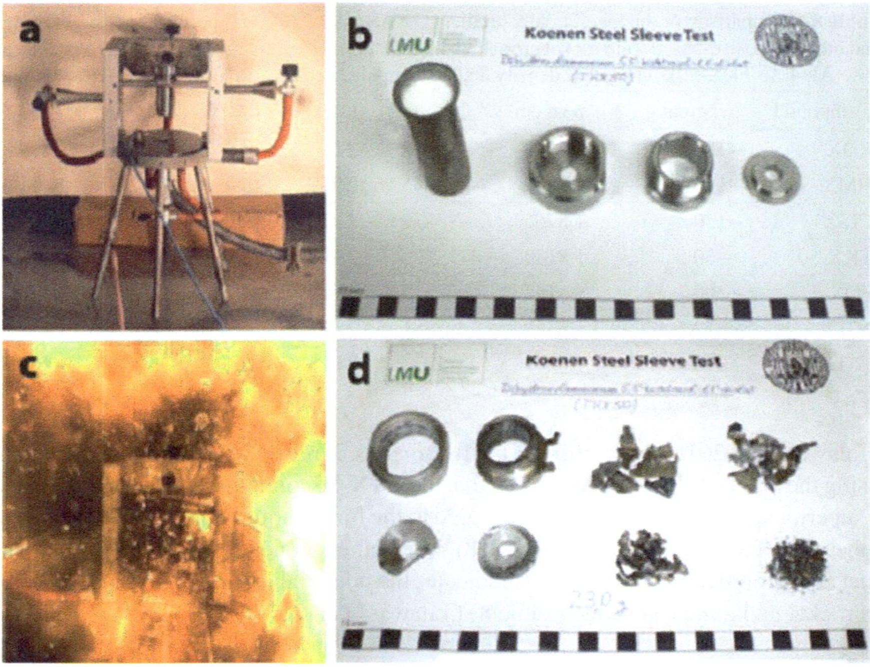

Fig. 34 Photographs showing the experimental test set-up for the Kcenen test in which 23.0 g TKX-50 was used (above) and the fragments of the steel tube present after the test arranged by size (Fischer et al. 2012). Reproduced with kind permission from the publisher from (Fischer et al. 2012)

safety of an explosive material, it requires a large quantity of explosive—in this test with TKX-50, 23 g were required. Using 23.0 g TKX-50 and a 10 mm orifice, the steel container was ruptured into approximately 100 pieces (Fig. 34). This can be compared with TNT which causes destruction of the steel container with an orifice of 6 mm and RDX which destroys the steel tube up to 8 mm (Fischer et al. 2012). To complete these series of tests, a fast cook-off test was also performed on TKX-50 in which no explosion was observed and controlled deflagration was observed instead (Fischer et al. 2012). Depending on the medium in which the pressure wave generated by an explosive expands into its surroundings, one can distinguish between

(i) underwater explosions (UNDEX) and
(ii) air explosions (AIREX)

The unique properties of underwater explosions are due to the high velocity of sound in water meaning that the pressure wave travels approximately four times faster in water than it does in air. Furthermore, due to the high density and low compressibility of water, the destructive energy (from the explosion) can be efficiently transferred over relatively large distances. The most important effects caused by an underwater explosion are the corresponding shock wave and the gas bubble

Table 9 Values for the shock wave (E_{SW}) and bubble wave (E_{BW}) energies for TKX-50 and other secondary explosives (Klapötke et al. 2016)

Explosive	Mass (g)	E_{SW} (J)	E_{BW} (J)	E(total) (J)
TKX-50	0.2	167	428	595
	0.5	367	780	1147
	0.7	488	1020	1508
RDX	0.2	217	499	716
	0.5	446	901	1347
	0.7	591	1185	1775
HMX	0.2	221	492	713
	0.5	450	890	1340
	0.7	594	1154	1748
PETN	0.2	225	507	732
	0.5	460	926	1386
	0.7	629	1223	1852

pulsations. During detonation, an explosive undergoes rapid conversion into gaseous products with high temperature and pressure. Taking into account the damage which is caused by the detonation, this process (underwater explosion) can be considered as resulting from a shock wave and the subsequent gas bubble pulsations of the detonation products. The underwater explosion test is a widely used method to compare the effectiveness of various explosives and is a convenient method for measuring the brisance and heaving power. This approach is based on following hypothesis:

- the shock energy from an underwater explosion measures the explosive's shattering action in other materials (μs time scale);
- the bubble energy from the underwater explosion measures the heaving action of the explosive (ms time scale).

The experimentally obtained shock wave and bubble wave energies of small charges (0.2, 0.5 and 0.7 g) of TKX-50 and other secondary explosives for comparison are listed in Table 9 (Klapötke et al. 2016).

The fact that TKX-50 shows slightly lower performance in the reported small-scale UNDEX tests may be attributed to the fact that the charges used were under the critical diameter of TKX-50.

3.5 *Thermal Stability and Aging*

Since thermal stability is a highly important factor in deciding whether an energetic material will find application, experimental and theoretical investigations of the thermal stability and decomposition mechanism of TKX-50 are of great interest with respect to its future. Since the temperatures that explosives are exposed to in real-life situations can exceed 100 °C, it is important that during storage, transportation, and manipulation, that possible thermal initiation sources can be avoided, since thermal decomposition of large quantities of an energetic material could have extremely serious consequences. Therefore, it is unsurprising that the mechanism of the initial stages of the thermal decomposition of TKX-50 has been of interest to several research groups around the world. A recent report in the literature used quantum mechanics-based reaction studies in order to determine the initial decomposition reaction mechanism at the atomic level (An et al. 2014). Using DFTB-MD simulations (SCC-DFTB method), the optimized unit cell parameters for the solid-state structure at 0 K were calculated (a = 5.413 Å, b = 11.586 Å, c = 6.625 Å, β = 95.40°) (An et al. 2014) and compared with the experimentally determined single-crystal X-ray diffraction unit cell parameters at 100 K (a = 5.487 Å, b = 11.547 Å, c = 6.483 Å, β = 95.402°) (An et al. 2014) in order to make sure that there was good agreement between the two sets of results. The structure was then equilibrated using DFTB-MD at 300 K using the NVT ensemble for 500 fs (0.25 fs time steps), and finally, DFTB-MD temperature-programmed cook-off simulations were performed in which the temperature was continuously increased from 300 to 300 K at a heating rate of 180 K/ps with a total run time of 15 ps (An et al. 2014). It was found that the initial step in the thermal decomposition occurring at 491 K (1.05 ps, pressure = 0.01GPa) and corresponds to proton transfer from one NH_3OH^+ cation to the $C_2N_8O_2^{2-}$ anion to form hydroxylamine, NH_2OH and the corresponding (monoprotonated diolate anion) $C_2HN_8O_2^-$ anion. The $C_2HN_8O_2^-$ anion which is generated was observed not to react <1500 K, while further heating to 1700 K (1.06 GPa) results in the formation of N_2O and at 1900 K (1.68 GPa) N_2, which is presumably due to break-up of the bicyclic $C_2HN_8O_2^-$ anion with the formation of N_2 and N_2O (Fig. 35) (An et al. 2014). The break-up of the bicyclic anion is considered to be important in the initial stages of TKX-50 decomposition, since it provides a release of energy which is important for further decomposition steps. At temperatures >1900 K (1.68 GPa), formation of the secondary products H_2O and NH_3 is observed which is attributed to the decomposition and recombination of the NH_3OH^+ cation (Fig. 35) (An et al. 2014). The authors state that one or two protons always attach to the bicyclic anion before break-up of the anion occurs which they state indicates that proton transfer is catalyzing the ring-breaking reactions which are observed (Fig. 35) (An et al. 2014).

In a second approach, a lower maximum temperature was used. Using this approach, the structure equilibrated at 300 K as mentioned above was again used, but this time it was heated in the simulation very quickly to 1750 K (within 0.5 ps) and a NVT simulation at 1750 K for 20 ps was undertaken. Also in this simulation, it was found that proton transfer was the initial decomposition step (starting at 0.2 ps, 0.01 GPa). The later formation of H_2O and NH_3 was attributed to the decomposition of the NH_2OH molecule which was formed from deprotonation of the NH_3OH^+ cation. In contrast to the cook-off simulation described above, the formation of H_2O and NH_3 occurred before the break-up reactions of the bicyclic anion (An et al. 2014). In this simulations, the formation of N_2—indicative of ring breakage) occurred at

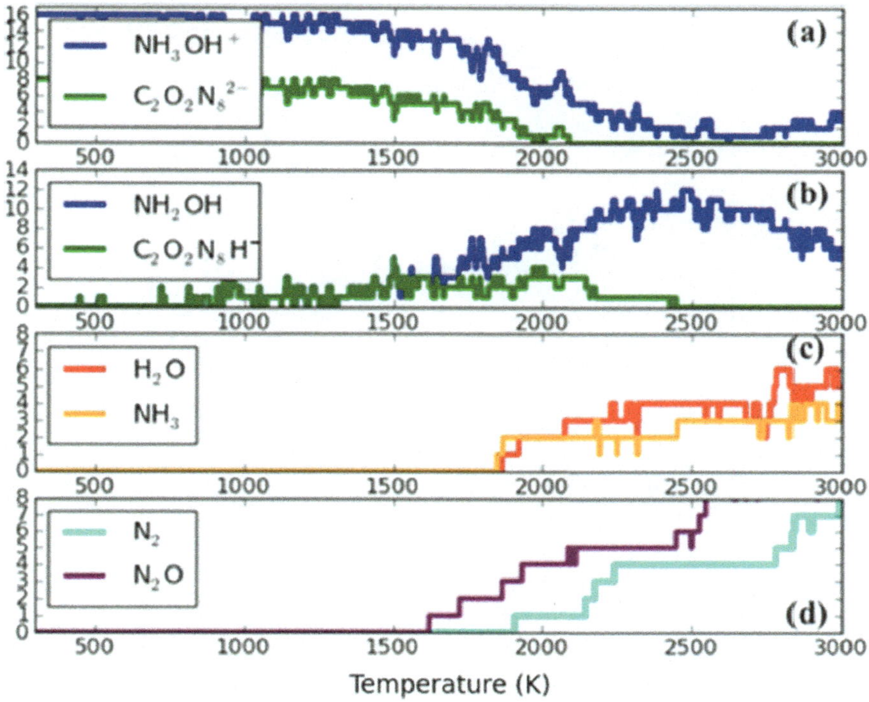

Fig. 35 DFTB-MD temperature-programmed cook-off simulations in which the temperature was continuously increased from 300 to 3000 K at a heating rate of 180 K/ps with a total run time of 15 ps predicted the following initial stages of decomposition for TKX-50 (An et al. 2014). Reproduced with kind permission from the publisher from An et al. (2014)

about 2.5 ps (1.39 GPa), with the release of N_2O then occurring from further ring decomposition (An et al. 2014). It is worth pointing out that the thermal decomposition of most energetic salts containing the NH_3OH^+ cation has been proposed as occurring via the dissociation of the NH_3OH^+ cation to form the corresponding B–H and NH_2OH products (Muravyev et al. 2017).

The computational results obtained using the DFTB-MD simulations (SCC-DFTB method) were then refined with the use of finite cluster quantum mechanics studies (B3LYP/6-311++G(d,p), Jaguar) for the thermal decomposition reaction mechanism of the bistetrazole dianion and found that the energy barrier for the release of N_2 was 45.1 kcal/mol through a concerted reaction (via TS3) (Fig. 36), which was lower than the mechanism via TS1, therefore making it the easier path for break-up of the bicyclic ring (An et al. 2014). The formation of N_2O can occur via TS4 in which N_2O dissociates via TS6 with a 72.2 kcal/mol energy barrier (Fig. 36); however, both of these barriers are reduced in the monoprotonated $C_2HN_8O_2^-$ anion, for which the energy barrier to release of N_2 decreased to 37.2 kcal/mol and for N_2O 59.5 kcal/mol (Fig. 36) (An et al. 2014). These results led to the prediction that the initial decomposition step in the thermal decomposition of TKX-50 is the release of N_2, and this

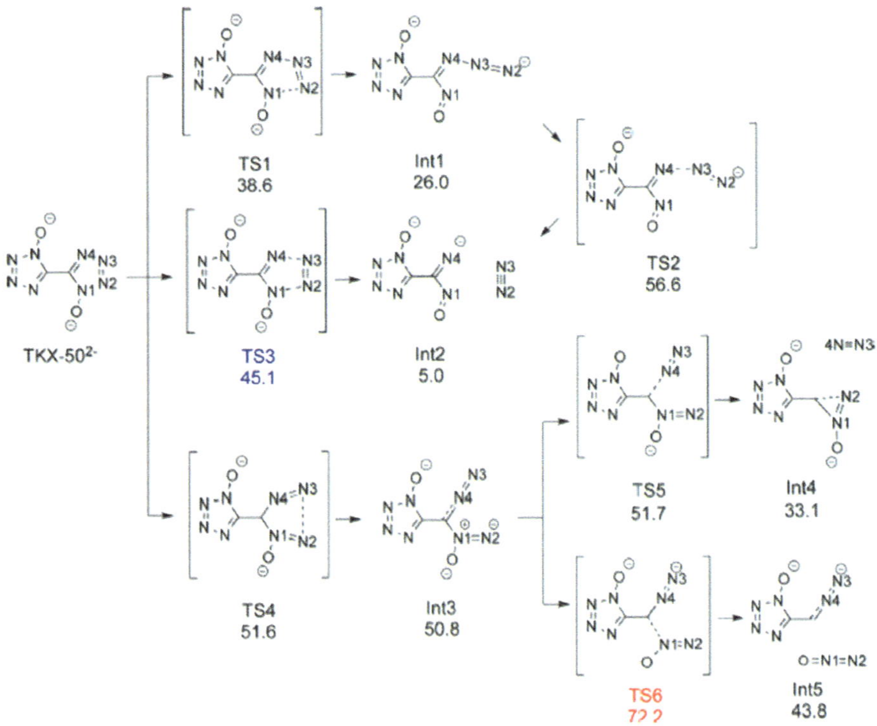

Fig. 36 Summary of the unimolecular decomposition reaction of the TKX-50 dianion $C_2N_8O_2{}^{2-}$ (reaction enthalpies are given in kcal/mol) from finite QM (B3LYP) calculations. The rate determining steps producing N_2 are indicated in blue while those for N_2O are indicated in red (An et al. 2014). Reproduced with kind permission from the publisher from An et al. (2014)

provides energy for further decomposition (Fig. 36) (An et al. 2014). Since it has already been discussed that proton transfer from the NH_3OH^+ cation to the $C_2N_8O_2{}^{2-}$ is predicted to be highly influential on the initial reaction stages in the thermal decomposition of TKX-50, the enthalpies of the decomposition pathways were calculated for each of the four prototopic tautomers, and it was found that the easiest fragmentation pathway for the formation of N_2 occurs via Int10 followed by formation of Int12 with an energy barrier of 37.2 kcal/mol (Fig. 37), or starting from a different tautomer, namely Int13 via Int16 with an energy barrier of 39.2 kcal/mol, which is considerably more exothermic releasing −14.7 kcal/mol (Fig. 37) (An et al. 2014). Important here is to emphasize that either of these mechanisms have for the monoprotonated anion $C_2HN_8O_2{}^-$ show reaction barrier heights which are 6–7 kcal/mol lower than that for the non-protonated dianion $C_2N_8O_2{}^{2-}$ which supports the proposal that the proton transfer reaction from the NH_3OH^+ cation to the $C_2N_3O_2{}^{2-}$ anion promotes the decomposition of TKX-50 by reducing the energy barrier for the release of N_2 and N_2O by approximately 10 kcal/mol (An et al. 2014). Interestingly, the formation of N_2O is also facilitated in terms of energy by monoprotonation of the anion, since

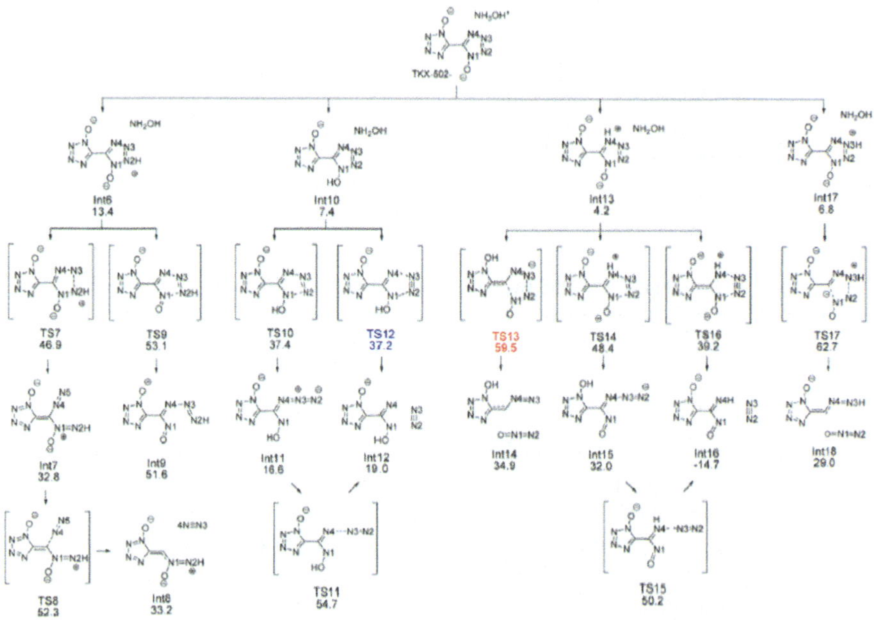

Fig. 37 Summary of the unimolecular decomposition reaction of the various prototopic tautomers of the TKX-50 dianion $C_2N_8O_2^{2-}$ (reaction enthalpies are given in kcal/mol) from finite QM (B3LYP) calculations. The rate determining steps producing N_2 are indicated in blue while those for N_2O are indicated in red (An et al. 2014). Reproduced with kind permission from the publisher from An et al. (2014)

the barrier height in the dianion with respect to the generation of N_2O are higher (An et al. 2014). The importance of the release of N_2 is due to the its exothermic nature, in contrast to the reactions in this systems which release N_2O which are endothermic (An et al. 2014). Further simulations were able to show that double protonation of the $C_2N_8O_2^{2-}$ anion to form the $C_2H_2N_8O_2$ diol molecule doesn't change the rate of N_2 release (An et al. 2014).

The main conclusions with respect to the initial stages of thermal decomposition which were elucidated by this investigative computational work were that the proton transfer from the NH_3OH^+ cation to the $C_2N_8O_2^{2-}$ anion occurs at considerably lower temperature than the initial stages of the ring break-up of the anion. The initial ring-breaking reactions release N_2 rather than N_2O and that the above-mentioned proton transfer results in a decrease in the energy barriers for the break-up of the ring to form N_2 and N_2O by approximately 10 kcal/mol (An et al. 2014).

Recently, *Muravyev* et al. have pointed out that the decomposition of the neutral hydroxylamine molecule (NH_2OH) in these simulations occurred at higher temperatures than BTO, despite the poor thermal stability of NH_2OH (Muravyev et al. 2017). On the basis of DSC/TGA analysis and experimentally detected intermediates, these authors have suggested that the thermal decomposition of TKX-50 occurs via two distinct stages, namely via the deprotonation of two NH_3OH^+ cations forming the

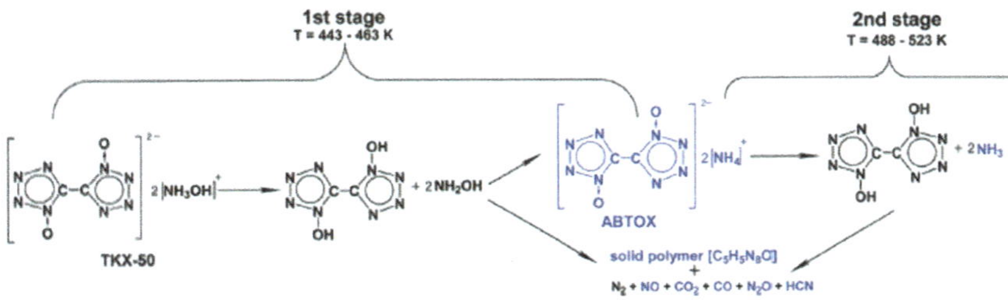

Fig. 38 Proposed mechanism for the two-stage thermal decomposition of TKX-50 by *Muravyev* et al.in which ABTOX is formed as an experimentally detected intermediate (indicted in blue) (Muravyev et al. 2017). Reproduced with kind permission from the publisher from Muravyev et al. (2017)

corresponding neutral base NH_2OH and diol acid $C_2H_2N_8O_2$ (the doubly protonated TKX-50 dianion) in the first stage, which occurs at 443–463 K and which is followed by the formation of ABTOX—which is simply the bisammonium salt of the TKX-50 dianion—as an experimentally detected intermediate (75% weight) as well as [$C_5H_8N_8O$] polymer (~25% weight). The ABTOX formed as an intermediate is then proposed to decompose in the second stage to form NH_3 and the doubly protonated form of the TKX-50 dianion, namely $C_2H_2N_8O_2$ and also gaseous decomposition products such as CO_2, N_2O. (Fig. 38) (Muravyev et al. 2017). The mass loss of the two main stages of decomposition were determined to be 49% and 38%, respectively (Muravyev et al. 2017). The ABTOX formed in the first stage of decomposition was identified using X-ray powder diffraction, and its formation rationalized by the fact that NH_2OH is known to decompose forming NH_3, N_2O, and H_2O at temperatures below 420 K; therefore, the source of NH_3 to form the NH_4^+ cations was postulated (Muravyev et al. 2017). Support for the formation of the solid polymer and gaseous products was obtained from TGA experiments in which the mass loss was directly detected (Muravyev et al. 2017). In contrast to the proposition by *Sinditskii* et al. that the decomposition of NH_2OH is the key process in the thermal decomposition of TKX-50 (Sinditskii et al. 2015), *Muravyev* et al. state that the decomposition of BTO is a key process (Muravyev et al. 2017). The assertion that ABTOX is the main component of the primary thermal decomposition product TKX-50-M of TKX-50 was also supported by recent investigations using ARC under adiabatic conditions (Wang et al. 2020). These authors reported that TKX-50-M appeared after an adiabatic temperature increase during the first stage of decomposition starting at 200.2 °C (Wang et al. 2020). Using SEM, the resulting TKX-50-M could be described as a khaki-colored solid in which crystals of 30–40 μm size (ABTOX decomposition product, confirmed using X-ray powder diffraction and single-crystal X-ray diffraction) were agglomerated by solid residue (proposed to be the decomposition product of tetrazole derivatives) and which differed from white, tabular crystals (100–200 μm size) of crystalline TKX-50 (Wang et al. 2020). The second stage of decomposition starting at 240.26 °C produced a dark brown powder which SEM showed to be a porous foam

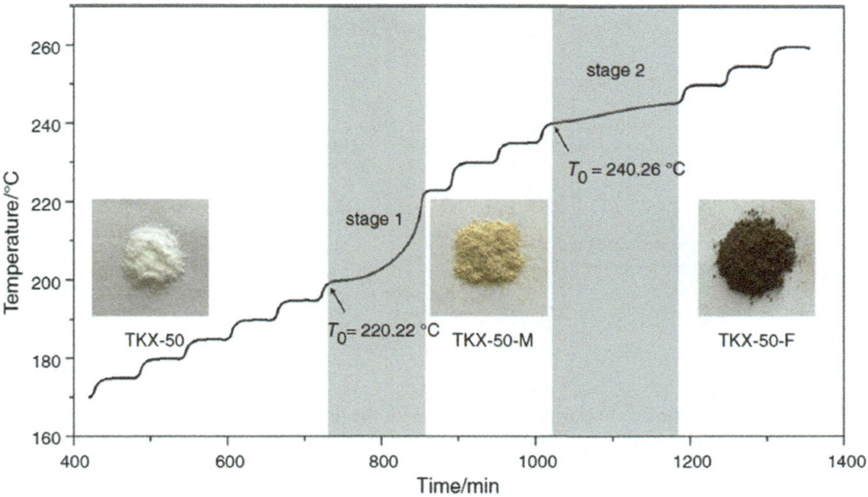

Fig. 39 Summary of the thermal decomposition process of TKX-50 under adiabatic conditions using ARC (Wang et al. 2020). Reproduced with kind permission from the publisher from Wang et al. (2020)

designated TKX-50-F (Wang et al. 2020). A summary of the thermal decomposition process of TKX-50 under adiabatic conditions using ARC is shown in Fig. 39 (Wang et al. 2020). The mass loss of TKX-50 in the first stage of decomposition in these ARC experiments in a confined space was found to be 35–40% (Wang et al. 2020) which was lower than that reported previously; however, this was explained as being a result of the NH_3 formed from decomposition of NH_2OH remaining in the confined space and being able to react with the BTO also produced meaning that a larger amount of solid ABTOX was produced than in previous reports in which non-confined spaces were used (Wang et al. 2020). A summary of the proposed decomposition mechanism for the first stage of decomposition showing the origin of BTO and NH_3 and subsequently ABTOX is given in Fig. 40 (Wang et al. 2020). However, the authors also state that a polymeric residue was also formed in the first stage of thermal decomposition and that TKX-50-M showed a self-decomposition reaction that beginning at 235.3 °C and finished at 253.9 °C with an adiabatic temperature rise of 8.59 °C, which is much lower than that of ABTOX (41.76 °C) which shows an initial self-heat temperature of 265.34 °C (Wang et al. 2020). The authors attributed the larger heat release of ABTOX compared to TKX-50-M as resulting from TKX-50-M containing polymeric residue which didn't undergo an exothermic reaction (Wang et al. 2020). The calculated heat capacity of TKX-50 with constant pressure, $C_p = 295$ kJ/mol (An et al. 2015) and at constant volume, $C_v = 292$ kJ/mol at room temperature (An et al. 2015). Since explosives must be able to be stored for longer periods of time, sometimes under extreme temperature conditions (munition in the desert can be exposed to temperatures of over 100 °C), it is important that the thermal stability and stability under common storage conditions can be

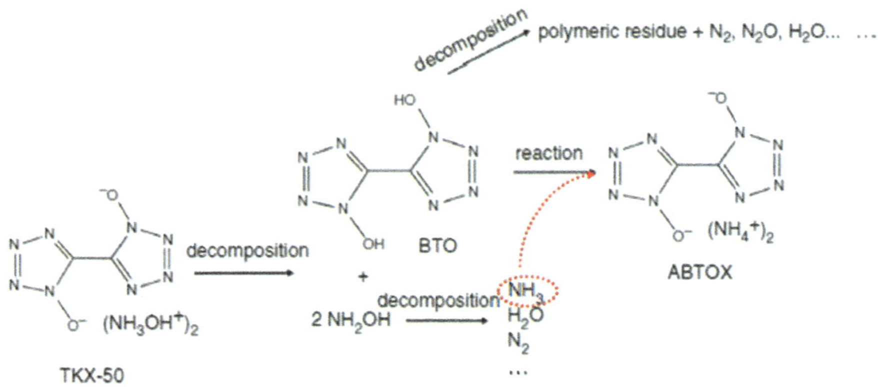

Fig. 40 Summary of the decomposition mechanism for the first stage of the thermal decomposition process of TKX-50 under confined conditions (Wang et al. 2020). Reproduced with kind permission from the publisher from Wang et al. (2020)

assessed. An energetic material even with outstanding performance which is only stable at temperatures below 100 °C would not be suitable for military applications. In addition to establishing the decomposition temperature using DSC among other methods, which TKX-50 was shown to decompose at 222 °C and therefore surpasses the 200 °C minimum decomposition threshold required for application, a long-term stability test was also performed (Fischer et al. 2012). Using this test, TKX-50 was heated in an open glass vessel for 48 h at 75 °C, and subsequently, a DSC was performed to see if it was identical to the same before heating at 75 °C (Fischer et al. 2012). No exothermic or endothermic events were observed for TKX-50 apart from the exothermic decomposition at 222 °C which is also observed for the raw sample before heating (Fischer et al. 2012). This showed that TKX-50 was unaffected by heating at 75 °C for 48 h under the conditions used, implying that it shows good safety with respect to temperature.

3.6 Compatibility

The compatibility of a promising new energetic material is a factor which must be established early on in the development of an energetic material, since energetic materials which show only poor compatibility will most likely find very limited possibilities for application. This is because energetic materials are often used in combination with other energetic and non-energetic compounds as components in energetic formulations. However, it is only possible to prepare formulations if there is high compatibility between the components. The compatibility is an important safety index, since it gives information about the potential of unwanted chemical reactions occurring between the components of the mixture. There are also other

Table 10 Criteria used for assessing the compatibility of an energetic material with other components using DSC

Criteris ΔTp (°C)	Rating	Note
≤2	A. Compatible or good compatibility	Safe for use in any explosive design
3–5	B. Slightly sensitized or fair compatibility	Safe for use in testing when the device will be used in a very short period of time; not to be used as a binder material or when long-term storage is desired
6–15	C. Sensitized or poor compatibility	Not recommended for use with explosive items
>15	D. Hazardous or poor compatibility	Hazardous. Do not use under any conditions

compatibility issues which have to be taken into account, for example, the energetic material must also be compatible with the metals used for production and storage equipment. One commonly used method to determine the compatibility of energetic materials is DSC.

Huang and co-workers looked at the different components of a mixture and determined using DSC which of the components possessed the lowest decomposition temperature. They then looked at the difference in the peak temperature (T_p) of the pure substance with the lowest T_p in a mixture, with the T_p of the mixture to obtain the ΔT_p value. If the difference in the T_p of the pure substance and the T_p of the mixture was less than or equal to 2 °C, then the mixture was concluded to have components which are compatible or showed good compatibility. The larger the value of ΔT_p, the poorer the compatibility was concluded to be. The criteria used to determine the level of compatibility between the components is as follows (Table 10) (Huang et al. 2015a, b): For example, the T_p for TKX-50 as a pure compound (not as part of a mixture) was found in this work to be 245.72 °C. In the 1:1 mixture of TKX-50 and HNE, it is HNE that has the lower T_p as a pure compound with a value of 136.61 °C. Therefore, it is the value of HNE which is used to calculate ΔT_p with respect to the mixture. Since the TKX-50/HNE mixture shows a T_p of 144.65 °C, this is a difference of −8.04 °C since $\Delta T_p = T_{p1}$ (pure compound—in this case HNE)—T_{p2} (mixture) (Table 11). This difference of −8.04 °C would categorize the TKX-50/HNE mixture as being compatible or showing good compatibility according to Table 11 (Huang et al. 2015a, b). Utilizing this procedure for the above-mentioned 1:1 mixtures, the following conclusion for the compatibilities of TKX-50 resulted: HNE > DNAN > HMX > GAP > Al > CL-20 > AP > NC > Centralite > TNT > boron > RDX > HTPB, with HTPB showing the poorest compatibility with TX-50 out of the compounds investigated under the conditions used (Huang et al. 2015a, b). However, it should be pointed out the difference in the maximum exothermic peak temperature for TKX-50 (single component system) which was reported as 221 °C (DSC @ 5 °C/min) in earlier work (Fischer et al. 2012) which is significantly lower

Table 11 Maximum T_{dec} peak temperatures observed for pure compounds and mixtures with TKX-50 obtained from DSC (10 °C/min heating rate, dynamic nitrogen atmosphere, 50 mL/min, 0.1 MPa pressure) and the compatibility rating (A–D) of TKX-50 with the component of the mixture (T_{p1} = max. exothermic peak T of a single system; T_{p2} = max. exothermic peak T of the mixture; $\Delta T_p = T_{p1} - T_{p2}$)

System		Peak temperature			Rating
Mixture system	Single system	T_{p1} (°C)	T_{p2} (°C)	ΔT_p (°C)	
TKX-50/DNAN	TKX-50	245.72	243.69	2.03	A–B
TKX-50/TNT	TKX-50	245.72	234.25	11.4	C
TKX-50/RDX	RDX	242.88	225.24	17.64	D
TKX-50/HMX	TKX-50	245.72	241.83	3.89	B
TKX-50/CL-20	CL-20	245.72	237.55	8.17	C
TKX-50/Centralite	TKX-50	245.72	234.90	10.82	C
TKX-50/NC	NC	202.15	192.30	9.85	C
TKX-50/NC + NG	NC + NG	206.32	189.42	16.9	D
TKX-50/AP	TKX-50	245.72	236.95	8.77	C
TKX-50/HNE	HNE	136.61	144.65	−8.04	A
TKX-50/Al	TKX-50	245.72	237.71	8.01	C
TKX-50/Boron	TKX-50	245.72	228.11	17.61	D
TKX-50/GAP	TKX-50	245.72	239.56	6.16	C
TKX-50/HTPB	TKX-50	245.72	17732	68.4	D

than that of 245.72 °C (DSC @ 10 °C/min) in these compatibility studies (Huang et al. 2015a, b). The compatibility of TKX-50 with TNT in melt-cast mixtures with different TNT:TKX-50 ratios was investigated using DSC (Table 12) (Li et al. 2019b). The aim of this work was to investigate the suitability of TKX-50 as a component of TNT melt-cast explosives as a replacement for well-known examples such as Comp B (mixture of 40% TNT and 60% RDX). It was found that TNT/TKX-50 mixtures with mass ratios of 50:50, 40:60, 30:70, 20:80, and 10:90 had good compatibility. However, interestingly, in contrast to TNT/RDX and TNT/Fox-7 mixtures which showed an increase in the decomposition activation energy as the content of RDX or Fox-7 increased, for the TNT/TKX-50 mixtures the opposite was true (Li et al. 2019b). To determine the compatibility of TKX-50 mixtures with TNT at different ratios, DSC was used in accordance with the STANAG 4147 standard which has the following classification criteria: <4 °C decrease in T_{dec} peak indicates acceptable compatibility; if the peak shift lies in the range >4 to <20 °C, then the compatibility is classified as being borderline, whereas a shift in T_{dec} > 20 °C indicates incompatibility of the components (Li et al. 2019b). In addition, the TGA technique was also used according to the STANAG 4147 standard which classifies the mass loss of the mixtures in comparison to that of the individual components as follows: If the observed mass loss for the mixture is greater than the total mass loss expected for each of the individual components of the mixture, then the components of the

Table 12 Summary of the assessment of compatibility of TKX-50/TNT melt-cast mixtures based on the DSC and TGA data (Li et al. 2019b)

Sample	OB (%)	DSC T_p (°C)			TGA mass loss (%)			Compatibility	Activation energy, E_a (kJ/mol)	
		Pure	Mixture	ΔT_p	Measured	Theoretical	ΔML		Kissinger	Ozawa
TNT	−74.0	313			96				113.8	117.1
TKX-50	−27.1	251 308	48						186.9 267.0	185.9 262.9
TNT:TKX-50 mixtures (mass ratio TNT:TKX-50)										
90:10	−69.3		251 295	0 18	97	87.6	9.4	Certain degree of incompatibility		
80:20	−64.6		240 284	11 29	92	83.2	8.8	Incompatible		
70:30	−59.9		240 281	11 32	83	78.8	4.8	Incompatible		
60:40	−55.2		240 276 308	11 37 0	79	74.4	4.5	Incompatible		
50:50	−50.6		248 308	3 0	71	70.0	1.0	Compatible	195.1 206.6	193.7 205.3
40:60	−45.9		249 307	2 1	69	65.6	3.4	Compatible	192.6 201.3	191.2 208.0
30:70	−41.2		250 309	1 −1	64	61.2	2.8	Compatible	167.1 196.6	166.9 195.7
20:80	−36.5		250 309	1 −1	55	56.8	−1.8	Compatible	166.5 170.9	166.4 171.3
10:90	−31.8		250 309	1 −1	50	52.4	−2.4	Compatible	150.8 170.5	151.5 171.1

mixture are considered incompatible (Li et al. 2019b). Furthermore, the greater the difference in the mass loss of the mixture with the total mass loss of the individual components is, the greater the incompatibility is. A mass loss difference of up to 4% is considered to indicate compatible components, a mass loss difference between 4 and 20% is considered to indicate an increase in the incompatibility and requires further testing, whereas a mass loss of >20% classifies the components as being incompatible with each other (Li et al. 2019b).

In this study, the TNT and TKX-50 were dried in an oven prior to mixing and ground to pass through a 270 mesh screen. The TNT and TKX-50 were then mixed together in the desired ratio (steps of 10%) in a mortar and ground to a fine mixture before heating at 90–95 °C for 2 h and subsequently allowing the mixture to cool to room temperature, before grinding the mixture so that the powder would pass through a 70 mesh, before performing DSC and TGA analyses (Li et al. 2019b). First of all, the DSC/TGA was analyzed of the two individual component explosives, namely TNT and TKX-50. The DSC of TNT shows an endothermic peak at 80 °C indicating the melting point of TNT, followed by a second endothermic peak at 258 °C (only) in the open system corresponding to the boiling point (Li et al. 2019b), and an exothermic peak at 313 °C (only) in the closed system which corresponds to the decomposition point of pure TNT (Li et al. 2019b). The DSC of TKX-50 ($\beta = 10$ °C/min) showed two exothermic peaks at 251 °C and 308 °C which correspond to the first and second stage of decomposition, respectively (Li et al. 2019b). Of particular importance is the exothermic peak corresponding to the first stage of decomposition which is shown in Fig. 19 to shift to lower temperature on decreasing the heating rate, β (Li et al. 2019b). The DSC curves of the TNT/TKX-50 mixtures with different mass ratios of the components (10% steps) is shown in Fig. 41. It can be seen that the three mixtures with the highest mass ratio of TNT (90, 80, and 70%) and consequently the lowest TKX-50 ratio (10, 20, and 30%, respectively) show two exothermic peaks which the authors assigned to the decomposition of TKX-50 (peak at lower temperature) and TNT (peak at higher temperature) (Li et al. 2019b). The decomposition peak of TNT (exothermic peak at higher temperature) in the TNT/TKX-50 90:10 mixture was found to be 18 °C lower than that of pure TNT; therefore, in alignment with the STANAG 4147 standard, the 90:10 TNT/TKX-50 mixture was classified as showing a degree of incompatibility (Li et al. 2019b) (Fig. 41). The TNT/TKX-50 mixtures with 80:20 and 70:30 mass ratios were deemed to be incompatible since the decomposition temperature of the peak assigned to TKX-50 at 251 °C in pure TKX-50 shifted 11 °C to lower temperature in both the 80:20 and 70:30 mixtures (Li et al. 2019b). And, in addition, the peak at 313 °C in pure TNT was shifted to 29 °C (80:20 mixture) and 32 °C (70:30 mixture) lower temperature (Table 12) (Li et al. 2019b). In the 60:40 mixture, for the first time in the DSC, three exothermic peaks are observed, which correspond to the decomposition of TKX-50, TNT, and ABTOX with ΔT_p values of 11 °C, 37 °C, and 0 °C, respectively (Li et al. 2019b). The latter compound being an intermediate decomposition product of TKX-50 (Li et al. 2019b). These values indicate that the 60:40 mixture is incompatible (Li et al. 2019b). However, in the mixtures with a higher TKX-50 mass ratio content, a higher degree of compatibility

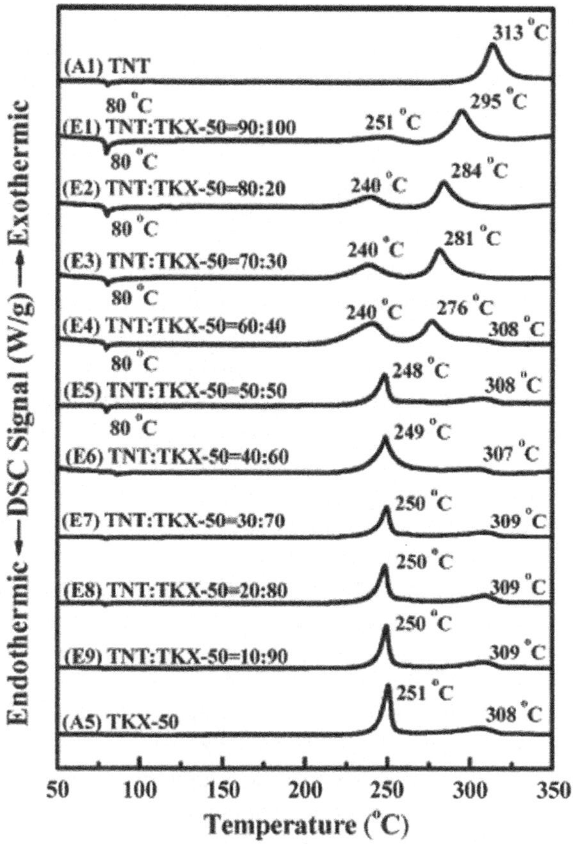

Fig. 41 DSC curves of different (varying % mass ratios of each of the components) TNT/TKX-50 mixtures at a heating rate of 10 °C/min (Li et al. 2019b). Reproduced with kind permission from the publisher from Li et al. (2019b)

is observed with TNT. Only two peaks are again observed in the DSC spectra of these mixtures, and the values of the T_{dec} peaks at these ratios are all within 4 °C of either pure TKX-50 or pure TNT (Li et al. 2019b), meaning that for TKX-50/TNT mass ratios of 50:50, 60:40, 70:30, 80:20, or 90:10 there is good compatibility (Table 12) (Li et al. 2019b). A similar conclusion is reached from the TGA data which also show that only TKX-50/TNT mixtures with mass ratios of 50:50 or higher TKX-50 values show good compatibility. The TNT-rich mixtures show a degree of incompatibility based on the mass loss (Table 12) (Li et al. 2019b). The decomposition activation energy for all of the mixtures discussed above has also been calculated using the Kissinger and Ozawa methods resulting in the conclusion that as the TKX-50 content increases, the activation energy to decomposition decreases (Table 12) (Li et al. 2019b). However, the criticism of applying the Kissinger and Ozawa methods based on DSC/TGA experiments for pure TKX-50 should be mentioned here (Muravyev et al. 2017). If the compatibilities of these melt-cast TNT/TKX-50 mixtures are compared with other TNT/energetic material mixtures that were investigated by the same authors, then the compatibility of the TNT/TKX-50 mixtures were classified with the same compatibility as the corresponding TNT/RDX mixture (Li et al. 2019b). Furthermore, the authors found that TNT/CL-20 mixtures were incompatible (Li et al. 2019b), whereas TNT/Fox-7 mixtures were compatible (Li et al. 2019b).

4 Nano-TKX-50

Interest in the synthesis of energetic materials with nano instead of the more commonly obtained milli- or micro-sized particles has started to receive considerable interest over recent times. In particular, the control and design of specific particle sizes and shapes of energetic materials is of particular importance, since these properties often have a considerable effect on the sensitivity of energetic materials. Very recently, a report appeared in the literature regarding the fabrication of 3-D network-like nanostructures of TKX-50 (Cao et al. 2019). In this work, the TKX-50 nanostructures were prepared using the facile liquid nitrogen-assisted spray-freeze-drying method using different concentrations of aqueous TKX-50 solutions (Cao et al. 2019). In this process, at ambient temperature a very small quantity (10 mg) of TKX-50 was added to a certain amount of deionized water in a beaker and dissolved by stirring at 40 °C using ultrasound. Subsequently, the aqueous TKX-50 solution was then sprayed rapidly into a beaker containing liquid nitrogen during which the rapid icing of the TKX-50 solution was observed. The samples then underwent vacuum freeze drying for 48 h which resulted in the formation of white, opaque TKX-50 material (Cao et al. 2019). Using X-ray powder diffraction, it could be established that the spray-dried TKX-50 samples while showing poor crystallinity showed chemical stability and no decomposition of TKX-50 had occurred during the process (Cao et al. 2019). The poor crystallinity suggested the formation of nano-particle TKX-50 (Cao et al. 2019). Furthermore, using SEM, the size and morphology of the samples was determined and showed that the size of the nano-scale networks increased with increasing TKX-50 concentration of the aqueous solution used in the preparation (Cao et al. 2019). The formation of more perfect 3-D nanostructures was promoted by lower TKX-50 concentration of the aqueous solution (e.g., 0.3 mg/ml TKX-50 solution) (Cao et al. 2019) and the stacking density of the nanostructures decreased from 1.196 g cm^{-3} in the raw material prior to treatment to a minimum value of 0.1272 g cm^{-3} when a 0.3 mg/ml TKX-50 solution was used (Cao et al. 2019). The concentration of the aqueous TKX-50 solution was highly influential on the nano-structure which resulted, with solutions of lower concentration producing more perfect network-like frameworks, whereas more concentrated solutions resulted in overgrowth occurring and even the formation of crystals at the highest concentration (Cao et al. 2019). Based on these results, the authors proposed a growth and formation mechanism for the nano-networks of TKX-50 as follows (Cao et al. 2019): The raw TKX-50 solid was dissolved in aqueous solution and subsequently rapidly injected into liquid nitrogen which resulted in the icing process occurring (Cao et al. 2019). Since the temperature of the aqueous TKX-50 solutions suddenly drops by a large amount, the TKX-50 molecules begin to nucleate and grow due to super-saturation of the solution (Cao et al. 2019). Crucially, since this process occurs in a very short space of time, the TKX-50 molecules are able to form a very large number of primary crystal nuclei throughout the solution before icing-formation occurs (Cao et al. 2019). However, since the TKX-50 crystal nucleus can't grow continuously due to limited space as a consequence of the freezing of the solution. The nano-sized TKX-50 particles can grow to a limited extent from the residual TKX-50 molecules remaining in solution and random agglomeration begins which results in self-assembling formation of the TKX-50 3-D network-like nanostructures (Fig. 42) (Cao et al. 2019). The ice which was produced during the process subsequently sublimed. The properties of nano-TKX-50 were of particular interest in comparison to those of raw non-nano-sized TKX-50 which is usually prepared (Cao et al. 2019). Recently, investigations of the properties of energetic materials with milli-sized particles compared to nano-sized particles have received considerable attention because of the higher stability toward atmospheric conditions. If the thermal decomposition temperature of TKX-50 with different particle sizes is investigated, then the two exotherm peak maxima indicating decomposition observed at 250.6 and 274.3 °C in the DSC ($\beta = 10$ °C/min) for raw TKX-50 (micro-sized particles) shifted to lower temperatures, for example to 236.0 and 259.7 °C for non-particle-sized TKX-50 (Fig. 43) (Cao et al. 2019). Therefore, nano-TKX-50 is reported to decompose at lower temperature (Cao et al. 2019).

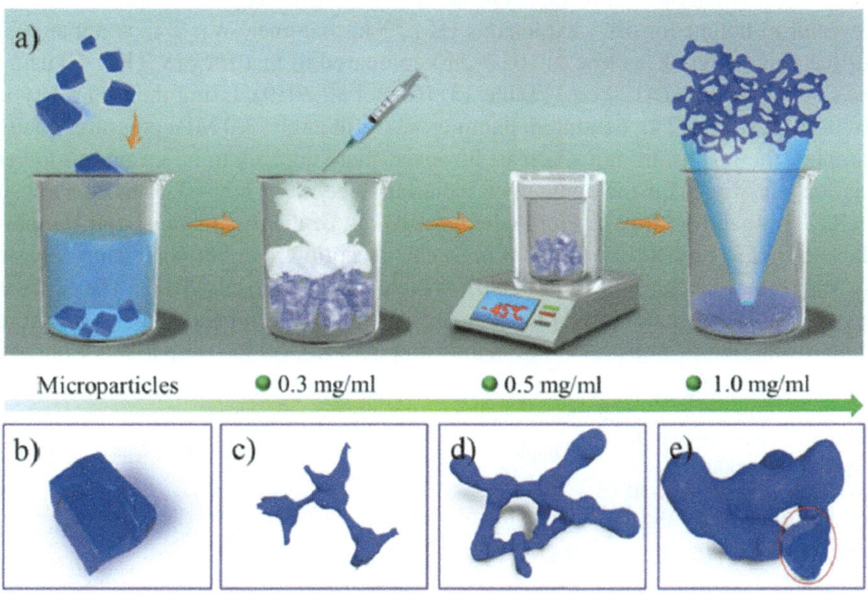

Fig. 42 Schematic representation of the formation of the TKX-50 3-D network-like nanostructures in a self-assembly process using the liquid nitrogen-assisted spray-freeze-drying method (Cao et al. 2019). Reproduced with kind permission from the publisher from Cao et al. (2019)

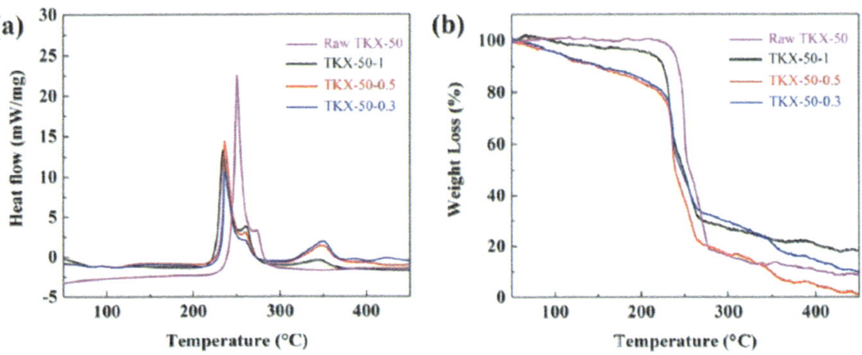

Fig. 43 DSC/TG curves of raw TKX-50 (pink) and three samples of TKX-50 consisting of 3-D network-like nanostructures, prepared by the liquid nitrogen-assisted spray-freeze-drying method (Cao et al. 2019). Reproduced with kind permission from the publisher from Cao et al. (2019)

Another important aspect which has also been investigated in other work for other energetic materials on the nano-scale is the sensitivity to external stimuli such as impact and friction. In both the impact and friction sensitivity tests of both the raw TKX-50 material and the nano-structure TKX-50 material, it was found that the latter material showed a higher sensitivity to both friction and impact than the raw TKX-50 material. With respect to the impact sensitivity, this is indicated by the lower drop height for 50% explosions (H_{50}, 2 kg hammer, WL-1 type apparatus) for the nanomaterial (as low as 103.3 cm) compared to that for raw TKX-50 using the same apparatus (>120 cm) (Table 13) (Cao et al. 2019). Using the MGY-1 type friction apparatus (2 kg pendulum hammer, <90° tilt angle, 3.0 MPa pressure, 20 mg sample), the results are not so clear; however, there appears to be a trend to lower friction sensitivity for the nano-sized material since raw TKX-50 material showed an 85% probability of explosion, whereas two out of the three nano-sized samples showed a 75% probability of explosion—the exception being the one nano-sample with a probability of explosion of 90% (Table 12) (Cao et al. 2019). The authors suggest the lower sensitivity of the nanostructures toward impact is due to the more facile formation of hotspots in the nanostructure TKX-50 in comparison with raw

Table 13 Comparison of the impact and friction sensitivities of raw TKX-50 and nano-sized TKX-50v (Cao et al. 2019)

Sample	Impact sensitivity (H_{50}/cm)	Friction sensitivity (%)
Raw TKX-50	>120	85
TKX-50-1	103.3	90
TKX-50-0.5	106.6	75
TKX-50-0.3	104.3	75

TKX-50 (Cao et al. 2019). The effect of TKX-50 particle size on the thermal decomposition and thermal stability of TKX-50 has also been investigated using three particle sizes: milliparticles (MMPs), microparticles (MPs), and nanoparticles (NPs) (Wang et al. 2018). The focus of this work was not the thermal decomposition of TKX-50 as much as the difference the particle size has on the thermal decomposition. It has been shown for other energetic materials that particle size can have a large influence on thermal properties which is attributed to the difference in the size of the surface area, surface energy, mass transfer energy, and heat transfer efficiency (Wang et al. 2018). The adiabatic ARC experiments again confirmed the two-stage decomposition of TKX-50; however, only the first stage of thermal decomposition will be discussed here in more detail. Using SEM, the differences in the particles could be described: TKX-50 NPs showed reticular-structured agglomerates with 100–200 nm pore diameters (Fig. 44), TKX-50 MPs exhibited tabular crystals with a high draw ratio (Fig. 44), and TKX-50 MMPs formed ellipsoids with a particle size of approximately 1 mm (Fig. 44) (Wang et al. 2018). It was found that all three samples showed a two-stage decomposition process which was shifted to lower temperature for nano-TKX-50 in comparison with milli- and micro-TKX-50 (Wang et al. 2018). The self-decomposition of NP TKX-50 started at 185.8 °C and had a duration of 189.8 min and an adiabatic temperature increase of 15.09 °C (Wang et al. 2018). The self-decomposition of MP and MMP TKX-50 started at 200.22 °C and 201.16 °C, respectively, with adiabatic temperature increases of 22.66 °C and 21.24 °C, respectively, and both showed a faster acceleration reaction in comparison with that of NP TKX-50 (Wang et al. 2018). Using ARC, it could be determined that the first-stage decomposition of NP TKX-50 was the mildest out of the three particle sizes tested with MP TKX-50 showing the most violent reaction (Wang et al. 2018). It was furthermore concluded by the authors that the thermal safety of TKX-50 could be increased by reducing the particle size of TKX-50 to the nano-level from the micro-size (Wang et al. 2018), since the E_a value for first-stage decomposition was higher for nano-TKX-50 (503.29 kJ/mol) than for micro-TKX-50 (287.72 kJ/mol) (Wang et al. 2018). The authors, however, continued that depending on the cooling rate and control of the crystal growth process, the milli-particle size TKX-50 in this work showed less interior defects in the crystals which the authors attribute to being the reason behind the higher E_a value for MMP TKX-50 (563.82 kJ/mol) than for MP or even NP TKX-50 (Wang et al. 2018). For the first stage in the thermal decomposition of TKX-50, the data were summarized as is shown in Table 14 (Wang et al. 2018). An analysis of the second-stage decomposition of TKX-50 was also discussed in the same work (Wang et al. 2018); however, it is not included in this review. That the NP TKX-50 showed the mildest decomposition in the first-stage decomposition process was also supported by examination of the time-dependent pressure and pressure change rate, which showed that the pressure rise of the NPs (357.26 kPa) was less and much slower than those of the MMPs and MPs in the first stage of decomposition (Wang et al. 2018). Although the thermal decomposition obtained from ARC is important, it is not the only relevant approach to considering the thermal stability of an explosive. The SADT is a useful value which can be calculated and used to estimate the thermal hazard that an explosive poses during storage and transport. The

Fig. 44 SEM images of the three particle types with different particle sizes: TKX-50 NPs (nano), TKX-50 MPs (micro), and TKX-50 MMPs (milli) (Wang et al. 2018). Reproduced with kind permission from the publisher from Wang et al. (2018)

SADT is defined as the lowest temperature at which self-accelerating decomposition may occur for a substance in the packaging used in transportation. The simulated temperature of no-return, T_{NR}, and the SADT values for NP, MP, and MMP TKX-50 are shown in Table 15 for 50 kg TKX-50 stored in wooden drums (Wang et al. 2018). From the data given in Table 14, it can be concluded that since the simulated SADT value for MMP TKX-50 is the highest, that MMP TKX-50 shows the highest thermal stability and not nano-TKX-50. Nano-TKX-50 had a higher SADT value than the MP

Table 14 Data for the first stage in the thermal decomposition of TKX-50 from ARC under real adiabatic and near adiabatic conditions

Parameters	TKX-50 NPs		TKX-50 MPs		TKX-50 MMPs	
	Measured data	Corrected data	Measured data	Corrected data	Measured data	Corrected data
T_o (°C)	185.77	183.38	200.22	195.42	201.16	197.50
T_f (°C)	200.86	1184.60	222.88	1698.91	222.40	1673.12
ΔT_{ad} (°C)	15.09	1001.22	22.66	1503.49	21.24	1475.62
m_o (°C/min)	0.029	1.924	0.028	1.858	0.028	1.858
m_m (°C/min)	0.166	11.014	1.301	86.321	0.358	23.753
T_m (°C)	198.23	1010.10	219.04	1448.93	214.90	1187.44

T_0 = initial self-heat temperature, T_f = final decomposition temperature, ΔT_{ad} = adiabatic temperature rise, m_o = initial temperature rise rate, m_m = maximum temperature rise rate, T_m = temperature of maximum temperature rise rate (Wang et al. 2018)

Table 15 Values for T_{NR} and SADT for NP, MP, and MMP TKX-50 (Wang et al. 2018)

Parameters	T_{NR} (°C)	SADT (°C)
TKX-50 NPs	195.14	191.52
TKX-50 MPs	188.69	182.53
TKX-50 MMPs	208.73	205.31

TKX-50, which described as being untreated crystals (Wang et al. 2018). Overall, the authors concluded that the high SADT value indicates that TKX-50 should show good thermal stability during the storage and transportation process (Wang et al. 2018).

5 Formulations Containing TKX-50

5.1 PBXs Containing TKX-50

Polymer-bonded explosives (PBXs) are mixed explosives—that is they are a blend of one or more explosives which constitute the main component and contain one or more polymeric binders as the minor component. PBXs usually consist of a high percentage of the energetic material (e.g., 90%). Since the main component of PBXs is the explosive(s), the performance of the PBX is largely dependent on the explosive compound(s) which it contains, thus making the choice of explosive and its explosive performance of utmost importance. The addition of a polymer binder to an explosive can result in advantageous properties being observed since the polymer can hold brittle explosive crystals together resulting in an improvement of the brittleness and also result in an improvement of the compressibility and response to

Table 16 E_{bind} (binding energies) of the fluoro-polymers PVDF and PCTFE with the different crystal surfaces of TKX-50 (Ma et al. 2016)

Interface	E_{bind} (kJ/mol)
TKX-50 (010)/PVDF	520.97
TKX-50 (011)/PVDF	829.47
TKX-50 (100)/PVDF	776.02
TKX-50 (010)/PCTFE	458.06
TKX-50 (011)/PCTFE	580.62
TKX-50 (100)/PCTFE	556.89

external stimuli such that the energetic material in the PBX is desensitized with respect to the pure energetic material. Preliminary investigations have been undertaken with respect to the formation of plastic bonded explosives (PBXs) using TKX-50 (Ma et al. 2016). The interest in such work is to improve the safe-handling and practical applications of an energetic material. For example, computational work on TKX-50 containing PBXs with polyvinylidene difluoride (PVDF) and polychlorotrifluoroethylene (PCTFE) both showed that the binding energies (E_{bind}) between the fluoro-polymers followed the following order: (011) > (100) > (010) (Ma et al. 2016), and furthermore, shorter interatomic distances and higher E_{bind} were observed for the TKX-50/PVDF interfaces compared to the TKX-50/PCTFE interfaces (Ma et al. 2016). The E_{bind} describes the interaction between the energetic material and the polymer in a PBX and is defined as: $E_{bind} = -E_{inter} = -(E_{total} - E_{TKX\text{-}50} - E_{polymer})$ (whereby E_{total} is the energy of the explosive surface and the polymer) (Ma et al. 2016). A summary of the calculated E_{bind} values for TKX-50/fluoro-polymer on different crystal surfaces of TKX-50 is given in Table 16 (Ma et al. 2016). Computational results using MD showed that the fluoro-polymers were found to desensitize the TKX-50 in the PBXs in comparison with pure TKX-50 and also showed improved ductility in comparison with pure TKX-50 (Ma et al. 2016). The calculated and reported mechanical properties of TKX-50 and its PBXs are shown in Table 17 (Ma et al. 2016). Importantly, a comparison of RDX, HMX, and TKX-50 shows that in the pure materials, the higher K and G values of TKX-50 compared to those of RDX and HMX indicates that TKX-50 shows a better load bearing and stiffness. The ductility can be estimated by the ratio of K/G, and from Table 17, it can be concluded that the average ductility of TKX-50 is higher than that of RDX, and is only slightly lower than that of HMX, meaning that especially in comparison with RDX, TKX-50 shows low brittleness and a higher ductility and more plastic deformation rather than fracture on exposure to external stress (Ma et al. 2016). In pure TKX-50, the *K*, *G*, and *E* values increase in the order (011) < (100) < (011), but these values show a dramatic decrease for the TKX-50/fluoro-polymer interfaces which means that less strain is required to produce the same strain in the PBXs than in pure TKX-50, which is attributed to the flexibility and amorphous properties of the fluoropolymer chains (Ma et al. 2016).

The influence of external stimuli such as impact or friction on explosives is of utmost importance, since accidental detonation during handling or storage must be avoided. External stimuli such as impact or friction can result in compression or

Table 17 Calculated and reported mechanical properties of TKX-50 in comparison with RDX and HMX as well as in comparison with the TKX-50-based PBXs containing the fluoro-polymers (Ma et al. 2016)

Interface	Bulk modulus K (GPa)	Shear modulus G (GPa)	Poisson's ratio $\upsilon \times 10^{-1}$	Tensile modulus E (GPa)	K/G
RDX	11.70	6.00	–	–	2.00
HMX	12.50	3.40	–	–	3.70
TKX-50	31.58	9.07	3.69	–	–
TKX-50 (010)	32.50	9.55	3.66	26.09	3.40
TKX-50 (011)	24.43	7.66	3.58	20.81	3.19
TKX-50 (100)	28.23	8.36	3.65	22.80	3.38
TKX-50 (010)/PVDF	23.27	6.14	3.79	16.94	3.79
TKX-50 (011)/PVDF	17.03	4.83	3.70	13.24	3.53
TKX-50 (100)/PVDF	19.31	5.27	3.75	14.48	3.67
TKX-50 (010)/PCTFE	15.78	4.10	3.80	11.32	3.85
TKX-50 (011)/PCTFE	11.84	3.32	3.72	9.11	3.57
TKX-50 (100)/PCTFE	13.96	3.80	3.75	10.46	3.67

shear slide which results in internal stress or high sensitivity of the energetic material, since the generation of hotspots has been related to the amount of internal stress an explosive experiences (Ma et al. 2016). This is another advantage that the formation of a PBX can bring—it can result in a lowering of the sensitivity of an energetic material, thereby facilitating the safe handling and storage of the material. Computational work was again able to show that the internal stress caused by compression decreased in the following order: TKX-50 > TKX-50/PVDF > TKX-50/PCTFE > PVDF > PCTFE meaning that the addition of the fluoro-polymers PVDF or PCTFE results in a decrease in the internal stress and therefore also have a desensitizing effect on TKX-50 (Ma et al. 2016). A desensitizing effect of the fluoro-polymers with respect to TKX-50 is also reported from MD simulations for the bulk shear process (Ma et al. 2016). The PCTFE polymer shows a lower E_{bind} with TKX-50 which is reported to make the PCTFE chains easier to slide on the TKX-50 surface compared to the PVDF fluoro-polymer and TKX-50 (Ma et al. 2016). The first experimental investigation of TKX-50 containing PBXs was recently published (Niu et al. 2018). The PBX was prepared by the solution-water suspension method, in which approximately 90% (by mass) of TKX-50 was dissolved in water under stirring, and to this, 10% (by mass) binder $[(C_3H_4O_5N)_n]$ dissolved in ethyl acetate was slowly added while maintaining the temperature of the mixture at 70 °C, resulting in evaporation of the ethyl acetate and the sticking of the binder to the surface of TKX-50 to form the PBX abbreviated

PBX-T (Fig. 45) (Niu et al. 2018). Using thermal analysis (DSC, TG-DTG, β = 10 °C/min in both), it was found that there are three main mass loss processes in PBX-T (Niu et al. 2018). The first stage is due to decomposition of the binder at 184.3–205.2 °C (peak maximum at 195.2 °C in DTG) and corresponds to a 9.97% mass loss (Niu et al. 2018). The second-stage decomposition corresponds to the thermal decomposition of TKX-50 and occurs at 205.2 °C – 243.72 °C (peak maximum at 232.8 °C in DTG), and the third stage also corresponds to the thermal decomposition of TKX-50 and occurs at 243.7–271.96 °C (peak maximum at 258.9 °C in DTG), and the combined mass loss of stages 2 and 3 is 89.98% (Fig. 46) (Niu et al. 2018). The authors state that the main decomposition stage of PBX-T (stage 2) is faster (5.05% °C^{-1}) than the stage 1 decomposition of binder-free TKX-50 (3.73% °C^{-1}), which the authors suggest may be due to an accelerating affect caused by decomposition of

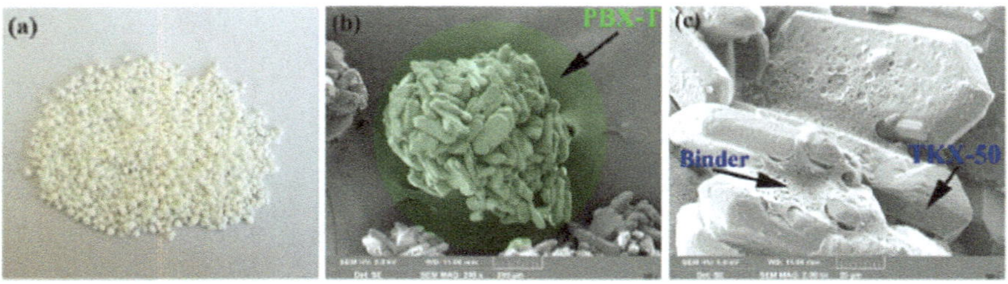

Fig. 45 Normal photograph of PBX-T showing the spherical shape of the powder (**a**), SEM images showing the effect of the binder on the TKX-50 crystals in PBX-T (**b**), (**c**) (Niu et al. 2018). Reproduced with kind permission from the publisher from Niu et al. (2018)

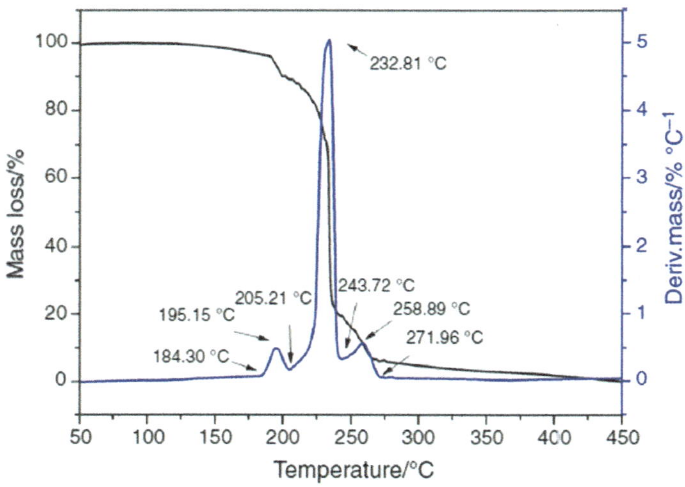

Fig. 46 TG-DTG curves for PBX-T at a heating rate of 10 °C/min showing the three stage decomposition process (Niu et al. 2018). Reproduced with kind permission from the publisher from Niu et al. (2018)

Table 18 Summary of the difference in the thermal performance between raw TKX-50 and the TKX-50-containing PBX, PBX-T, where T_{e10} = initial decomposition temperature at the heating rate of 10 °C/min; T_{p10} = maximum peak temperature at the heating rate of 10 °C/min; ΔH = heat generated by the decomposition reaction; $C\Theta$ p = specific heat capacity at 298 K: t_{TIad} = adiabatic time-to-explosion (Niu et al. 2018)

	T_{e10} (°C)	T_{p10} (°C)	ΔH (J g^{-1})	$C\Theta$ p (J g^{-1} K^{-1})	t_{TIad} (s)
PBX-T	225.23	234.67	2423.72	1.17	41.07
TKX-50	229.79	238.08	2199.61	1.13	112.65

the binder, which occurs at temperatures below that of stage 2 of PBX-T (Niu et al. 2018), and leads the authors to propose that the decomposition mechanism of PBX-T may be different from that of raw TKX-50 (Niu et al. 2018). The DSC tests also confirmed the three-stage decomposition process of PBX-T, with three exothermic peaks being observed at 196.5, 234.7, and 262.8 °C (Niu et al. 2018).The authors then applied the Kissinger and Ozawa methods to obtain an activation energy to thermal decomposition for PBX-T of 165.48 kJ/mol (Ozawa) and 165.68 kJ/mol (Kissinger) (Niu et al. 2018). A comparison of the thermal performance of TKX-50 and PBX-T is given in Table 18 (Niu et al. 2018).

The SADT of PBX-T was calculated to be 184.98 °C which indicates that PBX-T shows good thermal stability and the adiabatic time-to-explosion value was 41.0 s (Niu et al. 2018). Another important parameter is the 5 s delay explosion temperature, which for PBX-T was found through testing and calculation to be 287.43 °C (Niu et al. 2018). However, it is not just thermal aspects which are important for assessing the stability of a PBX, the sensitivity of a PBX (or explosive) to external mechanical stimuli such as impact or friction is also important. Using a 5 kg drop weight, the H_{50} value for PBX-T was found to be 56 cm. Unfortunately, the authors did not give the values of any other PBXs or explosives for comparison (Niu et al. 2018); however, they concluded that this test indicated that PBX-T is insensitive to impact (Niu et al. 2018). The thermal stabilities of TKX-50-based polymer-bonded explosives were investigated using DSC, TG-DTG, and ARC and the TMR_{ad} and SADT values determined (Niu et al. 2017). The TKX-50-containing PBX sample (PBX-T) was prepared using the solution-water suspension method, in which 90% TKX-50 was added to water under stirring to which 10% of the binder $(C_3H_5O_5N)_n$ dissolved in ethyl acetate was added while maintaining the solution at 70 °C (Niu et al. 2017). Due to the elevated temperature, the ethyl acetate evaporated and the binder coated the surface of the TKX-50 crystals which then stuck together to form the TKX-50-based PBX (Niu et al. 2017). A closed look at the samples using SEM showed that while the TKX-50 crystals are flat and spindle-shaped with smooth surfaces, the PBX-T powder was spherical; however, although the binder stuck the TKX-50 crystal together, the TKX-50 crystals were not enclosed inside the binder, meaning that a considerable amount of TX-50 crystals are still located on the surface of the PBX-T (Niu et al. 2017). The thermal stability was investigated using DSC and showed two clear exothermic peaks at 183.16 and at 215.34 °C, the latter of which corresponds to the first-stage decomposition process of TKX-50 (Niu et al. 2017).

Table 19 First-stage thermal decomposition temperatures of raw TKX-50 and TKX-50 in the PBX PBX-T at different heating rates using DSC where T_0 = initial decomposition temperature, T_p = maximum peak temperature, and ΔH = heat generated by decomposition reaction (Niu et al. 2017)

Sample	β (°C/min)	T_0 (°C)	T_p (°C)	ΔH (J g^{-1})
TKX-50	2	205.78	215.45	−2214.50
	4	216.44	225.40	−2161.03
	6	219.20	229.10	−2462.71
	8	225.45	234.25	−2249.73
	10	229.79	238.08	−1912.35
PBX-T	2	206.87	215.38	−2678.20
	4	215.34	222.44	−2392.59
	6	218.75	277.26	−2273.68
	8	221.77	230.70	−2471.54
	10	225.23	234.67	−2302.61

The peak at 183.16 °C was assigned to decomposition of the binder and using TG was found to correspond to approximately 10% mass loss, in agreement with the mass percent of binder present in the sample (Niu et al. 2017). A comparison of the temperatures of first-stage decomposition of TKX-50 and PBX-T at different heating rates using DSC showed that at the same heating rate, there was essentially no change in the decomposition temperature of TKX-50 in the raw material and in the PBX (PBX-T) (Table 19) (Niu et al. 2017). Therefore, the authors concluded that the binder only has a very weak effect on the decomposition temperature of TKX-50 (Niu et al. 2017). The authors found though, that the decomposition rate of TKX-50 in PBX-T was faster than that of raw TKX-50 which was proposed result from the decomposition of the binder accelerating the decomposition of TKX-50 in PBX-T (Niu et al. 2017). The SADT values of raw TKX-50 and of PBX-T in wooden drums were calculated using AKTS software and found to be 128.87 °C and 92.02 °C, respectively (Niu et al. 2017). Therefore, it was concluded that the thermal stability of TKX-50 decreases on addition of the binder. Figure 47 shows the calculated SADT profile for raw TKX-50 (Niu et al. 2017). Therefore, it can be concluded that during transportation and storage, the temperature raw TKX-50 is exposed to must be maintained below 128.87 °C, and for PBX-T, the temperature would have to be kept below 92.02 °C (Niu et al. 2017); otherwise, above these temperatures, the explosive materials will begin to decompose releasing heat which could result in a thermal explosion (Niu et al. 2017). Combining the advance method with finite element analysis, it could be shown that if the surrounding temperatures are much higher than the SADT temperatures given above, thermal explosions will happen very quickly and violently (Niu et al. 2017). At 50 and 100 °C, raw TKX-50 is predicted to be stable for a long period of time; however, when the surrounding temperature is 150 or 300 °C—which is much higher than the calculated SADT of raw TKX-50—explosions would occur in a very short period of time, i.e., in 1.3 days at 150 °C and in less than one hour at 300 °C surrounding temperature (Niu et al.

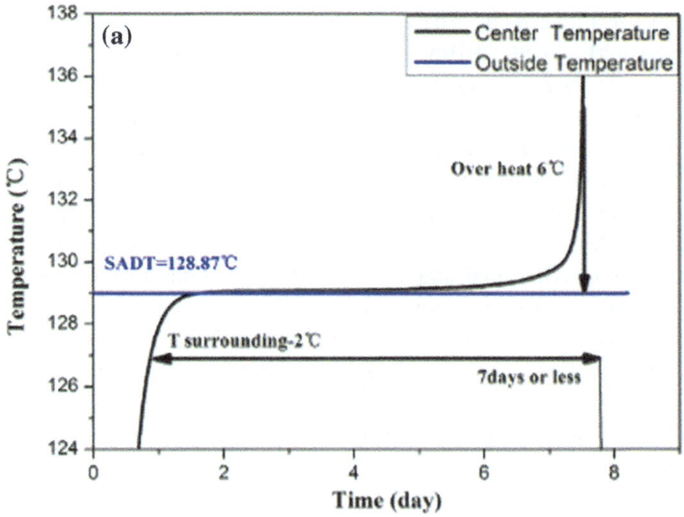

Fig. 47 SADT simulation for 25 kg of TKX-50 in a wooden drum (Niu et al. 2017). Reproduced with kind permission from the publisher from Niu et al. (2017)

2017). For PBX-T, these values are lower and correspond to 0.6 days at 150 °C and less than 1 h at 300 °C (Niu et al. 2017). Since the SADT of PBX-T is less than 100 °C, a weak thermal explosion of PBX-T would be expected after 3.8 days (Niu et al. 2017).

5.2 *Melt-Cast Explosives Containing TKX-50*

Melt-cast explosives are explosives which are normally used for formulations which contain TNT as the melt-castable component and RDX or HMX as the energetic filler and can be poured into molds of various shapes. Therefore, melt-cast explosives must have a low melting point. Melt-cast explosives are widely used for both industrial and military applications. One of the most important explosives used in melt-cast explosives is TNT, due to its low melting point, good thermal and chemical stability, inexpensive production costs, and safe-handling. Usually, in melt-cast explosives involving TNT, one (or more) of the other components are selected to increase the oxygen balance of the system, since the poor oxygen balance of TNT affects its detonation performance. One of the most famous melt-cast explosives is Composition B, which is a mixture of 39.5% TNT, 59.5% RDX, and 1% wax.

Recently, the suitability of TKX-50 as a replacement for RDX in a melt-cast explosive with TNT as the other component was investigated (Li et al. 2019b), by mixing different ratios of TNT with TKX-50 and TNT with other explosives for comparison and investigating the thermal stability and compatibility at different mass ratios of TNT:TKX-50. The explosive performance of the mixtures was not investigated yet (Li et al. 2019b). A further investigation into the possibility of using TKX-50 in melt-cast explosives was also reported in which TKX-50 was investigated as a possible replacement for RDX in TNT-containing melt-cast explosives with and without the presence of Aluminum (Badgujar and Talawar 2017). The low mechanical sensitivity of both RDX and TKX-50 and comparable performance justify this. The incorporation of Aluminum in melt-cast explosives results in an enhanced blast effect due to the secondary combustion of aluminum widening the detonation zone (Badgujar and Talawar 2017). Aluminum-containing energetic compositions undergo two reaction stages: In the first stage, the CHNO explosive undergoes detonation, whereas in the second stage, the detonation products which are produced in the first stage react at high temperature with the aluminum in a very exothermic reaction (Badgujar and Talawar 2017). The aluminized compositions (25% explosive (TKX-50 or RDX), 60% TNT, 15% Al) which were used in this study were prepared analogously to the non-aluminized compositions (30% explosive (TKX-50 or RDX), 70% TNT) that were used for comparison, by adding TKX-50 to molten TNT under continuous stirring followed by the addition of aluminum on a scale of up to 10 g of composite (Badgujar and Talawar 2017). DSC curves for raw TKX-50, and the TKX-50/TNT, and TKX-50/TNT/Al compositions show an endotherm at 81.6 and 79.27 °C for the latter two, which corresponds to the melting of TNT. The exothermic decomposition peak for TKX-50 was observed at 244.67 °C, whereas in TKX-50/TNT, the only exotherm observed was found at 285 and at 303.77 °C in TKX-50/TNT/Al (Badgujar and Talawar 2017). For comparison, the RDX/TNT and RDX/TNT/Al compositions show endotherms corresponding to the melting of TNT at 78.4 °C and 78.4 °C, respectively, whereas the exotherms corresponding to decomposition were observed at 231.6 °C and 231.3 °C, respectively, which were very similar to the decomposition exotherm for raw RDX which is observed at 238.0 °C (Badgujar and Talawar 2017). It must be pointed out here that for the TKX-50/TNT mixtures without Al, although the endotherm corresponding to the TNT melting showed only a range of approximately 5 °C, a considerable difference was observed for the first exothermic peak which is observed in the range of 172.92–197.34 °C depending on the heating rate, and an even larger difference in the temperatures observed for the second exotherm was observed depending on the heating rate, with a range of 266.2 °C–297.1 °C (Badgujar and Talawar 2017). Based on the DSC curves (see Sect. 2.2 for concerns about this approach), the authors calculated a value of E_a for the decomposition of TKX-50/TNT ($\beta = 5$ °C/min) at 266.2 °C of 223.04 kJ/mol, whereas for TKX-50/TNT/Al ($\beta = 5$ °C/min), it was 120.37 kJ/mol using the Kissinger method (Badgujar and Talawar 2017). The impact sensitivity, h_{50}, of the TKX-50/TNT composition (82 cm) was found to be slightly higher than that of the TKX-50/TNT/Al composition (98 cm) and higher than the corresponding RDX compositions: RDX/TNT = 104 cm, RDX/TNT//Al => 170 cm (c.f. TNT = 130 cm) (Badgujar and Talawar 2017). However, the friction sensitivity of TKX-50/TNT (36 kg) was lower than that od RDX/TNT (30 kg) and comparable with that of TKX-50/TNT/Al (36 kg) and RDX/TNT/Al (36 kg) (Badgujar and Talawar 2017). Calculation of the detonation performance of the compositions using the BKW equation of state for the non-aluminized composites predicted VoD of 7921 m/s and P_{C-J} of 2.92 Pa for TKX-50/TNT at 1.72 g/cm^3, in comparison with VoD of 7844 m/s and P_{C-J} of 2.66 Pa for RDX/TNT at 1.71 g/cm^3 (Badgujar and Talawar 2017). The authors concluded that the TKX-50/TNT composition showed a better thermal stability and FS than the RDX/TNT composition (Badgujar and Talawar 2017).

A comparison of the structure and properties of TKX-50 formulations involving TNT with those of HMX formulations containing TNT was reported in a combined experimental and computational report (Yu et al. 2017). In particular, melt-cast formulations based on TKX-50 using TNT as the dispersant were investigated and compared with the same formulation in which HMX was present instead of TKX-50 (Yu et al. 2017). For each combination of energetic materials, a formulation containing no wax and a second containing paraffin wax were prepared (Yu et al. 2017). The explosive formulations were all prepared in a similar manner regardless of whether TKX-50 or HMX was used as follows (Yu et al. 2017): TKX-50 or HMX were added to molten TNT under continuous stirring followed by the addition of paraffin wax (where appropriate) to the molten system. After stirring for *ca.* 10 min, the mixture was transferred to a mold, cooled to ambient temperature and transformed into the required size (Yu et al. 2017). First of all, using X-ray powder diffraction was used to establish that during the melt-cast process that the energetic material components had survived unchanged in crystal form (Yu et al. 2017). Furthermore, SEM showed that TNT coated the TKX-50 crystals with only few TKX-50 crystals exposed on the surface (Yu et al. 2017). In the same formulation but which also includes wax, almost no TKX-50 crystals were found on the outside surface, which is also the case for KMX which the authors conclude shows that coating can be improved by the addition of paraffin wax (Yu et al. 2017). The thermal profiles of the formulation containing TKX-50 without wax shows an endothermic peak at 79.0–86.8 °C which corresponds to the melting of TNT (Yu et al. 2017). No endotherm of exotherm is observed in the temperature range 87.0–213.8 °C, after which an exotherm is observed at 254.7 °C followed by a small exotherm at 309.3 °C (Yu et al. 2017). Moreover, in the formulation containing paraffin wax, the exotherms remain relatively unchanged (Yu et al. 2017). Using DSC in combination with the Kissinger or Ozawa approaches, the activation energy of pure TKX-50 (148 $\pm$ 0.5 kJ/mol, Kissinger; 149 $\pm$ 0.4 kJ/mol, Ozawa) was found to be relatively similar to that of TKX-50/TNT ($E_a = 144 \pm 0.3$ kJ/mol, Kissinger; 145 $\pm$ 0.1 kJ/mol, Ozawa), but in the formulation containing wax (TKX-50/TNT/paraffin wax), the E_a was found to be higher (161 $\pm$ 0.6 kJ/mol, Kissinger; 161 $\pm$ 0.3 kJ/mol, Ozawa) which the authors propose suggests that the addition of wax could improve the thermal stability of TKX-50 formulations, in contrast to its effect on the corresponding HMX-containing formulation (Yu et al. 2017). *Muravyev* has voiced concerns about using DSC in combination with the Kissing or Ozawa methods for assessing the activation energy to thermal decomposition for TKX-50 as is mentioned in Sect. XX (Muravyev et al. 2017).

The specific heat capacity of TKX-50/TNT was found to be 1.40 J (g^{-1} °C) and of the TKX-50/TNT/wax formulations 1.42 J (g^{-1} °C) at 25 °C which was higher than that of the corresponding HMX-based formulations (Yu et al. 2017). This was interpreted by the authors as providing evidence for the better heat absorbing capacity of TKX-50-based formulations compared to the HMX formulations, meaning that the TKX-50-based formulations can reduce the temperature of hotspots, thereby preventing the explosive self-accelerating reaction thereby reducing the likelihood of a chemical reaction occurring at the hotspot (Yu et al. 2017). This has the important consequence of making the TKX-50-based formulations less sensitive to impact (TKX-50/TNT H_{50} = 63 cm, TKX-50/TNT/wax H_{50} = 66 cm, 5 kg drop weight) than the corresponding HMX-based formulations which were studied (Yu et al. 2017). The stability of the TKX-50 formulations was compared to those of the corresponding HMX formulations also using VST, whereby it was found that EM/TNT and EM/TNT/wax (EM = TKX-50 or TNT) all showed good thermal stability (Yu et al. 2017) and also the 5 s explosion temperatures were 310 °C again indicating good thermal stability (Yu et al. 2017).

5.3 Composite Explosives Containing TKX-50

Composite explosives are mixtures of a crystalline energetic material (e.g., TKX-50) and a polymeric binder, which has been cured with isocyanate (e.g., hydroxy-terminated polybutadiene, HTPB/diisocyanate). Very recently TKX-50-based composite explosives were prepared with TNT or TNT/wax as the carrier and compared with NTO-, HMX-, or RDX-based formulations (Zhou et al. 2019). The properties which were selected to be investigated were the following: crystal stability, surface micro-topography, thermal stability, energetic performance and mechanical properties and sensitivities (Zhou et al. 2019). Possibly the most famous composite explosive is Composition B which consists of 60% (w/w) RDX in molten TNT, and which has been used for almost 60 years. Comp B shows good performance, processability and is convenient to produce however there are concerns about RDX due to its high toxicity, and therefore, currently much attention is being focused on TKX-50 as a possible replacement for RDX in composite explosives. In the investigations involving TKX-50, both wax-containing and non-wax formulations involving TKX-50 and TNT were investigated (Zhou et al. 2019). The non-wax formulations were prepared as follows: TNT was completely melted under stirring in a stainless-steel beaker in an oil bath at 100 °C. The TKX-50 was then added to the melt under stirring and subsequently cooled to ambient conditions and mechanically formed into the desired form (Zhou et al. 2019). The same process was also applied to generate the TKX/TNT formulation containing wax, with the only difference being the addition of the wax to the molten TNT before the addition of TKX-50 (Zhou et al. 2019). Overall, two formulations containing TKX-50 were studied in this work, namely TT with the composition TNT 40%, TKX-50 60%, and TTW with the composition TNT 30%, TKX-50 50%, wa × 10% (Zhou et al. 2019). Due to the addition of TKX-50 to molten TNT at temperatures which exceeded 85 °C, X-ray powder diffraction patterns were obtained of the TT and TTW formulations which showed that in both, the TKX-50 remained unchanged and had not undergone decomposition or phase change on addition to the molten TNT at 85 °C (Zhou et al. 2019). The SEM images of both formulations showed that the TKX-50 crystal were well coated by the TNT and that TNT was evenly spread out over the surface of the TKX-50 crystals, with only few TKX-50 crystals remaining exposed (Zhou et al. 2019). The addition of wax as in TTW resulted in better coating than in the formulation without wax (Zhou et al. 2019). The thermal decomposition of raw TKX-50 and of the formulation of TKX-50 with TNT (TT) and of TKX-50 with TNT and wax (TTW) were compared using DSC at different heating rates (1 mg sample, dynamic nitrogen atmosphere), and in all three samples, an endotherm was observed at approximately 80 °C corresponding to the melting of TNT (Zhou et al. 2019). Furthermore, in both TT and TTW, the exotherm occurring at 221–256 °C [depending on the heating rate) remained essentially unchanged (Zhou et al. 2019). A comparison of the E_a values (see Sect. 2.2 for a critical evaluation of this approach (Zhou et al. 2019)] indicated that the composite formulations (not just for TKX-50 formulations but also for the corresponding RDX/TNT, NTO/TNT and HMX/TNT formulations with and without wax) exhibited a higher E_a (157.17 kJ/mol (Kissinger, TT), 157.55 kJ/mol (Ozawa, TT), 161.26 J/mol (Kissinger, TTW), 161.23 kJ/mol (Ozawa, TTW)) in comparison with the corresponding raw explosive (i.e., TKX-5 (148.05 kJ/mol (Kissinger), 149.44 kJ/mol (Ozawa)), RDX, HMX or NTO) which the authors concluded showed a decrease in the stability of the formulations compared to the raw explosive (Zhou et al. 2019). Within the series of formulations studied it was concluded that the stability of the raw explosives decreased in the following order: TKX-50 > RDX > HMX > NTO (Zhou et al. 2019) while the formulations (with or without wax) decreased in the following order: TKX-50 > RDX > NTO > HMX (Zhou et al. 2019), and therefore that the TKX-50 formulations (with or without wax) showed a higher thermal stability in comparison with the corresponding formulations in which RDX, HMX, or NTO replaced the TKX-50 (Zhou et al. 2019).

The 5 s explosion temperatures of the two TKX-50 formulations, namely TT without wax (312.17 °C) and TTW with wax (313.29 °C), showed that the addition of wax had little effect on the 5 s explosion temperature (Zhou et al. 2019). However, the 5 s explosion temperatures of the TKX-50 formulations were lower than those of the NTO-containing formulations335.7 °C without wax, 340.5 °C with wax), higher than those of the RDX formulations (296.7 °C without wax, 292.6 °C with wax), and similar to those of the HMX formulations (311.6 °C without wax, 310.1 °C with wax) (Zhou et al. 2019). An important feature with respect to safety if of course the impact sensitivity, since safe-handling is a must. The impact sensitivity values given as H_{50} (height of 50% probability of explosion in cm, hammer) were not given for raw TKX-50 and only for the two formulations. The value of H_{50} = 63.7 cm for the wax-free TKX-50-formulation TT was lower than that of the wax-containing TKX-50 formulation TTW for which H_{50} = 68.0 cm, meaning that the addition of wax appeared to reduce the impact sensitivity of the formulation (Zhou et al. 2019). The same trend was observed for the other wax-free and wax-containing formulations involving NTO (H_{50} = 64.2 cm wax free, H_{50} = 79.3 cm with wax), RDX (H_{50} = 58.8 cm wax free, H_{50} = 66.2 cm with wax), or HMX (H_{50} = 40.2 cm wax free,

$H_{50} = 46.8$ cm with wax) instead of TKX-50 (Zhou et al. 2019). Overall, the impact sensitivity of the formulations was as follows: HMX > RDX > TKX-50 > NTO (Zhou et al. 2019).

Finally, the tensile strength σ_t of the formulations was investigated, and the authors concluded that the tensile strength of the TKX-50 and NTO formulations containing wax decreased, whereas for HMX and RDX formulations containing wax it increased (Zhou et al. 2019). The authors attributed these observations to the fact that the wax was a good binder for HMX and RDX, whereas for TKX-50 and NTO the opposite was the case (Zhou et al. 2019). Furthermore, the authors concluded that the mechanical properties of TKX-50 composites were in the medium range, lower than those of HMX and RDX, but higher than those of the NTO composites (Zhou et al. 2019). A composite of TKX-50/GO in which GO is graphite oxide was prepared and investigated using a variety of techniques in order to establish whether the surface active GO sheets form a desired TKX-50 morphology during crystallization (Wang et al. 2017). The potentially energetic properties of GO (can undergo violent exothermic decomposition) make it particularly interesting (Wang et al. 2017). The TKX-50/GO composite was prepared by dissolving recrystallized TKX-50 in water at 55 °C, to which an H_2O/GO was added under ultrasound. Afterward, ethanol was added on stirring to yield the TKX-50/GO composite which was filtered off and dried (Wang et al. 2017). In order to assess the coating efficiency of GO with respect to TKX-50, the mass of GO coated on TKX-50 could be determined by putting the TKX-50/GO composite in hot water and on stirring the mixture, the TKX-50 dissolved in the water, whereas the GO was dispersed on the water. After filtration and drying, the amount of solid GO could be determined (Wang et al. 2017).

The morphology of TKX-50 and TKX-50/GO was studied using SEM and showed that the tabular crystal morphology of TKX-50 with high draw ratio and smooth crystal faces was not maintained in the TKX-50/GO composite which adopted a polyhedron morphology almost spherical-like structure with a low draw ratio and many wrinkles on the surface (Wang et al. 2017). The authors concluded that TKX-50 crystals were well coated on the surface by sheets of GO and the composite had a uniform size distribution of approximately 20 μm (Wang et al. 2017). A further comparison of TKX-50, GO and the TKX-50/GO composite using X-ray powder diffraction showed that the characteristic peak corresponding to GO was not observed in the diffraction pattern of the composite because of the high degree of crystallinity of TKX-50 which was much higher than that of GO (Wang et al. 2017). Due to the different morphologies of TKX-50 in raw TKX-50 and the composite, the intensities of peaks in the powder diffraction patterns of raw TKX-50 and the composite were different, in agreement with the conclusions from SEM (Wang et al. 2017). The presence of both TKX-50 and GO in the composite was shown by Raman spectroscopy, in which peaks attributable to each of the components were observed for the composite (Wang et al. 2017).

The thermal properties of the components and the composite were investigated using DTA, which showed that the first-stage decomposition exothermic peak temperatures showed little change (241.51 °C TKX-50/GO, 243.04 °C TKX-50) (Wang et al. 2017), indicating that the thermal stability of TKX-50 appeared to have not been reduced after coating with GO in the TKX-50/GO composite (Wang et al. 2017). Using the results from DTA ($\beta = 2$ K/min), the authors calculated a E_a of 164.49 kJ/mol (Kissinger), 164.48 kJ/mol (Ozawa) for the TKX-50/GO composite and 157.72 kJ/mol (Kissinger), 158.07 kJ/mol (Ozawa) for raw TKX-50 (Wang et al. 2017). For concerns regarding the use of this approach for TKX-50, see Sect. 2.22. Powder X-ray diffraction and DTA methods were used to study the aging behavior of the TKX-50/GO composite. After approximately 6 months, the results from both techniques showed that no significant changes had occurred in comparison with the freshly prepared TKX-50/GO sample (Wang et al. 2017).

The coating of the TKX-50 particles by GO in the TKX-50/GO composite was shown to result in a reduction in the impact and friction sensitivities in comparison with raw TKX-50 (Wang et al. 2017). Not only that, but again it was confirmed that the TKX-50/GO composite had a lower IS and FS than HMX-containing composites of CL-20-containing composites with which it was compared. A summary is given in Table 20 (Wang et al. 2017).

The properties of composite propellants containing TKX-50 have been investigated experimentally with different mass fractions of TKX-50 particles—specifically on TKX-50/HTPB binder slurries and AP/HTPB/Al propellant slurries (Pang et al. 2018). It is possible with the inclusion of specially selected energetic materials to increase the gravimetric specific impulse or the density specific impulse of propellants; however, the energetic materials chosen for this purpose must show low sensitivity to external stimuli such as impact and friction. Since TKX-50 shows a high detonation velocity, but at the same time a low sensitivity to friction (comparable or lower than those of RDX, HMX, and CL-20) and impact (lower than those of HMX,

Table 20 Summary of the IS and FS of TKX-50/GO as well as of HMX-containing and CL-20-containing composites (Wang et al. 2017)

Sample	Content of carbon materials (wt%)	IS		FS
		P_I (%)	H_{50} (cm)	P_F (%)
HMX/GO-1	1.0	70		40
HMX/GO-2	2.0	10	43	32
HMX/CNT	2.0		45	24
HMX/Viton/G	1.0		49	
HMX/Viton/GO	1.0		66	
HMX/(Wang et al. 2018) Fullerene3	1.0	60		70
CL-20/551glue/G	0.5	20		22
CL-20/551glue/GO	0.5	25		30
CL-20/551glue/rGO	0.5	22		28
TKX-50/GO	0.5	8	84	1

IS was measured using a WL-1 apparatus, H50, 50 mg sample, 5 kg drop weight. FS was measured using a MGY-1 pendulum friction apparatus, 20 mg sample, steel anvils, 1.5 kg pendulum hammer, 90° tilt angle, 3.92 MPa pressure P_F is the explosion probability

RDX, and CL-20), it is an interesting candidate not only as an explosive, but also as an ingredient in propulsion compositions (Pang et al. 2018). In order to assess the potential of TKX-50 as a possible future component of solid rocket motors, composite solid propellants containing the mass fractions indicated in Table 21 were prepared (Pang et al. 2018). It was found that the I_{sp} and velocity increases as the TKX-50 content of the propellant composition increases up to 10% mass fraction of TKX-50, whereas at higher mass fraction of TKX-50, a decrease is observed (Pang et al. 2018). In contrast, the adiabatic flame temperature (T_c) decreases with increasing mass fraction of TKX-50 (Pang et al. 2018). Therefore, the following summary can be given: I_{sp}: PHT-3 (266.7 s) > PHT-2 (266.3 s) > PHT-1 (265.4 s) > PHT-4 (261.4 s) > PHT-5 (256.7 s) > PHT-6 (255.2 s) and T_c: PHT-1 (3393.5 K) > PHT-2 (333.7 K) > PHT-3 (3261.4 K) > PHT-4 (3042.9 K) > PHT-5 (2698.1 K) > PHT-6 (2648.2 K) (Pang et al. 2018). The effect of TKX-50 on the experimentally determined energetic properties of the composite solid propellants studied is summarized in Table 22, from which

Table 21 Mass fraction compositions of the main ingredients of the composite solid propellants investigated (Pang et al. 2018)

Sample	HTPB (%)	Al (%)	AP (%)	TKX-50 (%)	Additives (%)
PHT-1	12	18	68	0	2
PHT-2	12	18	63	5	2
PHT-3	12	18	58	10	2
PHT-4	12	18	48	20	2
PHT-5	12	18	38	30	2
PHT-6	12	18	28	40	2

Binder is HTPB (hydroxyl-terminated polybutadiene) plasticized with di-2-ethylhexyl sebacate and cured with 2,4-toluene diisocyanate; Al = micro-sized powder, $d_{50} = 5\ \mu m$; AP = tri-modal ammonium perchlorate: First mode is pure research grade, average particle size = 105–147 μm, second mode aver age particle size = 178–250 μm, third mode is AP ground to average particle size of 1–5 μm in fluid energy mill; part of first mode AP was replaced by TKX-50 in the propellant compositions PHT-1–PHT-6

Table 22 Experimentally determined energetic properties of the composite solid propellants PHT-1–PHT-6 containing different mass ratios of TKX-50 (Pang et al. 2018)

Sample	Heat of explosion ($J\ g^{-1}$)		Heat of explosion efficiency (%)	Density [$kg\ m^{-3}$ (× 10^3)]		Density efficiency (%)
	Theoretical	Measured		Theoretical	Measured	
PHT-1	6609	6528	98.77	1.767	1.762	99.72
PHT-2	6497	6441	99.14	1.764	1.760	99.77
PHT-3	6364	6312	99.18	1.761	1.756	99.72
PHT-4	6026	5838	96.88	1.756	1.750	99.66
PHT-5	5775	5531	95.78	1.750	1.740	99.43
PHT-6	5437	5158	94.87	1.744	1.734	99.42

Table 23 Comparison of the IS and FS of composite solid propellants containing different mass fractions of TKX-50 (Pang et al. 2018)

Sample	Friction (*P*) (%)	Confidence level of 95% believe level	Impact (Nm)	Standard deviation S (logarithmic value)
PHT-1	92	75%, 99%	4.29	0.04
PHT-2	84	64%, 96%	4.71	0.12
PHT-3	80	59%, 93%	5.39	0.10
PHT-4	80	59%, 93%	11.55	0.11
PHT-5	76	55%, 91%	20.05	0.04
PHT-6	72	51%, 88%	24.68	0.08

it can be seen that the density of the TKX-50 containing composites show lower densities than composition PHT-1 which contains no TKX-50 (Pang et al. 2018). The authors attribute this to the lower density of TKX-50 than AP, and therefore on increasing mass fraction of TKX-50, increasing amounts of AP are replaced by TKX-50 resulting in an overall lower density for the composite propellant (Pang et al. 2018). The highest measured heat of explosion was for the composite containing not TKX-50 (PHT-1) which the authors considered to be a result of the lower heat of formation of TKX-50 (Pang et al. 2018). As was mentioned previously, although the energetic performance is important, it should not be at the expense of safety. Therefore, the sensitivity of the compositions PHT-1–PHT-6 were also investigated toward impact and friction with the results shown in Table 23 (Pang et al. 2018). Important, it can be concluded that all of the composition containing TKX-50 were insensitive to impact and friction under the test conditions, whereas the composition PHT-1 which contains no TKX-50 was more sensitive (Pang et al. 2018). Therefore, the authors concluded that the inclusion of TKX-50 in these samples had resulted in a decrease in the sensitivity and increase in the safety compared to the sample which was TKX-50-free (Pang et al. 2018). Another relevant factor in considering a composition propellant for potential use is the viscosity and resulting classification of the flow properties of the slurry. The authors found that the composites PHT-1–PHT-4 showed flowing properties classed as A or B and should therefore be acceptable for large-scale production (Pang et al. 2018). Overall, in the compositions studied, the flowing properties of the slurry were found to decrease and become poorer as the mass fraction of TKX-50 increased (Pang et al. 2018). Finally, the authors considered the burning rate of the propellant compositions, since the burn rate of the propellant determines the rate of gas generation which in turn determined the pressure inside the motor and overall thrust (Pang et al. 2018). The addition of TKX-50 was found to affect the combustion behavior of the composite solid propellant and change the burning rate (Pang et al. 2018). Overall, it was found that the burning rate and pressure exponent of the composite solid propellants which were studied decreased with increasing TKX-50 content (Pang et al. 2018). Interestingly, it is worth pointing out that in contrast to several other reports in the literature concerned with the thermal stability of raw TKX-50, the authors observed three exothermic decomposition processes in the curves of the TKX-50-containing composites: The first exotherm (at lowest temperature, 232.5 °C for PHT-5) was assigned to the main (first) stage decomposition of TKX-50, the second exotherm (249.7 °C for PHT-5) was described as being a small decomposition process which is affiliated with the first-stage decomposition process and finally a third exotherm at considerably higher temperature (340.0 °C for PHT-5) (Pang et al. 2018).

6 TKX-50 Cocrystals

The formation of energetic cocrystals is a relatively new area of energetic materials research which has attracted a lot of interest. Using the cocrystal methodology, it is hoped that the advantageous properties of particular energetic materials can be enhanced, while the disadvantageous properties improved on, by the formation of cocrystals of carefully selected pairs (or more) of energetic materials. Traditionally, recrystallization to obtain different polymorphs, polymer coating, and the formation of composites have been the methodologies used to modify the properties of a particular explosive, the incorporation of a second or third energetic material into the crystal lattice is clearly a different approach. An interesting report recently appeared regarding the possibility of modifying the too-high impact sensitivity of PETN using TKX-50 through the formation of an energetic cocrystal (Xiao et al. 2019). It was hoped that the resulting cocrystals would show lower impact sensitivity due to the low impact sensitivity of TKX-50, despite the high sensitivity of PETN, but that the explosive performance of PETN would not be negatively affected. In addition to the high impact sensitivity of PETN, it also shows poor chemical stability in the vacuum stability test (VST) (Xiao et al. 2019). There is a foundation for optimism that energetic properties can be modified and tailored by the formation of energetic cocrystals, since it has already been shown that the impact sensitivity of a HMX/TATB cocrystal was considerably reduced in comparison with that of pure HMX (Shen et al. 2011). In the case of the PETN/TKX-50 cocrystal, PETN was considered to be the sensitive ingredient and TKX-50 selected as the insensitive ingredient (Xiao et al. 2019). The synthesis of the TKX-50/PETN cocrystal used the solvent/non-solvent method in which both PETN and TKX-50 were both dissolved in DMF (1:1 molar ratio) at 60 °C using ultrasound. The clear solution that resulted was then slowly dropped into chloroform under stirring. After filtration and undergoing a vacuum drying process, a powder corresponding to the PETN/TKX-50 cocrystal was obtained (Xiao et al. 2019). The presence of TKX-50 and PETN in the cocrystal was established using IR and Raman spectroscopy, in which characteristic vibrational bands of both PETN and TKX-50 could both be observed in the spectra of the cocrystal (Xiao et al. 2019). In the powder X-ray diffraction, however, the diffraction pattern of the cocrystal was distinctly different from those of raw PETN and raw TKX-50 indicating that a cocrystal had been formed and not just a mixture of two different compounds (Xiao et al. 2019). Furthermore, SEM showed that the cocrystal material was homogeneous (mean granularity of 1 μm) with distinct morphology

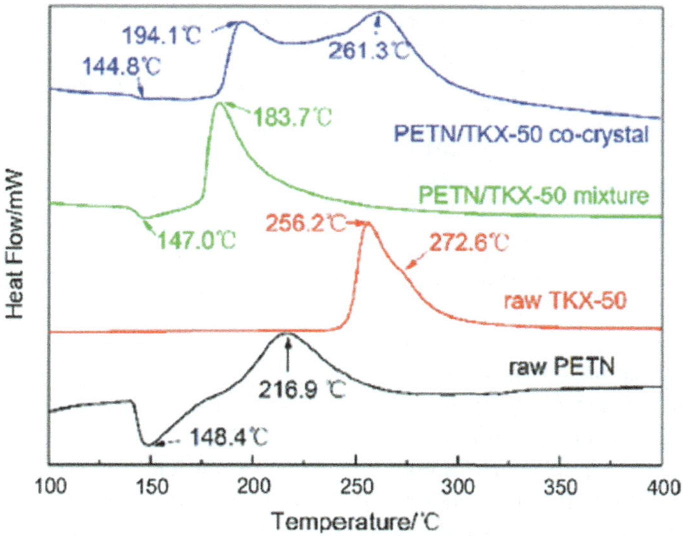

Fig. 48 Comparison of the DSC curves of the two raw energetic materials TKX-50 and PETN, as well as of a TKX-50/PETN mixture in comparison with the TKX-50/PETN cocrystal (β = 20 °C/min) (Xiao et al. 2019). Reproduced with kind permission from the publisher from Xiao et al. (2019)

from that of the PETN and TKX-50 components (Xiao et al. 2019). The thermal behavior of the cocrystal was also found to be different from that of the components. The DSC curve of the cocrystal showed a very weak endothermic peak at 148.4 °C followed by an exothermic peak maximum indicating decomposition at 194.1 and at 261.3 °C. In comparison, a mixture of TKX-50 and PETN showed an endothermic signal at 147.0 °C and exothermic peak maximum at 183.7 °C, whereas the raw components TKX-50 showed an exothermic peak maximum at 256.2 °C followed by a second at 272.6 °C and PETN an endothermic signal at 148.4 °C and exothermic peak maximum at 216.9 °C (Fig. 48) (Xiao et al. 2019). Interestingly, the same authors state that the compatibility of TKX-50 with PETN is poor and that on mixing the two components, an exothermic reaction occurs (Xiao et al. 2019). However, they state that in the cocrystal, there is good compatibility which they attribute to the formation of hydrogen bonds between the two components which are absent in the mixture (Xiao et al. 2019).

Despite the concerns of *Muravyev* et al. about the use of DSC/Kissinger/Ozawa methods to determined activation energies of decomposition for TKX-50 (see Sect. 2.2) (Gottfried and Witkowski 2017), the authors concluded that based on the results shown in Table 24, the higher E_a and ΔH values of the cocrystal in comparison with the two raw components indicated that the cocrystal required more energy to decompose (Xiao et al. 2019). Of fundamental importance to the TKX-50/PETN cocrystal was establishing its mechanical sensitivity (impact and friction) in comparison with the two components, in order to demonstrate the ability of the

Table 24 Summary of the DSC peak temperatures and kinetic parameters for the raw components, mixture, and cocrystal ($\beta = 5$ °C/min) (Xiao et al. 2019)

Sample	T_p (°C)	E_a (kJ/mol)	ΔH (kJ/mol)	ΔG (kJ/mol)	ΔS (J K^{-1} mol^{-1})
Raw PETN	194.7	104.4	100.3	137.9	−76.6
Raw TKX-50	229.7	108.4	104.0	150.5	−87.9
PETN/TKX-50 mixture	168.0	143.3	139.5	126.8	27.8
TKX-50/PETN cocrystal	175.1 241.6	123.2 152.0	119.4 147.6	130.6 150.5	−24.0 −5.4

cocrystal formation to reduce the sensitivity of PETN if combined with an energetic material of low sensitivity such as TKX-50 (Xiao et al. 2019). In fact, the IS of the cocrystal was decreased to a value of $H_{50} = 42.2$ cm in comparison with raw PETN ($H_{50} = 14.5$ cm) but was higher than that of raw TKX-50 ($H_{50} = 55.4$ cm) (Xiao et al. 2019). This indicated that the formation of the cocrystal resulted in a PETN-containing compound with lower impact sensitivity than raw PETN. A similar situation was also observed for the friction sensitivity. The probability of explosion of raw PETN (68%) was higher than that of the cocrystal (12%), whereas that of raw TKX-50 (28%) was lower than that of the cocrystal (12%) (Xiao et al. 2019). In both the IS and FS, the values for the PETN/TKX-50 mixture ($H_{50} = 20.7$ cm, $P =$ 60%) and PETN/TKX-50 cocrystal ($H_{50} = 42.2$ cm, $P = 12\%$) were significantly different (Xiao et al. 2019). Furthermore, the authors stated that the sensitivity of the cocrystal was lower than that of RDX ($H_{50} = 21.5$ cm, $P = 36\%$) (Xiao et al. 2019). The low sensitivity of the cocrystal was attributed by the authors to the likely formation of intermolecular hydrogen bonds between TKX-50 and PETN in the cocrystal (Xiao et al. 2019), which they suggest results in a closer packing of the molecules and reduces crystal defects (Xiao et al. 2019), in addition to the uniform particle size (~1 μm) and relatively smooth crystal surface (Xiao et al. 2019). In order to probe into more detail about the hydrogen bonding in the cocrystal, computational tools were used to simulate possible hydrogen bonds between the two components of the cocrystal (Xiao et al. 2019). The computational results were not discussed in detail by the authors; however, the simulated hydrogen bonds were shown in a figure (Fig. 49) (Xiao et al. 2019).

A theoretical (molecular dynamic simulations) report has been given of the TKX-50/HMX cocrystal, with the aim of determining whether a cocrystal of this type would result in an enhancement of the properties of the resulting energetic material in comparison with those of the individual components (Xiong et al. 2017). Since cocrystals by definition have different molecules incorporated into the same crystal lattice by intermolecular interactions and—crucially—in a fixed stoichiometric ratio, such systems can be simulated using molecular dynamics simulations. In order to estimate the strength of the interaction between components in a cocrystal, the binding energy E_{bind} must be calculated. The larger the E_{bind}, the stronger the strength of the interaction is and the higher the thermal stability of the cocrystal (Xiong et al. 2017).

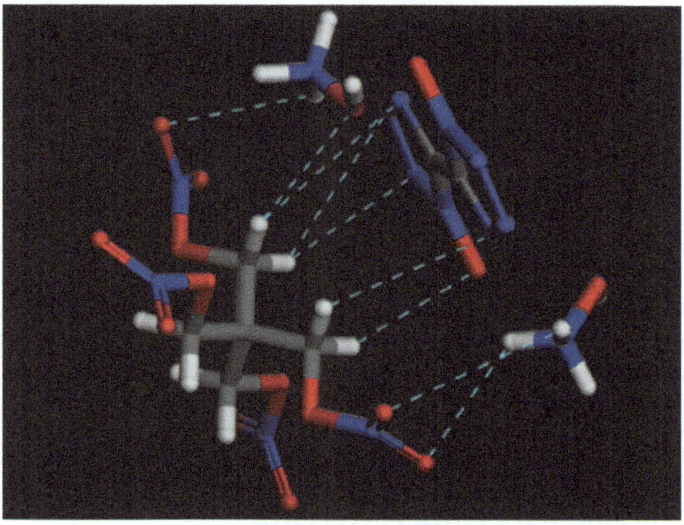

Fig. 49 Possible hydrogen bonds between PETN and TKX-50 in the cocrystal simulated using the computational tool material Studio (Xiao et al. 2019). Reproduced with kind permission from the publisher from Xiao et al. (2019)

Calculation of E_{bind} for the TKX-50/HMX cocrystal shows that substitution with HMX on the slow-growing facets of TKX-50 results in the more stable cocrystals (E_{bind} (kcal/mol): 2445.59 [011]/HMX; 2505.33 [100]/HMX; 2351.11 [022]/HMX; 2250.26 [110]/HMX; 2215.85 [11-1]/HMX; 1924.97 [12-1]/HMX) (Xiong et al. 2017).

Simulation of the powder X-ray diffraction spectrum indicated that the original characteristic peaks of the two components TKX-50 and HMX disappeared or decreased, whereas new peaks appeared in the TKX-50/HMX cocrystal which the authors concluded was indicative of destruction of the crystal structures of the individual components, and a new structure is formed in the cocrystal (Xiong et al. 2017). Finally, the authors concluded that the TKX-50/HMX cocrystal should show a lower sensitivity than raw HMX (Xiong et al. 2017).

A further cocrystal combination as examined, namely that of TKX-50 and CL-20 (Xiong et al. 2016). It is hoped that this multicomponent molecular crystal will have the advantageous properties of TKX-50 (high energy storage, low IS, and low toxicity) and ε-CL20 (very high energy density) without the disadvantages of ε-CL20, i.e., its high IS. The authors constructed models for the TKX-50/ε-CL-20 cocrystal using Material Studio software which were optimized by the polymer consistent force field (PCFF) and smart minimizer (SM) method. The equilibrium structures were then obtained after performing MD simulations (Xiong et al. 2016). The main growth facets of the TKX-50 crystal were established to consist of six dominant ones: [100], [011], [020], [110], [11-1], [12-1], with the [011] facet having the greatest morphological importance since it occupies the largest percentage area (Xiong et al.

2016). The [100], [011], and [020] facets are the slow-growing facets (Xiong et al. 2016). The calculated E_{att} values show that the values for the cocrystals of TKX-50 substituted with ε-CL-20 on its slow-growing facets are lower meaning that these cocrystal structures should be more stable (Xiong et al. 2016). Moreover, the lowest value for E_{att} was found for the cocrystal in which TKX-50 was substituted with ε-CL-20 on its [011] facet (Xiong et al. 2016).

As was discussed above for the TKX-50/HMX cocrystal, the calculated XRD diffraction peaks for the cocrystal do not simply correspond to a superimposition of those of the two components TKX-50 and ε-CL-20, but rather, the XRD pattern of the cocrystal is calculated to show new diffraction peaks, which the authors interpret as corresponding to the destruction of the crystal lattices of the TKX-50 and ε-CL-20 components, and the formation of a new structure corresponding to the cocrystal is formed (Xiong et al. 2016).

7 Conclusions and Future Perspective

This review has aimed to give an overview of the current standing of the research surrounding the energetic material TKX-50. Due to its low impact and friction sensitivity, facile synthesis, high density, and detonation performance, TKX-50 is currently one of the most promising candidates as a possible future replacement for standard secondary explosives such as RDX or HMX in many applications. Much of the current work on TKX-50 is computationally based; however, much remains to be investigated using experimental procedures. The decomposition mechanism of the first stage of the thermal decomposition of TKX-50 has been of considerable interest to various research groups and the main decomposition products have been identified. However, the kinetics of the thermal decomposition of TKX-50 is still being discussed by several research groups. Future investigations and scale-up tests will determine if TKX-50 has future applications and can fulfill its promise.

References

An Q, Cheng T, Goddard WA III, Zybin SV (2015) Anisotropic impact sensitivity and shock induced plasticity of TKX-50 (Dihydroxylammonium 5,5′-bis(tetrazole)-1,1′-diolate) single crystals: from large-scale molecular dynamics simulations. J Phys Chem C 119:2196–2207

An Q, Liu WG, Goddard WA, Cheng T, Zybin SV, Xiao H (2014) Initial steps of thermal decomposition of dihydroxylammonium 5, 5′-bistetrazole-1, 1′-diolate crystals from quantum mechanics. J Phys Chem C 118:27175–27181

Badgujar D, Talawar M (2017) Thermal and sensitivity study of dihydroxyl ammonium 5, 5′-Bistetrazole-1, 1′-diolate (TKX-50)based melt cast explosive formulations. Propellants, Explos, Pyrotech 42:883–888

Cao X, Shang Y, Meng K, Ye G, Yang L, Liu Y, Deng P, Hu L (2019) Fabrication of three-dimensional TKX-50 network-like nanostructures by liquid nitrogen-assisted spray freeze-drying method. J Energet Mater 37:356–364

Chen F, Zhou T, Li J, Wang X, Cao D, Wang J, Yang Z (2019) Crystal morphology of dihydroxylammonium 5, 5′-bistetrazole-1, 1′-diolate (TKX-50) under solvents system with different polarity using molecular dynamics. Comp Mater Sci 168:48–57

Cooper PW (1996) Explosives engineering. Wiley-VCH, New York

Dong W, Chen S, Jin S, Chen Y (2019) Effects of carboxymethylcellulose sodium on the morphology and properties of TKX-50, an insensitive high-energy explosive. J Energet Mater 37:199–211

Dreger ZA, Breshike CJ, Gupta YM (2017a) Chem Phys Letts 679:212–218

Dreger ZA, Stash AI, Yu ZC, Chen YS, Tao Y (2017b) High-pressure structural response of an insensitive energetic crystal: dihydroxylammonium 5, 5′-bistetrazole-1, 1′-diolate (TKX-50). J Phys Chem C 121:5761–5767

Dreger ZA, Tao Y, Averkiev BB, Gupta YM, Klapötke TM (2015) High-pressure stability of energetic crystal of dihydroxylammonium 5, 5′-bistetrazole-1, 1′-diolate: Raman spectroscopy and DFT calculations. J Phys Chem B 119:6836–6847

Eaton PE, Zhang MX, Gilardi R (2000) Hepta-and octanitrocubanes. Angew Chem Int Ed 39:401–404

Fischer D, Klapotke TM, Musanic SM, Stierstorfer J, Suceska M (2013) TKX-50. In: Pachman J, Selesovsky J, Matyas R (eds) Proceedings of the 16th seminar on new trends in research of energetic materials. University of Pardubice, Czech Republic, pp 566–577

Fischer N, Fischer D, Klapötke TM, Piercey DG, Stierstorfer J (2012) Pushing the limits of energetic materials–the synthesis and characterization of dihydroxylammonium 5, 5′-bistetrazole-1, 1′-diolate. J Mater Chem 22:20418–20422

Foltz MF, Coon CL, Garcia F, Nichols AL (1994a) The thermal stability of the polymorphs of hexanitrohexaazaisowurtzitane, Part I. Propellants, Explos, Pyrotech 19:19–25

Foltz MF, Coon CL, Garcia F, Nichols AL (1994b) The thermal stability of the polymorphs of hexanitrohexaazaisowurtzitane, Part II. Propellants, Explos, Pyrotech 19:133–144

Gao H, Shreeve JM (2011) Azole-based energetic salts. Chem Revs 111:7377–7436

Glasser L, Jenkins HDB, Klapötke TM, Anorg Z (2014) Allg Chem 640:1297–1299

Gottfried JL, Witkowski TMG (2017) Estimated detonation velocities for TKX-50, MAD- X1, BDNAPM, BTNPM, TKX-55, and DAAF using the Laser–induced air shock from energetic materials technique. Propellants, Explos, Pyrotech 42:353–359

Hammerl A, Holl G, Klapötke TM, Mayer P, Nöth H, Piotrowski H, Warchhold M (2002) Salts of 5, 5′-azotetrazolate. Eur J Inorg Chem 2002:834–845

He P, Zhang JG, Yin X, Wu JT, Wu L, Zhou ZN, Zhang TL (2016) Energetic salts based on tetrazole N-oxide. Chem Eur J 22:7670–7685

Huang H, Shi Y, Yang J, Li B (2015a) Compatibility study of dihydroxylammonium 5, 5′-bistetrazole-1, 1′-diolate (TKX-50) with some energetic materials and inert materials. J Energet Mater 33:66–72

Huang HF, Shi YM, Yang J (2015b) Thermal characterization of the promising energetic material TKX-50. J Therm Anal Calorim 121:705–709

Jia J, Liu Y, Huang S, Xu J, Li S, Zhang H, Cao X (2017) Crystal structure transformation and step-by-step thermal decomposition behavior of dihydroxylammonium 5, 5′-bistetrazole-1, 1′-diolate. RSC Adv 7:49105–49113

Jia J, Xu J, Cao X, Li S, Huang S, Liu Y, Li J (2019) Stability of dihydroxylammonium 5, 5′-bistetrazole-1, 1′-diolate (TKX-50) in solvents. Propellants, Explos, Pyrotech 44:989–999

Keshavarz MH, Abadi YH, Esmaeilpour K, Damiri S, Oftadeh M (2018) Centr Eur J Energet Mater 15:364–375

Keshavarz MH, Haz J (2009) Mat 166:762–769

Klapötke TM (vol ed), (2007) Structure and bonding. Mingos DMP (series ed) High energy density compounds, vol 125/2007. Springer, Berlin/Heidelberg

Klapötke TM (2017) Chemistry of high-energy materials, 4th edn. De Gruyter, p 179

Klapötke TM, Mayer P, Stierstorfer J, Weigand JJ (2008) Bistetrazolylamines—synthesis and characterization. J Mater Chem 18:5248–5258

Klapötke TM, Stierstorfer J (2009) The CN7—anion. J Am Chem Soc 131:1122–1134

Klapötke TM, Stierstorfer J (2014) Energetic tetrazole N-oxides. In: Brinck T (ed) Green energetic materials. Wiley, pp 133–177

Klapötke TM, Witkowski TG, Wilk Z, Hadzik (2016) Determination of the initiating capability of detonators containing TKX-50, MAD-X1, PETNC, DAAF, RDX, HMX or PETN as a base charge, by underwater explosion test. J Prop Expl Pyrotech 41:92–97

Klenov MS, Guskov AA, Anikin OV, Churakov AM, Strelenko YA, Fedyanin IV, Lyssenko KA, Tartakovsky VA (2016) Synthesis of Tetrazino-tetrazine 1, 3, 6, 8-Tetraoxide (TTTO). Angew Chem Int Ed 55:11472–11475

Klenov MS, Guskov AA, Anikin OV, Churakov AM, Tartakovskij VA (2015) Method for producing [1,2,3,4]-tetrazino-[5,6-e]-[1,2,3,4]-tetrazino-1,3,6,8-tetraoxide. Russian Patent RU2, 593,993 C1, 22nd

Lee WH, Kwon K (2019) Cover picture: safe synthesis of TKX-50 using an insensitive intermediate. Propellants, Explos, Pyrotech 44:1353–1353

Li JS, Chen JJ, Hwang CC, Lu KT, Yeh TF (2019a) Study on thermal characteristics of TNT based melt-cast explosives. Propellants, Explos, Pyrotech 44:1270–1281

Li M, Chen H, Xiao X, Yang L, Peng C, Qin Y, Wang T, Sun W, Wang C (2019b) Computational study of transition states for reaction path of energetic material TKX-50. J Energet Mater 37:240–250

Lu Z, Xue X, Meng L, Zeng Q, Chi Y, Fan G, Li H, Zheng Z, Nie F, Zhang C (2017a) Heat-induced solid-solid phase transformation of TKX-50. J Phys Chem C 121:8262–8271

Lu Z, Xue X, Zhang C (2017b) A theoretical prediction on the shear-induced phase transformation of TKX-50. Phys Chem Chem Phys 19:31054–31062

Ma S, Li Y, Li Y, Luo Y (2016) Research on structures, mechanical properties, and mechanical responses of TKX-50 and TKX-50 based PBX with molecular dynamics. J Mol Model 22:43

Meng L, Lu Z, Wei X, Xue X, Ma Y, Zeng Q, Fan G, Nie F, Zhang C (2016) Two-sided effects of strong hydrogen bonding on the stability of dihydroxylammonium 5, 5′-bistetrazole-1, 1′-diolate (TKX-50). Cryst Eng Comm 18:2258–2267

Muravyev NV, Monogarov KA, Asachenko AF, Nechaev MS, Ananyev IV, Fomenkov IV, Kiselev VG, Pivkina AN (2017) Pursuing reliable thermal analysis techniques for energetic materials: decomposition kinetics and thermal stability of dihydroxylammonium 5,5′-bistetrazole-1,1′-diolate (TKX-50). Phys Chem Chem Phys 19:436–449

Nair UR, Sivabalan R, Gore GM, Geetha M, Asthana SN, Singh H (2005) Combustion Explosion Shock Waves 41:121–132

Niu H, Chen S, Jin S, Li B, Li X, Wang J, Ma X, Bao F, Li L (2018) Preparation, nonisothermal decomposition kinetics, heat capacity, and safety parameters of TKX-50-based PBX. J Thermal Anal Calorim 131:3193–3199

Niu H, Chen S, Jin S, Li L, Jing B, Jiang Z, Ji J, Shu Q (2016) Thermolysis, nonisothermal decomposition kinetics, calculated detonation velocity and safety assessment of dihydroxylammonium 5, 5′-bistetrazole-1, 1′-diolate. J Therm Anal Calorim 126:473–480

Niu H, Chen S, Shu Q, Li L, Jin S (2017) Preparation, characterization and thermal risk evaluation of dihydroxylammonium 5, 5′-bistetrazole-1, 1′-diolate based polymer bonded explosive. J Hazard Mater 338:208–217

Pang W, Li J, Wang K, Fan X, De Luca LT, Bi F, Li H (2018) Propellants, Explos, Pyrotech 43:1012–1022

Shen JP, Duan XH, Luo QP, Zhou Y, Bao Q, Ma YJ, Pei CH (2011) Preparation and characterization of a novel cocrystal explosive. Cryst Growth Des 11:1759–1765

Simpson RL, Urtiew PA, Ornellas DL, Moody GL, Scribner KJ, Hoffman DM (1997) CL-20 performance exceeds that of HMX and its sensitivity is moderate. Propellants, Explos, Pyrotech 22:249–255

Sinditskii V, Filatov S, Kapranov K, Asachenko A, Nechaev M, Lunin V, Shishov N (2015) Thermochim Acta 614:85–92

Singh RP, Gao H, Meshri DT, Shreeve JM (2007) Nitrogen-Rich heterocycles. In: Klapötke TM (vol ed) High energy density materials. Springer-Verlag, pp 60–62

Sun J, Xu J, Liu Y, Zhang H, Shu Y (2012) Polymorphism of Hexanitrohexaazaisowurtzitane in antisolvent crystallization. ICT Karlsruhe, Germany, 95:1–95-12

Szwarc M (1949) The dissociation energy of the NN bond in hydrazine. Proc Royal Soc A 198:267–284

Talawar MB, Nandagopal S, Singh S, Majahan AP, Badgujar DM, Gupta M, Khan M (2018) AS Chemistry Select 3:12175–12182

Tidey JP, Zhurov VV, Gianopoulos CG, Zhurova EA, Pinkteron AA (2017) Experimental charge-density study of the intra-and intermolecular bonding in TKX-50. J Phys Chem A 121:8962–8972

Tselinskii IV, Mel'nikova SF, Romanova TV (2001) Synthesis and reactivity of carbohydroximoyl azides: I. aliphatic and aromatic carbohydroximoyl azides and 5-substituted. Russ J Org Chem 37:430–436

Wallis JD, Dunitz JD (1983) J Chem Soc Chem Comm 910–911

Wang J, Chen S, Jin S, Shi R, Yu Z, Su Q, Ma X, Zhang C, Shu Q (2020) J Thermal Anal Calorim 139:1771–1777

Wang J, Chen S, Jin S, Wang J, Niu H, Zhang G, Wang X, Wang D (2018) Size-dependent effect on thermal decomposition and hazard assessment of TKX-50 under adiabatic condition. Propellants, Explos, Pyrotech 43:488–495

Wang J, Chen S, Yao Q, Jin S, Zhao S, Yu Z, Li J, Shu Q (2017) Preparation, characterization, thermal evaluation and sensitivities of TKX-50/GO composite. Propellants, Explos, Pyrotech 42:1104–1110

Xiao L, Guo S, Su H, Guo B, Liu Q, Hao G, Hu Y, Wang X, Jiang W (2019) Preparation and characteristics of a novel PETN/TKX-50 co-crystal by a solvent/non-solvent method. RSC Adv 9:9204–9210

Xing X, Zhao S, Wang X, Zhang W, Diao X, Fang W, Li WX (2019) The detonation properties research on TKX-50 in high explosives. Propellants, Explos, Pyrotech 44:408–412

Xiong S, Chen S, Jin S, Zhang C (2016) Molecular dynamics simulations on dihydroxylammonium 5, 5′-bistetrazole-1, 1′-diolate/hexanitrohexaazaisowurtzitane cocrystal. RSC Adv 6:4221–4226

Xiong S, Chen S, Jin S, Zhang Z, Zhang Y, Li L (2017) Molecular dynamic simulations on TKX-50/HMX cocrystal. RSC Adv 7:6795–6799

Xu Y, Wang Q, Shen C, Lin Q, Wang P, Lu M (2017) A series of energetic metal pentazolate hydrates. Nature 549:78–81

Yu Y, Chen S, Li T, Jin S, Zhang G, Chen M, Li L (2017) Study on a novel high energetic and insensitive munitions formulation: TKX-50 based melt cast high explosive. RSC Adv 7:31485–31492

Zhai L, Bi F, Luo Y, Wang N, Zhang J, Wang B (2019) New strategy for enhancing energetic properties by regulating trifuroxan configuration: 3,4-Bis (3-nitrofuroxan-4-yl) furoxan. Nat Sci Rep 9:4321

Zhang C, Jin S, Chen S, Zhang Y, Qin L, Wei X, Shu Q (2016) Solubilities of dihydroxylammonium 5,5′-Bistetrazole-1,1′-diolate in various pure solvents at temperatures between 293.15 and 323.15 K. J Chem Eng Data 61:1873–1875

Zhao SX, Xia QY, Zhang C, Xing XL (2019) Theoretical design of bistetrazole diolate derivatives as novel non-nitro energetic salts with low sensitivity. Ju XH Struct Chem 30:1015–1022

Zhou MN, Chen SS, Wang DX, Yu YH, Wang JQ, Li JX, Wang N, Chen ML (2019) A comparative study of performance between TKX-50-based composite explosives and other composite explosives. J Energet Mater 37:162–173

Chapter 2
Stability Assessment for Double Base Rocket Propellant During Long Natural/Artificial Aging Using Various Methods and Kinetic Modeling

Salim Chelouche, Djalal Trache, Ahmed Fouzi Tarchoun, Kamel Khimeche, and Abderrahmane Mezroua

Abstract During storage of double base rocket propellant (DBRP) several chemical and physical processes occurred at normal storage conditions leading to a decrease in the physicochemical properties as well as the performance of these solid propellants. In this work, the effect of long natural aging (up to 31 years of storage) and artificial aging (for 12 months at 338.65 K) on the chemical stability of three DBRPs have been inspected by spectroscopic techniques, stability tests, kinetic modeling on Vacuum Stabilty Test (VST) data, and the prediction of the lifetimes at different storage temperatures. With natural/artificial aging progress, FTIR and XRD results show a decrease in the intensities of the characteristic frequencies of nitrocellulose (NC) and the crystallinity indexes, respectively. Bergmann & Junk (B&J) test results showed that the unaged samples are stable according to the AOP-7 standard. Nevertheless, the VST results present some issues when compared to the same standard. Furthermore, the VST finding revealed three degradation stages during the aging of the DBRP. Kinetic modeling indicates a decrease in the activation energy value with aging progress and a change in the most probable reaction model after 10 months of artificial aging. The predicated lifetimes showed that the DBRPs could not be stored after 10 months of artificial aging and became hazardous.

Keywords Double base rocket propellant · FTIR · XRD · Stability tests · Kinetics · Storage lifetime

---

S. Chelouche (✉) · D. Trache (✉) · A. F. Tarchoun · K. Khimeche · A. Mezroua
Energetic Materials Laboratory, Teaching and Research Unit of Energetic Processes, Ecole Militaire Polytechnique, BP 17, 16046 Bordj El-Bahri, Algiers, Algeria
e-mail: salim.chelouche@gmail.com

D. Trache
e-mail: djalaltrache@gmail.com

D. Trache et al. (eds.), *Materials Research and Applications*,
Materials Horizons: From Nature to Nanomaterials,
https://doi.org/10.1007/978-981-15-9223-2_2

1 Introduction

Despite their lower performance compared to liquid propellants, most of modern missiles and rockets use solid propellants due to their simplicity, safety, lower cost, and minimum maintenance requirement (Chelouche et al. 2019f; Kuentzmann 2004). Double base propellants, which are consisting mainly of two nitrate esters, *viz*, nitrocellulose (NC) and nitroglycerine (NG) are one of the oldest propellant categories. NG acts as an energetic plasticizer. It is incorporated into the NC fibrous structure to increase the energetic performance and therefore, considered more suitable for rocket motor propellants rather than gun propellants (Davenas 2012). Moreover, double base propellants are easier to manipulate because of their high homogeneity and rigidity compared to composite propellants (Herder et al. 2003). Additives for easier manufacture, stability and ballistic characteristic improvement, and operational features increase are usually incorporated into DBRP formulation in order to ensure a safe and reliable function of rocket motors.

However, various chemical and physical processes, particularly, the diffusion/migration/evaporation of NG and the chemical decomposition of NC occur in DBRP blocks over the time, even under ambient storage temperature and normal humidity level (Chelouche et al. 2020a). These phenomena are seriously aggravated when hard storage conditions are expected (very hot and/or humid regions) (Chelouche et al. 2020b).

The migration/diffusion processes of nitroglycerine from the solid propellant towards the surface of the inhibitor can seriously alter both chemical stability and physical integrity of the energetic materials. This may decrease significantly its overall performance, reduces its shelf life, and makes it vulnerable.

Actually, in a DBRP, NG forms weak bonds with NC, making it more mobile, then moving more easily to the surface (Bobić et al. 2017). Hence, much more attention should be and have to be paid by the scientific community to the phenomena connected to the presence of NG in a solid propellant. Some works have been, indeed, carried out to get more insights on both thermodynamic and kinetic aspects of such processes (Sanja Matečić Mušanić et al. 2010; Sućeska et al. 2010b). Furthermore, concepts explaining both nitroglycerin migration mechanism and the ways to prevent/reduce from the harmful effects of this process in DB propellants were presented and debated in the review performed by Agrawal and Singh (Agrawal and Singh 1993).

On the other hand, Nitrate Esters-based Energetic Materials (NEEMs) exhibit, even though stored at ambient temperature and acceptable humidity level, a risk for self-decomposition accompanied by a self-heating ultimately resulting in autocatalytic run-way reactions (Chelouche et al. 2019a, 2018b). The thermal degradation of NEEMs has been widely and deeply investigated and several mechanisms have been proposed (Bohn 2017; Trache and Tarchoun 2018). Otherwise, the uncontrolled chemical decomposition could have harmful fallouts on the safety, efficiency, and performance of propulsion systems. Unexpected explosions in storage warehouses can be achieved when the chemical degradation of the propellant reaches an advanced stage under extreme conditions (Chelouche et al. 2019c, d). In order to prevent such dangerous situations and to increase the safety level in storage warehouses and also during operations, chemical substances called stabilizers are commonly added to NEEM formulations (Chelouche et al. 2018a). Because of their chemical structures, stabilizers are capable to fix the products of the thermal decomposition just after being formed, and hence slow down meaningfully its rate. Ultimately, both storage and service lifetimes of these energetic materials are substantially increased due to the inhibition of the autocatalytic character of this chemical degradation (Zayed et al. 2010).

Early, the inspection of the combustion regime of double-base propellant in a rocket motor had attracted much more attention from the experts (Aoki and Kubota 1982; Beckstead 1980; Sharma et al. 1991; Singh and Rao 1977; Tseng and Yang 1994; Zenin 1995). Afterward, the impact of several factors on the mechanical/ballistic characteristics as well as the thermal decomposition of such propellants have been well investigated, (Abdel-Ghani et al. 2016; Herder et al. 2003; Matečić Mušanić and Sućeska 2009; Mušanić and Sućeska 2013; Shekhar 2011; Sućeska et al. 2010a; Trache and Khimeche 2013a, b). The inventory makes that expect some works devoted to the evaluation of the chemical stability of modified double base propellant (Elbasuney et al. 2019, 2018; Elghafour et al. 2018), less attention has been paid to examine the effect of all the chemical and physical processes occurring during the DBRP lifetime, including NG migration, diffusion, and evaporation as well as the chemical decomposition of NC on its overall stability, especially after a long natural and/or artificial aging. This can be related to the difficulty of procuring such materials directly from weapon systems but also to the high risks associated with NG manipulation. One has to keep in mind that the use of a DBRP is strongly related to the preservation of satisfactory chemical stability throughout its lifetime, basically, evaluated by the achievement of accelerated aging operations.

In this work, the chemical stability of three DBRPs stored in the same conditions during long natural aging up to 31 years and artificial aging carried out at 338.65 K for 12 months has been deeply investigated. Sampling was performed every 2 months (60 days). The long artificial aging was conducted with a view to better reproduce the severe conditions of storage in hot regions. The chemical stability was studied by two spectrometric techniques, namely, Fourier transform infrared (FTIR) and X-ray diffraction, conventional stability tests, *viz*, Bergmann-Junk test and Vacuum Stability Test (VST). Kinetic modeling on VST data obtained at five isothermal temperatures has been carried out to assess the kinetic triplets for the thermal degradation of all the investigated samples and consequently get more insights on the controlling mechanisms. Using the obtained kinetic parameters, the storage lifetimes for the investigated DBRPs have been estimated as well at three different storage temperatures (293.15 K, 298.15 K, and 303.15 K). Beforehand, the effect of both types of aging on the propellant density has been studied by an electronic densimeter.

2 Experimental and Methods

2.1 Materials

All the experiments have been accomplished on three similar DBRPs with a nitrogen content of NC of 12.8%. The propellants, stored in the same warehouses and presented identical storage conditions history, have different natural aging periods: DBRP1, DBRP2, and DBRP3 were naturally aged for 31 years, 29 years, and 11 years, respectively. Likewise, the solid propellants studied in this work present the same chemical composition with 52% of nitrocellulose (NC), 30% of nitroglycerine (NG), 2.5% of methyl centralite (MC), 5% dibutyl phthalate (DBP), and 10.5% of other additives. Each propellant consisted of a single hard block. Therefore, to be able to be analyzed, the big blocks were reduced into very small pieces using a special rasp. Furthermore, all samples have been subjected to drying processes to eliminate residual humidity content before being tested.

2.2 Artificial Aging

Depending on the storage conditions, the chemical degradation processes of NEEMs could be slow and may be extended for several years. Thus, the investigation of the different aspects of that reaction without the use of an efficient methodology for time reduction should be practically very difficult. Artificial aging appears to be a very effective tool to solve this problem (Elbasuney et al. 2018; Elghafour et al. 2018; Mušanić and Sućeska 2013; Trache and Khimeche 2013b). This treatment aims to reduce meaningfully the time scale and delivers after a heating cycle, samples with overall characteristics comparable to samples aged naturally for a long period. In the current work, the requirements for the artificial aging process were inspired by STANAG 4117 (Standardization Agreement STANAG 4117. 1998). All the DBRPs have been heated for 12 months in an oven set at 338.15 ± 0.1 K, with sampling every 2 months (60 days). In all the following, A0, A2, A4, A6, A8, A10, and A12 correspond to the samples which are not artificially aged but present in the same time different natural aging durations, and those artificially aged for 2, 4, 6, 8, 10, and 12 months, respectively.

2.3 Density Measurement

A Helium gas pycnometer (Micromeritics, AccuPyc II 1340) was used to determine the densities of the studied propellants. The measurements were done in triplicate for each sample. By substituting the void space of the sample with helium, the volume of the samples is evaluated. Then, the density is obtained using the sample mass measured in prior. Each analysis involved 10 measurement cycles, providing 10 data points.

2.4 *FTIR Analysis*

A Shimadzu spectrometer 8400S was used to analyze both aged and unaged DBRPs samples which were dried beforehand. The spectra were recorded at ambient temperatures in transmittance mode in the range of 4000–400 cm^{-1}. In order to obtain a satisfactory signal–to–noise ratio, 64 scans were co-added. Furthermore, the resolution of the spectra was kept at 4 cm^{-1}. In all measurements, the background was obtained and subtracted from the sample spectrum.

2.5 *XRD Analysis*

XRD measurements were carried out at ambient temperature using a PANalytical X'Pert Pro Multi-Purpose Diffractometer. Data were collected over an angular range of 5–50°/2θ using An X'Celerator detector. The measurements were carried out on powdered specimens prepared using an adequate kit provided by the manufacturer.

2.6 *Stability Tests*

2.6.1 Bergmann & Junk (B&J)

Based on a quantitative determination of the nitrogen oxides evolved by the sample during an isothermal decomposition, the B&J test is efficiently used for the assessment of the thermal stability of nitrocellulose based-energetic materials including single and double base propellants. Both the duration and the temperature of the test depend on the type of the energetic material involved.

Dried DBRP samples (5 ± 0.0002 g) are put into glass test tubes enclosed by special glass adapters. The test tubes are then inserted into holes in a heating block. This latter is maintained at a predefined temperature of 388.15 K with a tolerance of ± 0.1 K. The test operating conditions were inspired by AOP-7, 2nd Edition ("North Atlantic Treaty Organization, Manual of Data Requirements and Tests for the Qualification of Explosives Materials for Military Use" June 2003). Once the heating cycle is finished (16 h), the tubes are extracted and allowed to cool to room temperature. Finally, the hydrogen peroxide solution is recuperated then titrated against sodium hydroxide (0.01 N).

2.7 Vacuum Stability Test

The STABIL Czech vacuum stability was used to assess the chemical stability of all the studied samples. A sample mass of 2 g, a temperature and a time of 333.15 ± 0.1 K and 4 h, respectively were found optimal to effectively wick away moisture from double base rocket propellant samples studied by VST (Fidanovski et al. 2016). Hence, these operating conditions were kept for the drying process of all the unaged and aged samples.

Afterward, the samples were introduced in special glass tubes. Using a vacuum pump, the pressure inside all the test tubes is brought back to 0.2 kPa. Once verified that no loss of vacuum due to gas leakage problem has occurred, the test tubes were put into the heating block for 48 h. The test temperatures were set at the following values: 333.15, 334.15, 353.15, 363.15 and 373.15 K. The discussion of the propellant stability was performed based on the results obtained at the specific test temperature of 373.15 K whereas the kinetic modeling was carried out on the data obtained at all the test temperatures.

Silicon oil provided by the manufacturer has been used to calibrate the heating bloc following a well-defined experimental protocol. Moreover, the pressure increase inside the test tubes because of the continuous gases release is estimated by the pressure transducers and displayed on the STABIL software conjointly with the volume of the evolved gases by the sample.

2.8 Kinetic Modeling on VST Data for the DBRP Thermal Decomposition

The stability appraisal as well as the prediction of the shelf-life for nitrate ester-based solid propellants imply the precise evaluation of the kinetic parameters for the decomposition reaction (Roduit et al. 2008). The general expression of the kinetic equation in the case thermally stimulated reactions describes the variation in the amounts of the reactants with time progress. Two forms are commonly used to conceive the kinetic equation: differential form (Eq. 1) or integral form (Eq. 2).

$$\frac{d\alpha}{dt} = k(T)f(\alpha) \tag{1}$$

$$g(\alpha) = k(T)t \tag{2}$$

where α is the extent of conversion, t refers to the time, and T stands to the temperature. $f(\alpha)$ and $g(\alpha)$ are respectively the differential and the integral form of the reaction model, however, k (T) is the rate constant given by the Arrhenius equation (Eq. 3):

$$k(T) = A\exp\left(\frac{-E_a}{RT}\right) \tag{3}$$

where A stands for the pre-exponential factor, E_a is the apparent activation energy, R refers to the universal gas constant, and T is the reaction temperature. The 41 model reaction frequently employed for the estimation of the reaction rate can be found elsewhere (Trache et al. 2017).

The values of the extent of conversion (α) can be derived from the VST experimental data according to Eq. 4:

$$\alpha = \frac{P - P_0}{P_{max} - P_0} \tag{4}$$

where P is the pressure at a time t, P_{max} refers to the final pressure inside the test tube, and P_0 stands to the initial pressure.

In the present work, two different approaches have been employed to evaluate the kinetic triplets, viz, model-fitting and model-free. The different steps followed for the computation of these kinetic parameters by these two methodologies were explicitly given in our recent works (Chelouche et al. 2019d, f).

2.9 Storage Lifetime Prediction

The storage lifetime can be defined as the period in which the energetic material can be stored safely while maintaining the reproducibility of performance. For many countries, the accurate determination of the storage lifetime for weapon systems, including missile and rocket motors plays a decisive role in the establishment of defense plans. In order to achieve this objective, temperature–time data determining the quantities for the propellant degradation are requested. These data should cover a representative degree of property change which is in our case none other than the pressure of the released gases in VST (Chelouche et al. 2019d; Fraunhofer-Institut fur Chemische Technologie 1997).

Equation 5 has been used to estimate the storage lifetimes for both aged/unaged samples (Vyazovkin and Wight 1998).

$$t = \frac{g(\alpha)}{Ae^{\left(\frac{-E_a}{RT}\right)}} \tag{5}$$

where t (s) is the storage lifetime, $g(\alpha)$ is integral model reaction value at the conversion extent (α), $A(\mathrm{s}^{-1})$ is the pre-exponential factor, E_a (J/mol) stands for the activation energy, T (K) refers to the storage temperature and R (J/K·mol) to the universal gas constant.

Using the procedure proposed by our research group (Chelouche et al. 2019b, e), the extent of conversion corresponding to the limit of stability was assessed for each sample and found to be around 2% of the total conversion. Therefore, the value of $\alpha = 0.02$ was taken as the limit of stability in the prediction of the storage lifetime for the studied DBRPs.

3 Results and Discussion

3.1 Density Measurement

Density is a physical property of both fundamental and practical importance for the study of the stability of energetic materials. In fact, several stability tests depend on the true value of this physical property. The density values for all the investigated DBRPs samples (A0 to A12) are summarized in Table 1. The uncertainty associated with the density values has been evaluated based on repeated measurements (threefold) and was found to not exceed 0.003 g/cm^3.

Typically, the density of the different DBRPs was found to range from 1.55 to 1.66 g/cm^3 which results from the densities of their raw materials (NC + NG) as well as that of the additives (Davenas 2012). The density was found to decrease with the progress in both types of aging (natural and artificial), which may be explained by the diminution of the sample masse caused by the gaseous release during aging and in the same time, the volume increase of the samples induced by the hemolytic cleavage, under heating, of NC polymeric chain and NG to form free radicals. A similar trend has been verified for other types of materials (Blazhnov et al. 2004; Narender et al. 2013).

The density evolution with the accelerated aging progress for the three batches of DBRP was found to be best correlated by a nonlinear regression fit:

$$\rho_t = a - b * c^t \tag{6}$$

where ρ_t is the density value after t aging duration; a, b, and c are the correlation constants; and t (months) is the aging progress. The lowest adjusted regression coefficient (adj.R^2) is found equal to 0.9968.

Table 1 Density values for the aged and unaged samples

Aging (month)	0	2	4	6	8	10	12
DBRP1[a]	1.612	1.592	1.576	1.563	1.549	1.538	1.533
DBRP2[a]	1.617	1.598	1.582	1.569	1.556	1.547	1.539
DBRP3[a]	1.627	1.608	1.593	1.583	1.572	1.560	1.552

[a]Uncertainty u associated with the density is $u(\rho) = \pm 0.003$ g/cm^3

3.2 FTIR Analysis

FTIR as a well-informing spectroscopic technique is often used to highlight any structural changes induced by a given treatment, including thermal ones. To extract useful information, FTIR spectra of reference materials are compared with those obtained for treated ones regarding position translation, intensity variation, and shape modification of the characteristic frequencies (Chelouche et al. 2018a, b).

Plot (a) in Fig. 1 shows the FTIR spectra for DBRP1 samples (A0–A12) whereas the assignments of the characteristic frequencies are displayed in Table 2. It should be noted that the two remaining propellants presented similar spectra. The stretching vibration of the –OH group in nitrocellulose and the deformation vibration of the C–H bond in the linear chain—CH–CH_2) in nitroglycerine observed at 3458 cm^{-1} and 2920 cm^{-1}, respectively, are the main frequencies in the high frequencies region of the spectra. Furthermore, the peaks observed at 1650 and 1277 cm^{-1} are attributed to the antisymmetric and symmetric stretching vibration of the NO_2 group while the peak at 837 cm^{-1} is assigned to the stretching vibration of the O–NO_2 bond. These three peaks with the highest intensities are the characteristic frequencies of nitrocellulose (Ma et al. 2012). It is obvious that these three frequencies are the most impacted by the progress in both natural and artificial aging.

Moreover, the peaks with lower intensities in the range of 1300–690 cm^{-1} are attributed to the different vibrations of the C–X (C, O, H) bonds. With aging progress, attenuation in the intensities of several bands can be clearly observed which is more pronounced for the characteristic frequencies of nitrocellulose (1650, 1277, and 837 cm^{-1}). However, this attenuation for these characteristic bands which are related to the NO_2 group is more significant after 2 months of artificial aging for DBRP1 and DBRP2 and after 4 months for DBRP3, which can be connected to the differences in the of natural aging periods between the three studied propellants. Such attenuation is found to slow down for all propellants after these periods of heating. Broadly, this may be explained by the cleavage of the O-NO_2 bond in the nitrate esters groups during aging.

Otherwise, plot (b) in Fig. 1 highlights the effect of the natural aging duration on the chemical structures of the investigated DBRPs. It is obvious that the progress in the natural aging impacts the chemical functions is a similar way than that of artificial aging. However, the effect is found less significant since the higher temperature of heating accelerates the reaction rate of the nitrate esters decomposition.

3.3 XRD Analysis

The measurements were carried out to inspect the effect of the natural/artificial aging on the crystallinity of the investigated samples. Plots (a) in Fig. 2 displays the X-ray diffractograms for DBRP2 samples. The remaining propellants presented similar trends. Furthermore, plot (b) shows a comparison between the diffractograms of

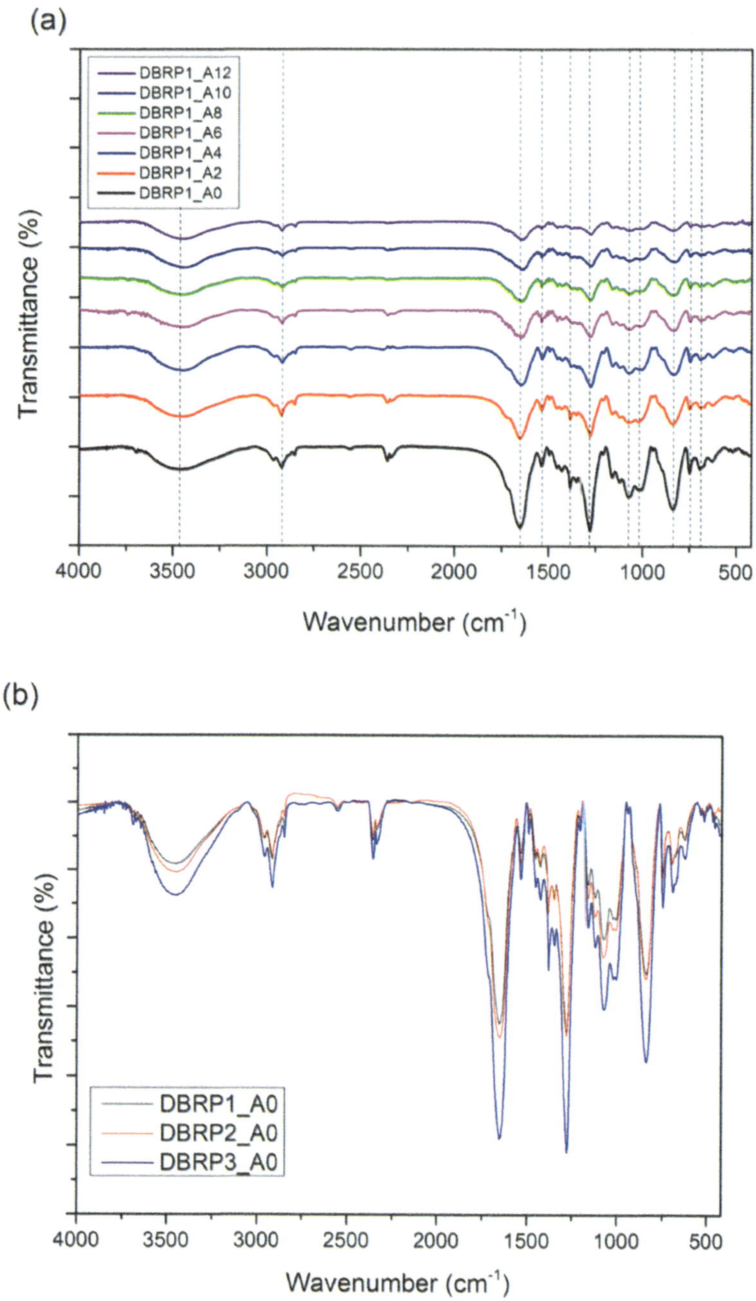

Fig. 1 FTIR spectra for **a** unaged and aged DBRP1; **b** effect of natural aging

Table 2 FTIR characteristic peaks for the investigated DBRPs

Wavenumber (cm^{-1})	Assignment
694 and 747	(C–H) aromatic out of plane deformation vibration
837	O–NO_2 stretching vibration
1008	C = C stretching vibration
1073	O–CH_3 stretching vibration
1277	NO_2 group symmetric stretching vibration
1384	N–O symmetric stretching vibration
1537	N–O asymmetric stretching vibration
1650	NO_2 group asymmetric stretching vibration
2920	C–H bond in the linear chain –(CH–CH_2) deformation vibration
3458	Linked OH group stretching vibration

naturally aged propellants. The raw data were treated by MATLAB software in order to eliminate the signal–noise (Bertacchini et al. 2012).

The crystallinity indexes of samples reported in Table 3 were calculated using Eq. 7. (Trache and Tarchoun 2019):

$$ka = \frac{\sum Scr}{\sum Scr + \sum Sam} \tag{7}$$

where ka is the crystallinity index, Scr and Sam are the peak area for the crystalline and amorphous parts, respectively. The two parts were estimated after the deconvolution of the XRD profiles using Pearson VII function (Chelouche et al. 2019c; Trache et al. 2016).

With aging progress, an increase in the intensity of the peaks at around 24.6°, 27.3°, 34.2°, and 43.1° was observed (Plot a and b in Fig. 2) leading to a raise in the crystallinity of all the studied samples. Furthermore, it can be inferred from Table 3 that for both types of aging (natural and artificial), the crystallinity indexes increase with aging evolution. This can be related to the acceleration of the hemolysis process with aging progress leading to an increase of the cleavage rate of the O–NO_2 bonds of the nitrate esters. On the other hand, the increased accessibility of hydroxyl (OH) groups likely to form intramolecular and intermolecular bonds, may be responsible for the rise in the crystallinity indexes with aging progress (Rodrigues Filho et al. 2000).

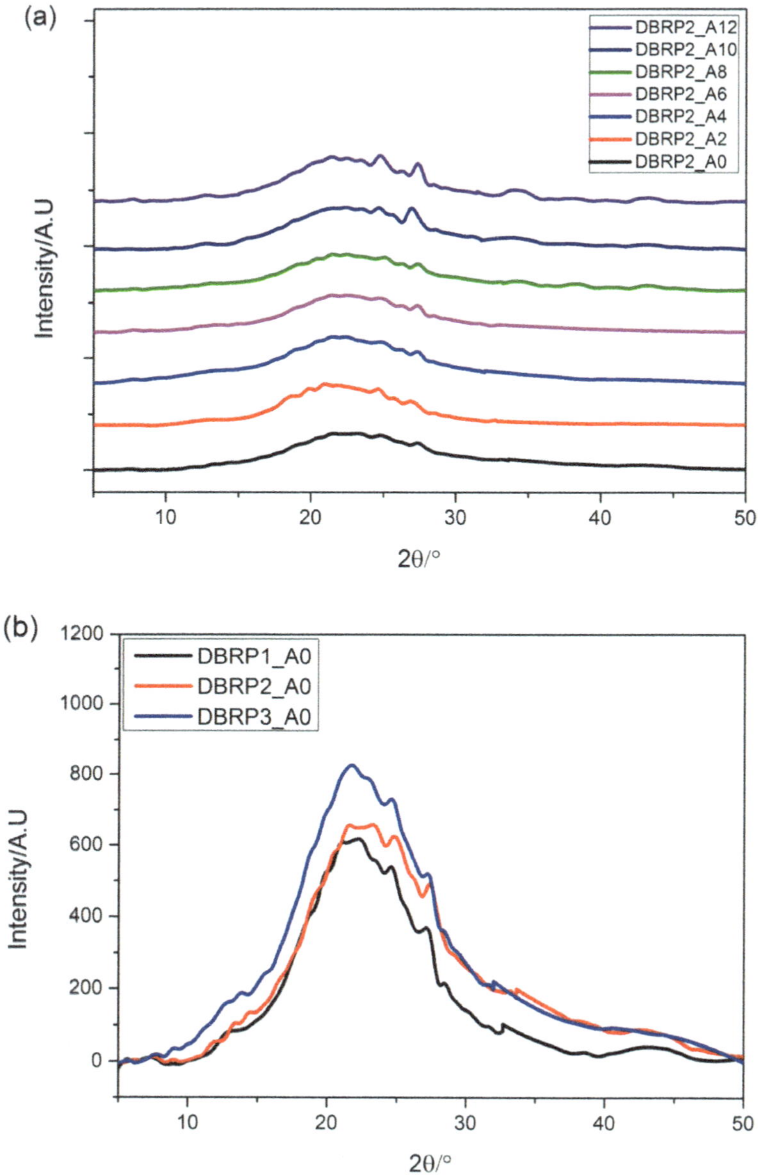

Fig. 2 X-ray diffractograms for **a** unaged and aged DBRP2; **b** effect of natural aging

Table 3 Crystallinity indexes for the DBRP samples

Sample	Crystallinity index		
	DBRP1	DBRP2	DBRP3
A0	22.60	22.24	20.60
A2	28.30	26.90	24.23
A4	30.77	29.77	25.14
A6	37.64	35.05	31.83
A8	40.22	39.53	38.07
A10	43.60	42.68	41.07
A12	45.09	44.19	42.82

3.4 Stability Tests

Conventional stability tests are often used to get accurate information about the current stability status of the NEEMs. Indeed, the masses of the samples are much more important and representative and the test duration allows a better approximation of the occurred decomposition reaction.

3.4.1 Bergmann & Junk

The volumes per propellant mass of the nitrous vapors are given in Table 4. The uncertainty associated with the evolved gases is found to be less than 0.04 ml/g (4%). According to AOP 7, 2nd edition ("North Atlantic Treaty Organization, Manual of Data Requirements and Tests for the Qualification of Explosives Materials for Military Use" June 2003), except the unaged propellants and the DBRP3 aged for two months, all the samples are unstable, which is in accordance with the FTIR results. Through the obtained results, the effect of natural aging is well elucidated. In fact, DBRP3 presents the lowest volume of the evolved nitrous vapors (0.31 ml/g), while the volume obtained for DBRP2 (0.41 ml/g) is lower than that of DBRP1 (0.45 ml/g). Similarly to what was obtained for single base propellants (Chelouche et al. 2019d), the artificial aging was found to negatively affect the chemical stability of double base rocket propellant since the volume of the released gases increases with aging progress. Similar trends were previously found for DBP and Modified DBP aged for 28 days at 353.15 K (Elbasuney et al. 2018).

3.4.2 Vacuum Stability Test

The volumes of the evolved gases per sample mass of the studied samples at the different test temperatures are given in Table 4 whereas 3D scatter plots (a and b) of the obtained results for DBRP1 are shown in Fig. 3 (similar evolution is observed for both DBRP2 and DBRB3). The uncertainty associated with the evolved gases volume

Table 4 Stability tests results

DBRP		Bergmann & Junk[a]		*VST*				
		V_{NaOH} (ml/g)	V^{b}_{NOx} (ml/g)	V^{c}_{gases} (ml/g)				
		$T = 388.15$ K		$T = 333.15$ K	$T = 343.15$ K	$T = 353.15$ K	$T = 363.15$ K	$T = 373.15$ K
DBRP1	A0	2.02	0.45	0.44	0.54	0.87	2.49	6.66
	A2	2.58	0.58	0.34	0.46	0.77	2.11	6.10
	A4	3.33	0.75	0.29	0.36	0.72	2.01	5.82
	A6	3.70	0.83	0.32	0.39	0.75	2.07	5.60
	A8	4.94	1.11	0.34	0.40	0.77	1.93	5.07
	A10	6.14	1.38	0.27	0.35	0.70	1.80	4.83
	A12	7.38	1.65	0.24	0.33	0.68	1.74	4.72
DBRP2	A0	1.85	0.41	0.43	0.56	0.87	2.52	6.79
	A2	2.30	0.52	0.35	0.48	0.81	2.13	6.29
	A4	3.18	0.71	0.30	0.38	0.75	2.01	6.02
	A6	3.50	0.78	0.33	0.41	0.77	2.13	5.72
	A8	4.58	1.03	0.35	0.43	0.78	1.95	5.21
	A10	5.58	1.25	0.30	0.35	0.72	1.83	5.01
	A12	6.82	1.53	0.26	0.33	0.69	1.74	4.85
DBRP3	A0	1.37	0.31	0.40	0.59	0.97	2.59	7.04
	A2	1.82	0.41	0.36	0.50	0.90	2.30	6.63
	A4	2.30	0.52	0.32	0.43	0.80	2.10	6.37
	A6	2.61	0.59	0.34	0.45	0.85	2.28	6.04

(continued)

Table 4 (continued)

DBRP		Bergmann & Junk[a]		*VST*				
		V_{NaOH} (ml/g)	V^{b}_{NOx} (ml/g)	V^{c}_{gases} (ml/g)				
		$T = 388.15$ K		$T = 333.15$ K	$T = 343.15$ K	$T = 353.15$ K	$T = 363.15$ K	$T = 373.15$ K
	A8	2.96	0.66	0.37	0.46	0.80	2.02	5.45
	A10	4.21	0.94	0.34	0.38	0.74	1.89	5.40
	A12	5.30	1.19	0.29	0.35	0.71	1.80	5.31

[a]Requirements: $\leq$ 6 to 20 ml 0.01n NaOH/5 g propellant, depending on the propellant type ("North Atlantic Treaty Organization, Manual of Data Requirements and Tests for the Qualification of Explosives Materials for Military Use" June 2003); $\leq$ 10 to 12 ml 0.01n NaOH/5 g double base rocket propellant (Bohn 2007)
[b]Uncertainty u associated with the volume of gases released in B&J is $u(V) = \pm 0.04$ ml/g (<4%)
[c]Uncertainty u associated with the volume of gases released in VST is $u(V) = \pm 0.08$ ml/g (<4%)

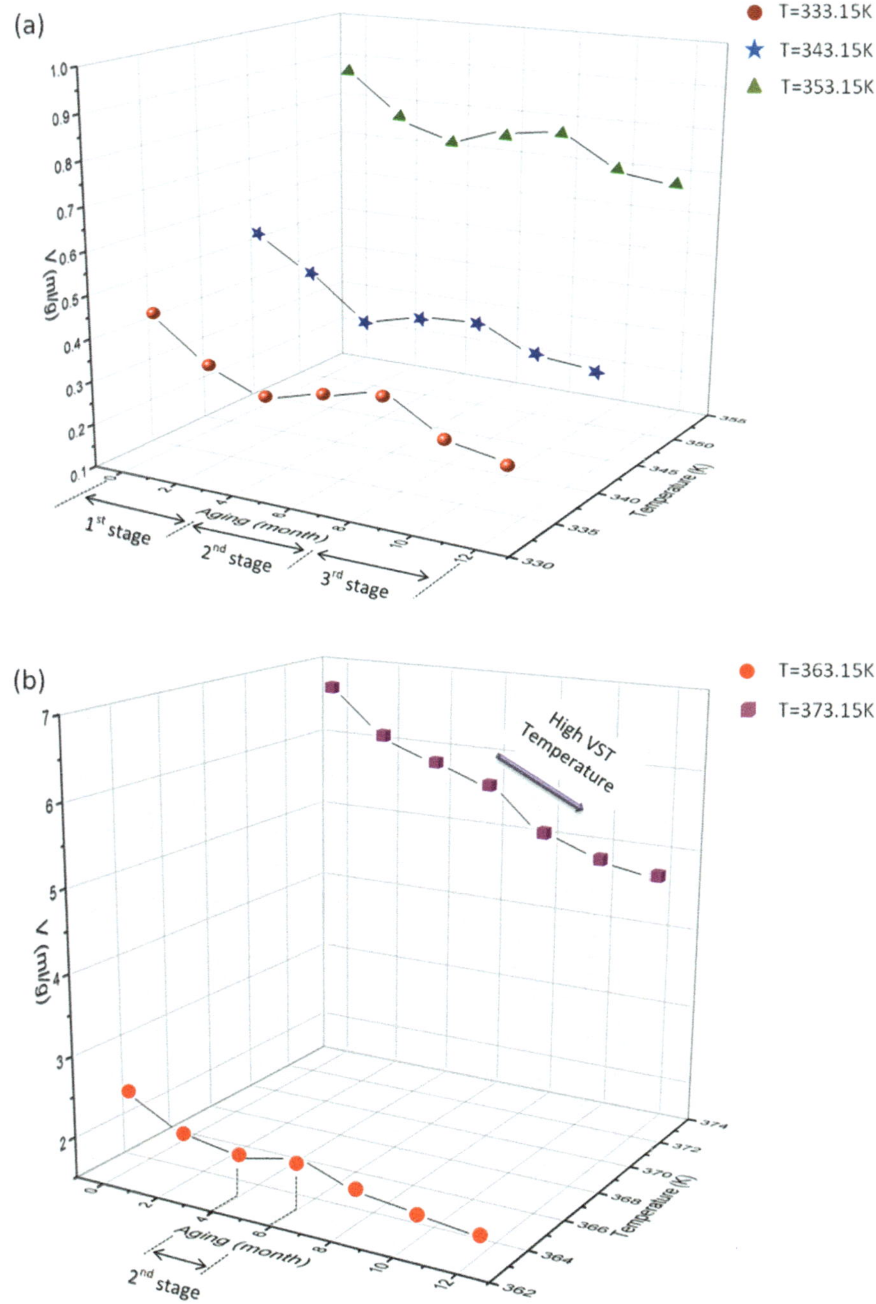

Fig. 3 3D scatter plot of the gas-evolved volumes for DBRP samples; **a** T = 333.15 K, 343.15 K and 353.15 K; (b) T = 363.15 K and 373.15 K

is evaluated based on repeated measurements of some chosen samples at different VST isothermal temperatures and found to be less than 0.08 ml/g (<4%). Moreover, plots a-e in Fig. 4 show the released gases pressure evolution versus the time test progress, by all the DBRP1 samples (A0-A12) at the different VST temperatures (333.15 K, 343.15 K, 353.15 K, 363.15 K, and 373.15 K). Likewise, plots a-g in Fig. 5 display the evolved gases pressure evolution against the time test progress, at the different VST temperatures by the DBRP3 samples (A0-A12). It should be pointed out that similar evolutions were obtained for the two other DBRPs in both previous cases.

In the first minutes of the test, the temperature elevation inside the heating block induces a sharp evolution of the pressure of the released gases by the samples. This behavior can be explained by the persistence of some residual moisture, even though a drying process was previously carried out (Zeman et al. 2013).

In the second stage, the pressure was found to increase gradually with test time evolution. The intensity of the pressure elevation is proportional to the isothermal temperature. It should be noted that such type of investigation is scarce in the literature.

According to AOP 7, 2nd Edition (US/202.01.001- samples heated for 48 h at 373.15 K), all unaged DBRPs are chemically unstable since the released volumes are higher than the typical value of 2 ml/g ("North Atlantic Treaty Organization, Manual of Data Requirements and Tests for the Qualification of Explosives Materials for Military Use" June 2003; Vogelsanger 2004). However, based on our results, such criterion seems to be less efficient to well investigate their stability, thus completing it with a kinetic analysis will provide better insights about the stability, as it will be shown later. Moreover, at the temperature test required by the AOP standard (T = 373.15 K), the volume of the released gases for the studied propellants decreases with natural aging progress, which can lead to an erroneous assessment of the chemical stability of DBRP. This could be connected to the fact that only processes that result in the gaseous release, are measured while processes of migration, diffusion, and adsorption are not taken into account. For the same reasons, VST method was previously mentioned by de Klerk to give no guarantee of compatibility between six propellants, including DBP, and several polymers, even if it is reported as an efficient tool in STANAG 4147 (de Klerk et al. 1995).

From Table 4 and Fig. 3, three degradation stages for the investigated DBRPs can be clearly outlined. In the first stage which extends until the fourth month of heating, the volumes of the evolved gases decrease with artificial aging progress. This behavior, valid at all the VST temperatures, is in opposition to what was previously found for NC-based propellants (Chelouche et al. 2019d). Such evolution can be connected to the NG evaporation during accelerated aging.

After 4 months of heating, the volumes of the gases evolved by the propellant samples were found to increase with artificial aging progress. This finding is maintained until 8 months of heating for the three first VST temperatures, and until 6 months of heating when a temperature of 363.15 K is used. However, this increase in the evolved gases cannot be observed for the test temperature of 373.15 K, as higher the temperature is, higher the rate of the degradation reaction is.

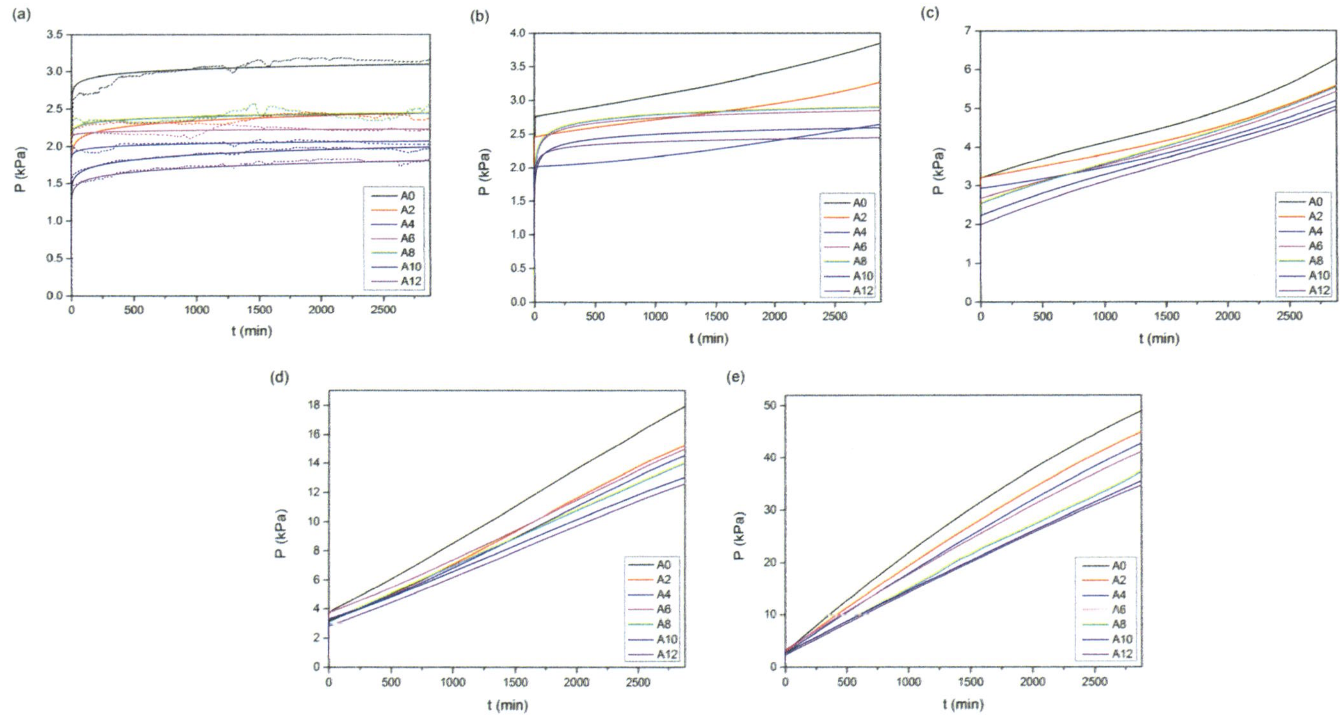

Fig. 4 Gases-evolved pressure evolution against VST time for the unaged and artificially aged DBRP1 samples at **a** $T = 333.15$ K, dashed and straight lines present raw and smoothed data, respectively; **b** $T = 343.15$ K; **c** $T = 353.15$ K; **d** $T = 363.15$ K; **e** $T = 373.15$ K

This opposite behavior can be the result of the predominance of one process, over the others occurring during the thermal degradation of the DBRP, in each step previously detected. In fact, in the first stage of degradation, the nitroglycerine evaporation is the dominant process. After a heating cycle of 4 months at 338.65 K, the rocket propellant will lose almost the total amount of the plasticizer (NG) and the decomposition of nitrocellulose becomes dominant justifying the increase in the volume of the gases released in the second degradation stage. Indeed, the NG depletion time was found by Mušanić and Sućeska (S Matečić Mušanić and Sućeska 2009) equal to 82 days for similar DBRP studied by isothermal thermogravimetry (TG) at a temperature of 373.15 K (S Matečić Mušanić and Sućeska 2009).

In the third degradation stage, the volumes of the evolved gases by all the DBRPs samples decreased again. This behavior, valid for all the VST temperatures, maybe the result of an eventual change in the mechanism of the decomposition reaction as propellants may under certain conditions adopt a different regime of degradation (Davenas 2012).

Otherwise, the catalytic effect of the higher temperature on the degradation reaction (Chelouche et al. 2019d) should be responsible for the increase of the evolved gases volumes by the propellant samples when the VST isothermal temperature

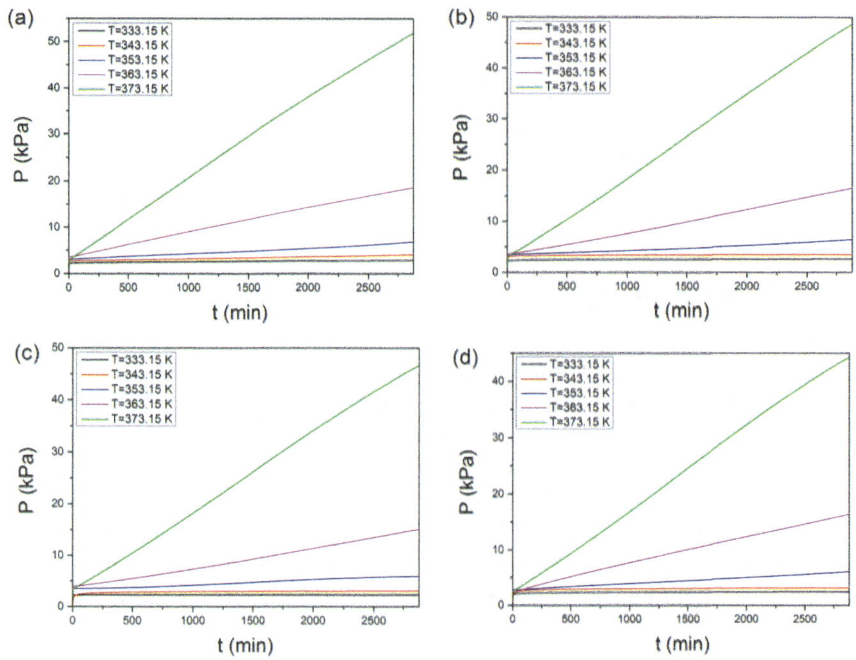

Fig. 5 Gases-evolved pressure evolution against VST time for DBRP3 samples at different isothermal temperatures; **a** Unaged sample; **b–g** samples aged artificially for 2, 4, 6, 8, 10, and 12 months, respectively

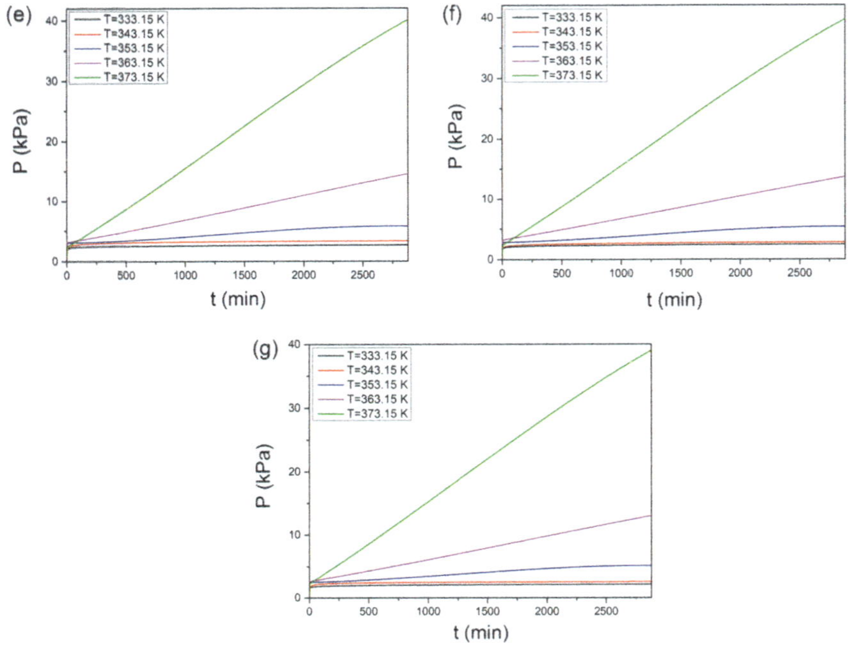

Fig. 5 (continued)

is augmented. This effect was found to be more important at higher temperatures (363.15 K and 373.15 K).

3.5 *Kinetic Modeling*

The use of multiple experimental techniques and approaches is highly recommended for an effective evaluation of the chemical stability of NEEMs (Benhammada et al. 2020). Recently, Kinetic modeling results were used efficiently to judge the stability and the compatibility status as well as the stabilizing potential of several organic stabilizers on energetic materials, including during accelerated aging (Chelouche et al. 2019b, c, d; Elbeih et al. 2017). Basically, kinetic investigation aims to parameterize the rate as a function of temperature and other variables as well as to gain mechanistic insights into the processes governing the chemical processes (Vyazovkin 2015).

VST pressure transducers are highly sensitive leading in some cases to instabilities in the VST signal. Without adequate treatment, fluctuations in the extent of conversion can be obtained during the computation which may affect the computation process of the kinetic parameters. That's why raw data have been smoothed using appropriate nonlinear fits as shown in plot (a) in Fig. 4 (Chelouche et al. 2019d).

Figures 6 and 7 show for all the studied DBRPs the evolution of the Arrhenius parameters (activation energy and the neperian logarithm of the pre-exponential factor) obtained by both kinetic procedures (model-fitting and model-free) versus the accelerated aging durations, including unaged samples. Moreover, the uncertainties associated with the computed values for both Arrhenius parameters (Ea and Ln A) have been evaluated and presented in Fig. 6 and 7 as well. The obtained activation energy values for the unaged DBRPs are in a good agreement with the small set of data available in the literature (Aoki and Kubota 1982; Tseng and Yang 1994). Furthermore, the reaction models provided by the model-fitting are given in Table 5.

The estimated parameters and the reaction model correspond to the best fitting results approved on the base of a statistical approach (${R^2}_{max}$ and F_{min}). Indeed, the linear correlation coefficients (R^2) from the plot of the neperian logarithm of the decomposition constant rate against the inverse of the temperature were found to range between 0.9241 and 0.9993.

For model-free, which is based on the isoconversional theory, the Arrhenius parameters have been evaluated after each advancement of 0.05 in the conversion extent. Then, the average values are computed conjointly with their standard deviations on the mean.

The compensation effect stating that a linear relationship should be established between the two Arrhenius parameters could be used efficiently to assess the quality of the results from the kinetic modeling. Figure 8 displays for both kinetic methods, the evolution of the Ln(A) with respect to Ea values. The linear regression coefficients R^2 have been found equal to 0.9987 and 0.9994 when fitting-model and model-free are used, respectively, allowing appreciating the accuracy of followed approaches for the kinetic modeling.

The Arrhenius parameters obtained by the two approaches for the unaged samples and those aged until 6 months were found closer. The relative deviation between the Ea values estimated by fitting-model and those by free-model were found to range between 2 and 9%. These deviations were ranging from 1 to 16% in the case of Ln(A). Exceeding this duration of heating, the deviations become much more important.

By definition, the activation energy is the energetic barrier to surpass to be able to initiate the DBRP degradation. For the investigated DBRP, the Ea values decrease with natural aging due to the loss of the chemical stability. Likewise, the values of the activation energies estimated by both models were found to decrease with artificial aging progress, which can be connected to the accelerating effect of high temperatures on nitroglycerine evaporation and nitrocellulose decomposition. Moreover, for the three studied DBRPs, it is obvious that the same degradation processes are occurring after each step of heating since the rate of decrease in the Ea values was found to be similar for the three batches of propellants after every artificial aging period.

Otherwise, the pre-exponential factor (A) indicates how quickly and frequently the decomposition reaction occurs. From Table 5, it is obvious that the highest value of Ea corresponds to the highest value of A due to the compensation effect (Vyazovkin and Linert 1995). On the other hand, larger pre-exponential factor means a faster process and, therefore, lower thermal stability. Therefore, an increase in the pre-exponential factor diminishes the thermal degradation deceleration if associated with

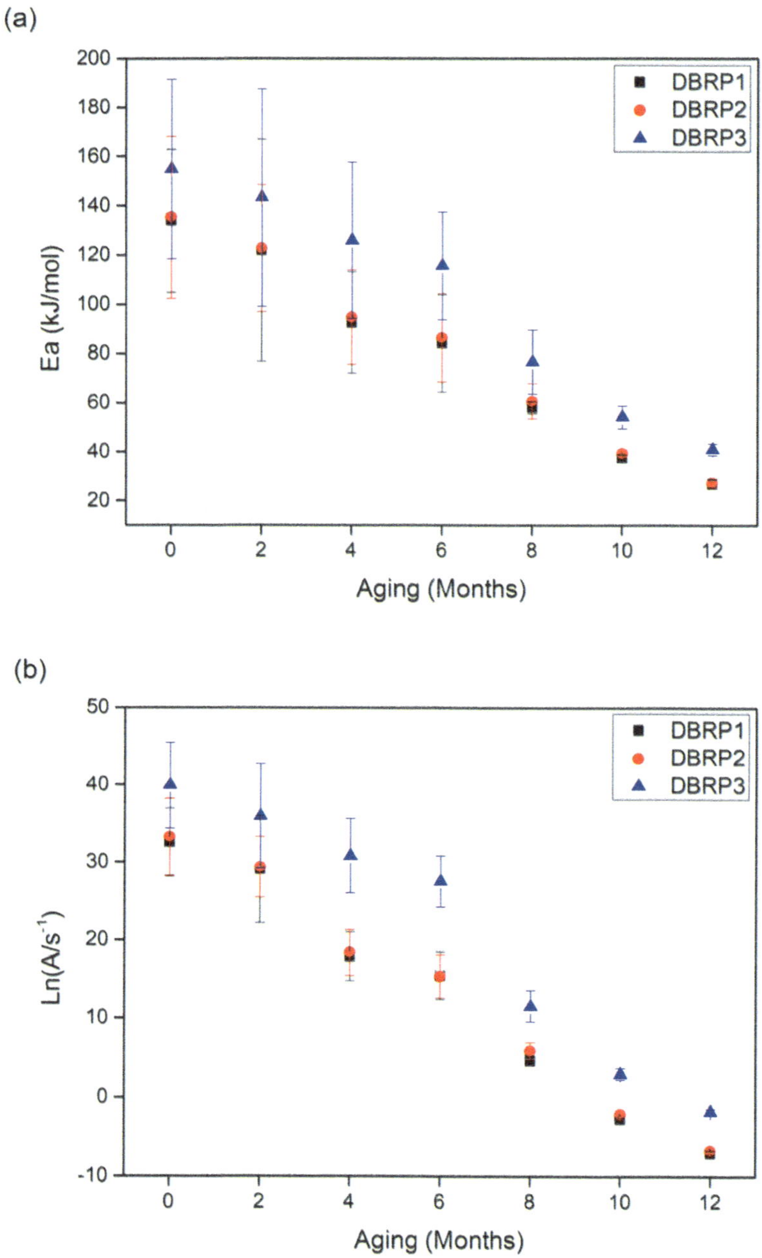

Fig. 6 Arrhenius parameters with their confidence intervals against aging duration for all the investigated DBRP by model-fitting: **a** activation energy; **b** neperian logarithm of the pre-exponential factor

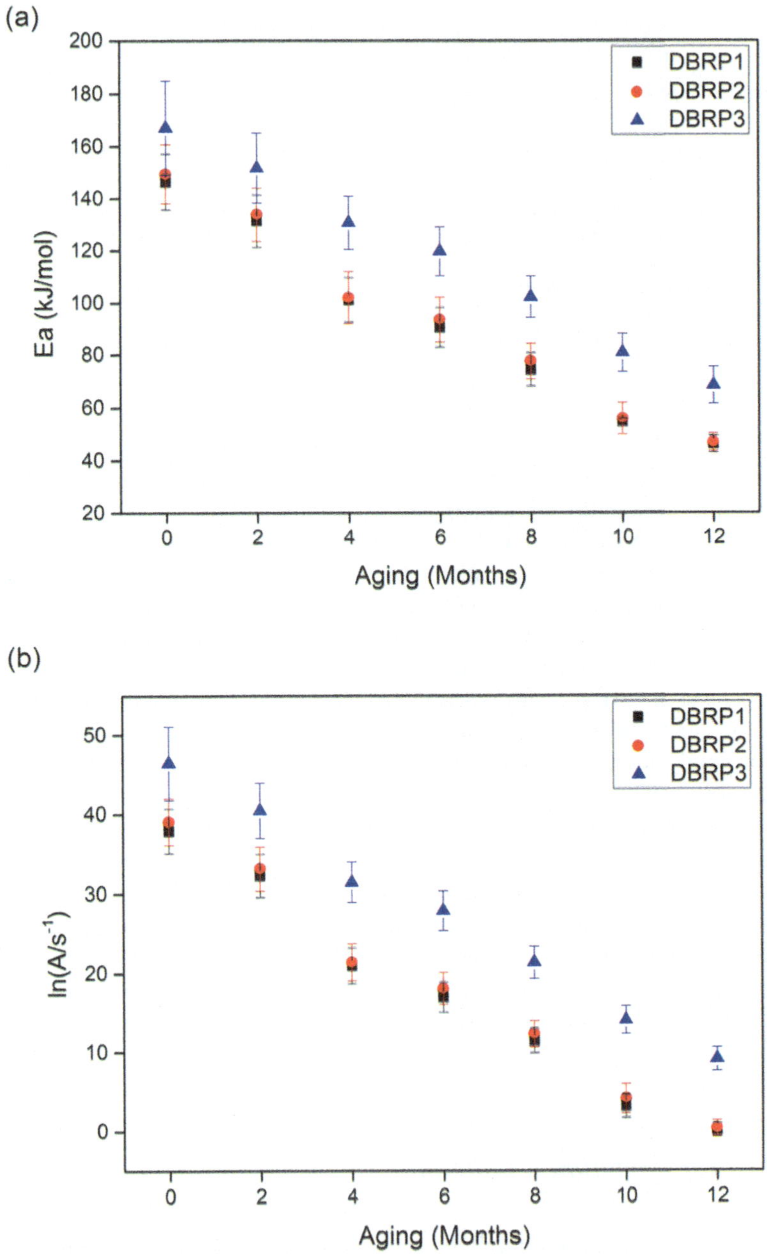

Fig. 7 Arrhenius parameters with their confidence intervals against aging for all the investigated DBRPs by model-free: **a** activation energy; **b** neperian logarithm of the pre-exponential factor

Table 5 Model reactions for DBRPs degradation reaction obtained by model-fitting on VST data

DBRP	A0	A2	A4	A6	A8	A10	A12
DBRP1	9/G_3	8/G_2	7/G_1	7/G_1	7/G_1	39/D_8	39/D_8
DBRP2	9/G_3	9/G_3	7/G_1	7/G_1	7/G_1	39/D_8	39/D_8
DBRP3	9/G_3	8/G_2	8/G_2	8/G_2	7/G_1	38/D_7	39/D_8

9/G_3 Chemical process or mechanism non-involving equations/chemical reaction
8/G_2 Chemical process or mechanism non-involving equations/chemical reaction
7/G_1 Chemical process or mechanism non-involving equations/chemical reaction
38/D_7 Deceleratory rate equations (equations based on the diffusion mechanism)/Three-dimensional diffusion
39/D_8 Deceleratory rate equations (equations based on the diffusion mechanism)/Three-dimensional diffusion

an increase in the activation energy (Vyazovkin 2015). From Table 5 it is obvious that natural/artificial aging diminish the thermal stability of the studied DBRPs that corroborates the stability tests findings.

The reaction models for the degradation reaction of the three studied DBRP are given in Table 5. These reaction models were provided by the fitting-model and correspond to the models that best fit the VST data. The unaged samples were found, indeed, to decompose according to chemical process or mechanism non-involving equations. The same reaction model was obtained for unaged NC-based propellant (Chelouche et al. 2019d). This reaction mechanism was maintained for the samples artificially aged until 8 months with a change in the mathematical model for DBRP1 aged for 2 and 4 months, after 4 months of aging for DBRP2 and after 2 and 8 months of aging for DBRP3. Furthermore, for all propellants, a change in the mechanism of the decomposition reaction is observed after 10 months of accelerated aging. In fact, deceleratory rate equations based on the diffusion mechanism (three-dimensional diffusion) took place. It was, indeed, reported that the mechanisms governing a thermal degradation might change with the aging of the material (Vyazovkin 2017). Furthermore, this result corroborates the hypothesis made before when a decrease in the amount of the released gases in VST is observed in the third degradation stage. It can be inferred that the kinetic modeling results are in good agreement with VST results.

3.6 Storage Lifetime Prediction

Lifetime, as the most commonly predicted parameter, is the time beyond which the material loses its properties and characteristics to the point of no longer being able to effectively ensure its intended purpose (Vyazovkin 2015). The lifetime of a material can be estimated by thermal analysis techniques, as long as the decline in the property of interest can be connected to a change in the mass or heat of the sample.

The storage lifetimes for the investigated DBRPs are given in Table 6. These

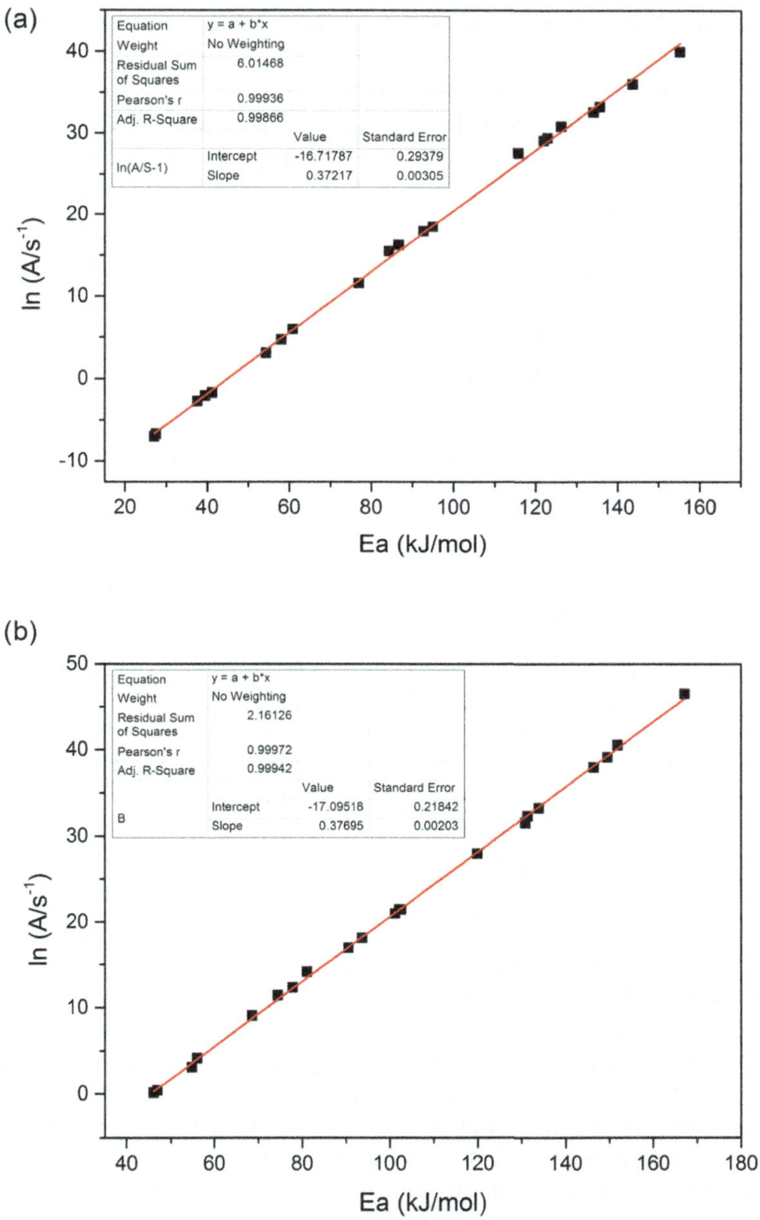

Fig. 8 Linear dependence of ln(A) with Ea by **a** model-fitting; **b** model-free

Table 6 Storage lifetimes (years) for the investigated DBRPs at $T_{storage}$ = 293.15 K, 298.15 K and 303.15 K

DBRP	Aging	Storage lifetime (years)		
		T = 293.15 K	T = 298.15 K	T = 303.15 K
DBRP1	A0	11.82	4.72	1.95
	A2	2.28	1.00	0.45
	A4	0.67	0.36	0.19
	A6	0.22	0.14	0.09
	A8	0.22	0.13	0.08
	A10	0.00	0.00	0.00
	A12	0.00	0.00	0.00
DBRP2	A0	12.25	4.83	1.97
	A2	3.18	1.37	0.61
	A4	0.95	0.51	0.27
	A6	0.25	0.17	0.12
	A8	0.26	0.14	0.09
	A10	0.00	0.00	0.00
	A12	0.00	0.00	0.00
DBRP3	A0	46.56	16.01	5.71
	A2	15.42	5.75	2.21
	A4	1.98	0.92	0.61
	A6	1.94	0.58	0.40
	A8	0.87	0.34	0.30
	A10	0.00	0.00	0.00
	A12	0.00	0.00	0.00

lifetimes were predicted at three storage temperatures (T = 293.15 K, 298.15 K, and 303.15 K) using the kinetic parameters provided by model-fitting method. As expected, the storage lifetime decreases with natural aging and storage temperature increase. All the unaged propellants are stable with the lowest lifetime found for DBRP1, which confirms the B&J test results. Furthermore, the storage lifetime for unaged DBRP3 was found to be about 3 folds higher than the two other propellants at the three storage temperatures used for the prediction.

With artificial aging progress, the storage lifetimes decrease because of the acceleration of the degradation reaction at the higher temperature of heating. Besides, after 10 months of accelerated heating, all the DBRPs did not present any storage lifetimes. They became hazardous and cannot be used under any conditions.

4 Conclusion

The assessment of the chemical stability of Double Base Rocket Propellant is of paramount importance during synthesis, handling, storage, and use as well as in the establishment of defense plans. The results obtained in this work showed that the use of several complementary approaches and methodologies allows deciding accurately on the stability status of DBRP.

The degradation of the DBRP during aging was found to be strongly connected to the migration, diffusion, and evaporation of nitroglycerine as well as the nitrocellulose chemical decomposition. FTIR and XRD analyses were found to provide valuable information on the current status of the chemical stability for DBRP. Therefore, these two non-thermal techniques can be used as guidance tools in energetic materials investigations.

Bergmann-Junk test gives efficient conclusions on the chemical stability of DBRP if compared to well-known standards. Nevertheless, the VST carried out according to AOP 7 requirements presents few issues concerning the stability assessment when only the released gas volume is taken into account. On the other hand, VST performed at several isothermal temperatures can be used efficiently to evaluate the kinetic parameters for the overall decomposition reactions occurring during natural and/or artificial aging. The activation energies computed using two different methodologies, *viz*, model-fitting and model-free are found to decrease with both types of aging (natural and aging). Until 6 months of accelerated aging, the application of the two methodologies to VST data provides close values of the Arrhenius parameters. Furthermore, the obtained kinetic parameters can be successfully used to predict the storage lifetimes for DBRP at different storage temperatures, which became hazardous after 10 months of accelerated heating at 338.15 K.

References

Abdel-Ghani N, Elbeih A, Helal F (2016) The effect of different copper salts on the mechanical and ballistic characteristics of double base rocket propellants. Central Eur J Energ Mater 13

Agrawal J, Singh H (1993) Qualitative assessment of nitroglycerin migration from double-base and composite modified double-base rocket propellants: Concepts and methods of prevention. Propellants, Explos, Pyrotech 18:106–110

Aoki I, Kubota N (1982) Combustion wave structures of high-and low-energy double-base propellants. AIAA J 20:100–105

Beckstead M (1980) Model for double-base propellant combustion. AIAA J 18:980–985

Benhammada A, Trache D, Kesraoui M, Tarchoun AF, Chelouche S, Mezroua A (2020) Synthesis and characterization of α-Fe_2O_3 nanoparticles from different precursors and their catalytic effect on the thermal decomposition of nitrocellulose. Thermochim Acta 178570

Bertacchini L, Durante C, Marchetti A, Sighinolfi S, Silvestri M, Cocchi M (2012) Use of X-ray diffraction technique and chemometrics to aid soil sampling strategies in traceability studies. Talanta 98:178–184

Blazhnov IV, Malomuzh NP, Lishchuk SV (2004) Temperature dependence of density, thermal expansion coefficient and shear viscosity of supercooled glycerol as a reflection of its structure. J Chem Phys 121:6435–6441

Bobić NR, Simonović RS, Drmanić SŽ, Milić SZ, Nikolić JB, Stoiljković SD, Terzić NS (2017) The influence of migration processes in gunpowder charge on the quality of mortar ammunition. Hemijska Industrija 71:231–240

Bohn MA (2007) NC-based energetic materials-stability, decomposition, and ageing. Paper presented at the Presentation on the meeting Nitrocellulose supply, Ageing and Characterization, Aldermaston, England

Bohn MA (2017) Principles of ageing of double base propellants and its assessment by several methods following propellant properties: STO-MP-AVT-268

Chelouche S, Trache D, Benayachi ZI, Tarchoun AF, Khimeche K, Mezroua A (2020a) A new procedure for stability control of homogeneous solid propellants by combination of vacuum stability test, FTIR and PCA. Propellants, Explosives, Pyrotechnics

Chelouche S, Trache D, Maamache I, Tarchoun AF, Khimeche K, Mezroua A (2020b) A new experimental way for the monitoring of the real/equivalent in-service-time of double base rocket propellant by coupling VST and PCA. Defence Technol

Chelouche S, Trache D, Neves CM, Pinho SP, Khimeche K, Benziane M (2018a) Solid+ liquid equilibria and molecular structure studies of binary mixtures for nitrate ester's stabilizers: Measurement and modeling. Thermochim Acta 666:197–207

Chelouche S, Trache D, Pinho SP, Khimeche K (2019a) Experimental and modeling studies of binary organic eutectic systems to be used as stabilizers for nitrate esters-based energetic materials. Fluid Phase Equilib 498:104–115

Chelouche S, Trache D, Pinho SP, Khimeche K, Mezroua A, Benziane M (2018b) Solid–liquid phase equilibria, molecular interaction and microstructural studies on (N-(2-ethanol)-p-nitroaniline+ N-(2-acetoxyethyl)-p-nitroaniline) Binary Mixtures. Int J Thermophys 39:129

Chelouche S, Trache D, Tarchoun AF, Abdelaziz A, Khimeche K (2019b) Compatibility assessment and decomposition kinetics of nitrocellulose with eutectic mixture of organic stabilizers. Journal of Energetic Materials1–20

Chelouche S, Trache D, Tarchoun AF, Abdelaziz A, Khimeche K, Mezroua A (2019c) Organic eutectic mixture as efficient stabilizer for nitrocellulose: kinetic modeling and stability assessment. Thermochim Acta

Chelouche S, Trache D, Tarchoun AF, Khimeche K (2019d) Effect of organic eutectic on nitrocellulose stability during artificial aging. J Energ Mater 1–20

Chelouche S, Trache D, Tarchoun AF, Khimeche K (2019e) Effect of organic eutectic on nitrocellulose stability during artificial aging. J Energ Mater 37:387–406

Chelouche S, Trache D, Tarchoun AF, Khimeche K, Mezroua A (2019f) Compatibility of nitrocellulose with aniline-based compounds and their eutectic mixtures. J Therm Anal Calorim 1–15

Davenas A (2012) Solid rocket propulsion technology: Newnes.

de Klerk W, van der Meer N, Eerligh R (1995) Microcalorimetric study applied to the comparison of compatibility tests (VST and IST) of polymers and propellants. Thermochim Acta 269:231–243

Elbasuney S, Elghafour AM, Radwan M, Fahd A, Mostafa H, Sadek R, Motaz A (2019) Novel aspects for thermal stability studies and shelf life assessment of modified double-base propellants. Defence Technol 15:300–305

Elbasuney S, Fahd A, Mostafa HE, Mostafa SF, Sadek R (2018) Chemical stability, thermal behavior, and shelf life assessment of extruded modified double-base propellants. Defence Technol 14:70–76

Elbeih A, Abd-Elghany M, Elshenawy T (2017) Application of vacuum stability test to determine thermal decomposition kinetics of nitramines bonded by polyurethane matrix. Acta Astronaut 132:124–130

Elghafour AM, Radwan M, Fahd A, Mostafa HE, Elbasuney S (2018) Novel approach to quantify the chemical stability and shelf life of modified double-base propellants. Defence Technol 14:720–724

Fidanovski B, Dimić M, Milojković A, Rodić V (2016) Determination of chemical stability of propellants using the vacuum stability test method. Sci Technical Rev 66:18–22

Fraunhofer-Institut fur Chemische Technologie I (1997) Kinetic description of the ageing of gun and rocket propellants for the prediction of their service lifetime. subgroup W1

Herder G, Weterings F, de Klerk W (2003) Mechanical analysis on rocket propellants. J Therm Anal Calorim 72:921–929

Kuentzmann P (2004) Introduction to solid rocket propulsion: Office national d'etudes et de recherches aerospatiales chatillion cedex

Ma S, Song G, Feng N (2012) Preparation and characterization of self-emulsified waterborne nitrocellulose. Carbohyd Polym 89:36–40

Matečić Mušanić S, Fiamengo Houra I, SućeSkA M (2010) Applicability of non-isothermal DSC and Ozawa method for studying kinetics of double base propellant decomposition. Central Eur J Energ Mater 7:233–251

Matečić Mušanić S, Sućeska M (2009) Artificial ageing of double base rocket propellant: effect on dynamic mechanical properties. J Therm Anal Calorim 96:523–529

Mušanić SM, Sućeska M (2013) Dynamic mechanical properties of artificially aged double base rocket propellant and the possibilities for the prediction of their service lifetime. Central Eur J Energ Mater 10:225–244

Narender K, Rao ASM, Rao KGK, Krishna NG (2013) Temperature dependence of density and thermal expansion of wrought aluminum alloys 7041, 7075 and 7095 by gamma ray attenuation method. J Mod Phys 4:331–336

NATO Standardisation Agreement (STANAG) 4147, Chemical compatibility of ammunition components with explosives (non nuclear applications); AC/310 (SG1) D/15 (Draft edition 2) I-96 NAVY/ARMY/AIR

North Atlantic Treaty Organization (2003) Manual of data requirements and tests for the qualification of explosives materials for military use (June 2003)

Rodrigues Filho G, da Cruz SF, Pasquini D, Cerqueira DA, de Souza PV, de Assunção RMN (2000) Water flux through cellulose triacetate films produced from heterogeneous acetylation of sugar cane bagasse. J Membr Sci 177:225–231

Roduit B, Guillaume P, Wilker S, Folly P, Sarbach A, Berger B, Mathieu J, Ramin M, Vogelsanger B (2008) Advanced simulation of the lifetime of energetic materials based on HFC signals. HFCS-EM 37:39

Sharma J, Wilmot G, Campolattaro A, Santiago F (1991) XPS study of condensed phase combustion in double-base rocket propellant with and without lead salt-burning rate modifier. Combust Flame 85:416–426

Shekhar H (2011) Effect of temperature on mechanical properties of solid rocket propellants. Defence Sci J 61:529–533

Singh H, Rao K (1977) Platonization in double-base rocket propellants. AIAA J 15:1545–1549

Standardization Agreement STANAG 4117 (1998) Explosives, stability test procedures and requirements for propellants stabilized with diphenylamine, ethyl centralite or mixtures of both. 3rd Edition. North Atlantic Treaty Organization (NATO)

Sućeska M, Mušanić SM, Fiamengo I, Bakija S, Bakić A, Kodvanj J (2010a) Study of mechanical properties of naturally aged double base rocket propellants. Central Eur J Energ Mater 7:47

Sućeska M,Mušanić SM, Houra IF (2010b) Kinetics and enthalpy of nitroglycerin evaporation from double base propellants by isothermal thermogravimetry. Thermochim Acta 510:9–16

Trache D, Abdelaziz A, Siouani B (2017) A simple and linear isoconversional method to determine the pre-exponential factors and the mathematical reaction mechanism functions. J Therm Anal Calorim 128:335–348

Trache D, Khimeche K (2013a) Study on the influence of ageing on chemical and mechanical properties of N, N′-dimethyl-N, N′-diphenylcarbamide stabilized propellants. J Therm Anal Calorim 111:305–312
Trache D, Khimeche K (2013b) Study on the influence of ageing on thermal decomposition of double-base propellants and prediction of their in-use time. Fire Mater 37:328–336
Trache D, Khimeche K, Mezroua A, Benziane M (2016) Physicochemical properties of microcrystalline nitrocellulose from Alfa grass fibres and its thermal stability. J Therm Anal Calorim 124:1485–1496
Trache D, Tarchoun AF (2018) Stabilizers for nitrate ester-based energetic materials and their mechanism of action: a state-of-the-art review. J Mater Sci 53:100–123
Trache D, Tarchoun AF (2019) Differentiation of stabilized nitrocellulose during artificial aging: spectroscopy methods coupled with principal component analysis. J Chemom 33:e3163
Tseng IS, Yang V (1994) Combustion of a double-base homogeneous propellant in a rocket motor. Combust Flame 96:325–342
Vogelsanger B (2004) Chemical stability, compatibility and shelf life of explosives. CHIMIA Int J Chem 58:401–408
Vyazovkin S (2015) Isoconversional kinetics of thermally stimulated processes. hiedelberg, Springer, Berlin
Vyazovkin S (2017) Isoconversional kinetics of polymers: the decade past. Macromol Rapid Commun 38:1600615
Vyazovkin S, Linert W (1995) Thermally induced reactions of solids: Isokinetic relationships of non-isothermal systems. Int Rev Phys Chem 14:355–369
Vyazovkin S, Wight CA (1998) Isothermal and non-isothermal kinetics of thermally stimulated reactions of solids. Int Rev Phys Chem 17:407–433
Zayed M, Mohamed AA, Hassan M (2010) Stability studies of double-base propellants with centralite and malonanilide stabilizers using MO calculations in comparison to thermal studies. J Hazard Mater 179:453–461
Zeman S, Elbeih A, Yan Q-L (2013) Notes on the use of the vacuum stability test in the study of initiation reactivity of attractive cyclic nitramines in the C4 matrix. J Therm Anal Calorim 112:1433–1437
Zenin A (1995) HMX and RDX-Combustion mechanism and influence on modern double-base propellant combustion. J Propul Power 11:752–758

Chapter 3
Assessment of the Migration of Combustion Moderator in Nitrocellulose-Based Propellant

Moulai Karim Boulkadid, Michel H. Lefebvre, Laurence Jeunieau, and Alain Dejeaifve

Abstract The impregnation in surface with combustion moderator (also called deterrent) is usually applied to improve the ballistic properties, for example in the case of deterred spherical propellant which is largely employed in defense applications. The disadvantage of this surface impregnation is the migration of the deterrent toward the center of the grains. The knowledge of the diffusion coefficient of deterrent into grain core during the lifetime of the propellant is of significant importance because it is a fitting factor for the estimation of the shelf life. Diffusion coefficient has been determined by various techniques. In the most recent articles, we find that the most employed method is infrared (IR) microscopy; however, this method is highly time consuming. Therefore, the aim of this work is to explore the possibility to obtain an efficient procedure needing less time to be achieved. In this paper, we present an improved procedure based on well-known experimental technique namely high pressure liquid chromatography (HPLC). It is composed of the following steps: (1) the preparation of propellant grains that have cylindrical shape and double base chemical composition (around 11% nitroglycerin) using extrusion process, (2) known quantities of deterrent (dibutylphthalate; DBP) are deposited on the surfaces perpendicular to the axis of the grains before the conditioning at constant temperature (85 °C) for 04 h. After that, each grain was sliced by a microtome into little pieces of a thickness equal to 20 μm. The last step comprises the determination of the concentration in DBP of each slice by HPLC. The results permit the establishment of the concentration profile and the migration coefficient of DBP in grain matrix by both infrared

M. K. Boulkadid (✉)
Laboratoire de Propulsion Energétique, UER. Procédés Energétiques, Ecole Militaire Polytechnique (EMP) Chahid Abderrahmane Taleb, BP 17 Bordj El-Bahri, 16046 Algiers, Algeria
e-mail: karim.boulkadid@gmail.com

M. H. Lefebvre · L. Jeunieau
Laboratory of Energetic Materials, Royal Military Academy, Av. de la Renaissance 30, 1000 Brussels, Belgium

A. Dejeaifve
PB Clermont SA, Rue de Clermont 176, 4480 Engis, Belgium

D. Trache et al. (eds.), *Materials Research and Applications*,
Materials Horizons: From Nature to Nanomaterials,
https://doi.org/10.1007/978-981-15-9223-2_3

(IR) microscopy improved procedure. The obtained values of the diffusion coefficient according to infrared (IR) microscopy and improved procedure are equal to 6.9×10^{-16} m²/s and 8.7×10^{-12} m²/s, respectively.

Keywords Deterrent concentration profile diffusion · Coefficient · DBP · Ball propellant · HPLC

1 Introduction

Small arms propellant grains are smokeless, double base powders containing nitrocellulose, and nitroglycerin as energetic ingredients. Nitrocellulose (NC) and nitroglycerin (NG) are usual energetic ingredients of solid gun propellants. In the production of propellants, additional elements are included to improve their chemical stability and plasticity, to control their burn rate, and to diminish the combustion temperature (Boulkadid et al. 2019, 2020; Benhammada and Trache 2019; Hanafi et al. 2019). NC and NG are unstable at ambient conditions, and stabilizing agents should be included into the energetic formulations to restrain and retard the decomposition reactions that can take place (Trache and Tarchoun 2018, 2019). Moderators are chemicals that are present down to a certain depth into small arms propellant particles to decrease their early combustion properties (when the combustion surface is high). The impregnation in surface with deterrent is employed to improve the ballistic properties. The inconvenience of this surface treatment is the diffusion of the deterrents inside the grains. When the diffusion process occurs, the combustion rate of the first propellant layers increases accounting for increasing maximum chamber pressure, which may destruct the gun or confer arise to malfunction of the armaments. Therefore, the knowledge of the migration rate of the deterrent into the propellant core during the lifetime of the propellant is of major importance and a relevant factor for the determination of the shelf life of the propellant (Jeunieau et al. 2007a, b). Diverse procedures of analysis have been used to determine the deterrent concentration profile (Boulkadid et al. 2015, 2016, 2017; Jeunieau et al. 2007a, b; Stiefel and Devine 1979; Fong and Cooke 1982; Brodman and Devine 1978; Brodman et al. 1974; Pesce-Rodriguez et al. 1992; Rat and Lacroix 1994; Vogelsanger et al. 1996; Louden et al. 1985, 1989; Varriano-Marston 1987; Louden and Kelly 1991). As an example ^{14}C-labeled and autoradiographs, but this method has the disadvantage that the propellant must be specially synthesized for this purpose; in fact a scaled-down production procedure was used to deter spherical nitrocellulose propellant grains (containing nitroglycerin) with ^{14}C-labeled DBP. Staining and optical methods have also been used but are not able of directly establishing the concentration profile in the deterred region of the propellant grain. Infrared (IR) microscopy has been proved to be a powerful and efficient method for the determination of qualitative and quantitative concentration profiles of deterrents in propellants; however, these methods are highly time consuming (Jeunieau et al. 2007a, b). In fact, IR microscopy measurements need calibration and showed some difficulties such as cutting problems of the ball propellant grains. Therefore, the aim of this work is to explore the possibility to obtain an efficient procedure needing less time to be achieved.

In this chapter, an improved method has been proposed for the assessment of the diffusion of combustion moderator (DBP) for gun propellant. It is based on HPLC and allowed the determination of the deterrent concentration profile in nitrocellulose/nitroglycerin matrix grain. In addition, the diffusion coefficients have been calculated. For comparison, the diffusion was investigated by infrared (IR) microscopy.

2 Diffusion Models

The deterrents diffuse through the nitrocellulose-based matrix by one of two diffusion models, either Fickian (case I) or non-Fickian (case II) (Vogelsanger et al. 1996). Fickian migration is identified by an exponential decrease in concentration from the surface inward while case II is distinguished by a fixed concentration followed by a sharp drop. A first examination of our results suggested that the simple Fickian diffusion model is well suited to fit our concentration profiles. The Fick's diffusion model for a one-dimensional diffusion in anisotropic media may be represented by (Vogelsanger et al. 1996)

$$\frac{\partial c}{\partial t} = D\frac{\partial^2 c}{\partial x^2} \tag{1}$$

where c is the diffusant concentration (kg/m^3), t the diffusion time (s), D the diffusion coefficient (m^2/s), and x the Cartesian space coordinate (m). Providing constant diffusion coefficient within the segment where diffusion takes place, general solutions of Eq. (1) can be obtained for a variety of initial and boundary conditions. Equation (2) describes the concentration profile as function of time and space, when started by an instantaneous plane source:

$$c(x,t) = \frac{s}{\sqrt{\pi Dt}}\mathrm{e}^{\frac{-x^2}{4Dt}} \tag{2}$$

where s is the surface concentration of diffusant (kg/m^2).

By using the deterrent concentration at the particle surface, Eq. (2) can be rewritten as

$$c(x,t) = c(0,t)\mathrm{e}^{\frac{-x^2}{4Dt}} \tag{3}$$

In this equation, the entire deterrent is supposed to be at the surface of the propellant at the initial time ($t = 0$ s).

3 Experimental Procedure

3.1 HPLC Measurement

A spherical nitrocellulose-based propellant with 10.0% of nitroglycerin is mixed with tetrahydrofuran to ease the mixing process. The spherical propellant does not contain deterrent, and it is stabilized with diphenylamine (DPA). The content of the propellant is summarized in Table 1. Mixing process occurs at ambient temperature. This mixture has been extruded to obtain cylindrical propellant strands of about 10 mm in diameter. To increase homogeneity of the samples, extrusion is repeated twice. The extrusion chamber is at 80 °C. The manufactured samples are stored at 40 °C during a week to ensure complete evaporation of THF. The DBP is deposited on the surfaces perpendicular to the axis of the grains. It is ensured that the lateral surfaces of the cylindrical strands are isolated to prevent the DBP from diffusing through it. Then the strands with DBP are introduced in oven at 85 °C during 24 h (accelerated aging). After removing the grains from the oven, they are sliced by a microtome into little pieces of a thickness equal to 20 μm. The concentration of DBP in each slice is determined by means of HPLC.

Figure 1 summarizes the experimental procedure (after preparation of the double base cylindrical grains) followed in this work to study the diffusion of the combustion moderator (DBP) in the grain matrix.

In order to correct the slice thickness, the following equation is used:

$$e_s = \frac{m_s * 0.01}{\pi * r_s^2}\left(\frac{\%\text{NC}}{\rho_{\text{NC}}} + \frac{\%\text{NGL}}{\rho_{\text{NGL}}} + \frac{\%\text{DBP}}{\rho_{\text{DBP}}} + \frac{\%\text{KN}}{\rho_{\text{KN}}} + \frac{\%\text{DPA}}{\rho_{\text{DPA}}}\right) \quad (4)$$

where

e_s slice thickness
m_s mass of slice
ρ_i density of compound i
$\% X_i$ mass fraction of compound i
r_s radius of the slice.

Table 1 Contents of the tested propellants

Component	Mass fraction (%)
NC	86.94
NG	11.20
KNO_3	0.61
DPA	1.25

NC Nitrocellulose, *NG* Nitroglycerin, *DBP* Dibutyl phthalate, *DPA* Diphenylamine

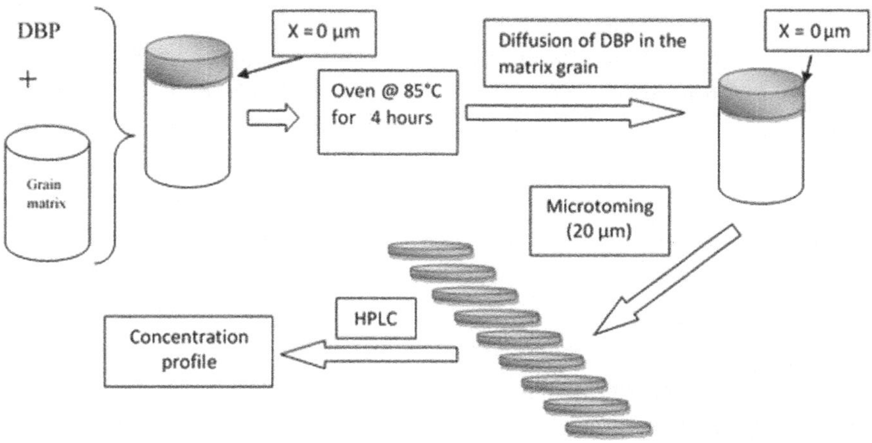

Fig. 1 Diagram describing the experimental procedure to study the diffusion of the deterrent (DBP) in the grain matrix

3.2 *Infrared Microscopy (IR) Measurement*

A spherical deterred propellant was used for this study. The spherical propellant had a mean diameter of 553 µm and comprise around 10% of nitroglycerin stabilized with diphenylamine (DPA). The mean content of the moderator dibutyl phthalate (DBP) was 4.8%. The composition of the investigated propellants is given in Table 2. The propellant grains were embedded in a polyepoxy resin. 7-µm-thick sections were prepared using a Leica microtome and a tungsten carbide knife.

A Bruker Hyperion infrared (IR) mounted on a Vector 33 Fourier transform spectrometer was used in this study. A medium-band MCT (HgCdTe) detector in the microscope gives high sensitivity in the 4000–600 cm^{-1} range. Propellant grain cross sections were placed on a KBr pellet, and IR spectra were taken at 10 µm intervals. A15X Cassegrain mirror objective was employed to obtain the infrared

Table 2 Composition (wt.%) of the spherical deterred propellant

Component	Mass fraction (%)
NC	82.32
H_2O	0.49
KNO_3	0.86
NGL	10.40
DBP	4.80
DPA	0.59
NDPA	0.54

NC Nitrocellulose, *NGL* Nitroglycerin, *DBP* Dibutyl phthalate, *DPA* Diphenylamine, *NDPA* Nitrodiphenylamine spectra. The IR spectrometer was operating at a resolution of 4 cm^{-1}, and 32 scans were acquired for each measurement position. The size of the aperture was 30 μm × 30 μm. For this investigation, unaged and aged propellants (10.91 days at 80 °C) samples were tested.

4 Results

4.1 HPLC

The diffusion profiles of DBP in grain matrix (cylindrical strand) are given in Fig. 2. The deterrent concentration profile shows an exponential diminution of moderator from the surface inward. Such concentration distribution agrees with a classical Fickian migration model [See Eq. (2)]. To estimate the best diffusion coefficient (Dt) fitting the experimental data points, we employ an adequate nonlinear regression.

In this part of the work, at $t = 0$ s the entire deterrent is at the surface of the propellant (see Fig. 1). By consequence, the time in Eq. (3) is equal to the time of aging at 85 °C ($t = 4$ h).

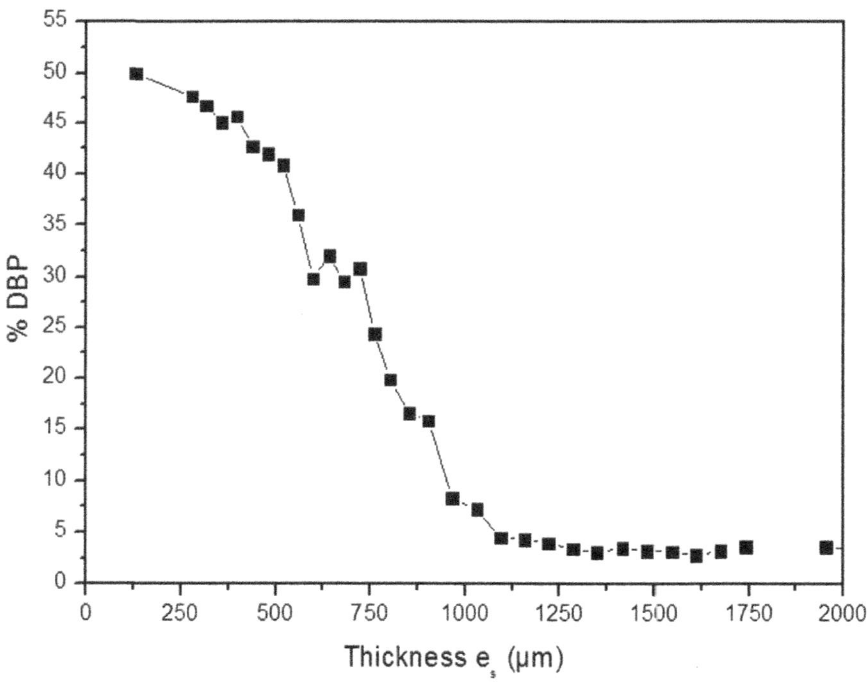

Fig. 2 Diffusion profiles of DBP in grain matrix (cylindrical strand)

The diffusion coefficient D can then be calculated using the Eq. (5):

$$D = \frac{Dt}{t} \quad (5)$$

The calculated value of the diffusion coefficient D according to Eq. (5) is equal to 8.7×10^{-12} m^2/s.

4.2 *IR Microscopy*

Figure 3 exhibits some moderator concentration distributions of the investigated propellant, unaged and aged 10.91 days at 80 °C. It is noticed that the moderator concentration diminishes in surface with aging and augments in the center part of the propellant with aging. This is due to the diffusion of the DBP from surface toward the center of the grain.

In Eq. (3), the entire deterrent is supposed to be concentrated at the surface of the propellant at the initial time ($t = 0$). In our specific case, a deterrent concentration profile is already present before the aging of the propellant. By consequence, the

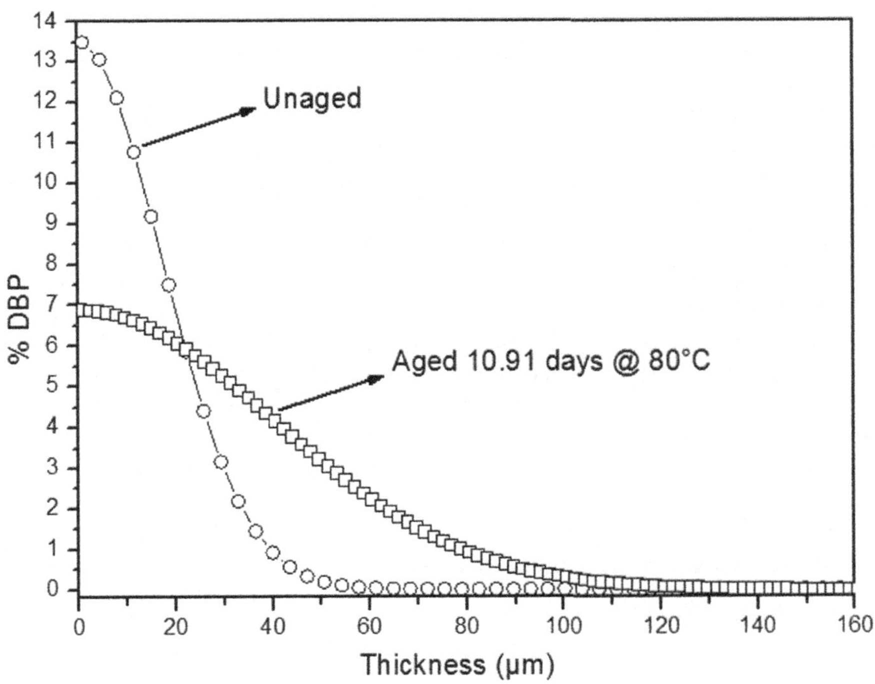

Fig. 3 Moderator concentration distributions of DBP in the unaged and aged at 80 °C investigated propellant

time in Eq. (3) must be equal to the summation of the time needed to achieve the deterrent concentration profile observed before aging and to the time of aging. The experimental curves (Fig. 3) are numerically fitted to obtain the value of $c(0, t)$ and of the corresponding parameter Dt. The diffusion coefficient can then be calculated using the set of Eq. (6):

$$\begin{cases} D & \text{from initial profile, before aging} \\ Dt_a & \text{from analyzed profiles, after aging time} \\ t_a - t_0 & \text{aging duration} \end{cases} \tag{6}$$

The obtained value of the diffusion coefficient D according to Eq. (6) is equal to 6.9×10^{-16} m^2/s.

5 Discussion

In this work, the diffusion process of DBP (deterrent) in double base grain propellant was investigated by both infrared microscopy and an improved procedure based on HPLC. According to IR microscopy measurements, the deterrent concentration profiles show an exponential reduction of moderator from the surface inwards; this is consistent with the observed diffusion models in the literature (Boulkadid et al. 2017, Jeunieau et al. 2007a, b). The obtained value of the diffusion coefficient is equal to 6.9×10^{-16} m^2/s.

In the case of the improved procedure measurements, the diffusion profile of the moderator in the grain reveals also an exponential concentration distribution, similar to the one obtained by IR microscopy. The calculated value of the diffusion coefficient D is equal to 8.7×10^{-12} m^2/s.

The obtained D-value by the improved method based on HPLC is four times greater in magnitude than the D-value found by IR measurements. The observed difference between the obtained D-values could be explained by the sensitivity of the D-value to the temperature of aging. The temperature sensitivity of the D-values is defined well by the Arrhenius equation $D = D_o \exp(E_a/RT)$. Furthermore, migration demonstrated to be very sensitive to the grain matrix. Two propellant samples fabricated under similar conditions except for one single step diverged in their D-values by a factor of two (Vogelsanger et al. 1996). Knowing that the steps followed for the manufacturing of the grains (cylinder) for HPLC measurements are completely different from that employed for IR measurements, this is an additional explanation for the observed difference between the found D-values.

6 Conclusions

An improved experimental methodology for assessment of the migration of combustion moderator (deterrent) in nitrocellulose-based propellant has been shown in this chapter. It is based on well-known technique, namely HPLC. The concentration distribution of deterrent (DBP) in nitrocellulose/nitroglycerin grain matrix has been determined, and the diffusion coefficient was found equal to 8.7×10^{-12} m^2/s. The deterrent concentration profiles show an exponential diminution of moderator from the surface inward; this was consistent with the observed diffusion models in the literature. The comparison of the obtained D-values based on HPLC and IR microscopy showed a divergence; indeed, the calculated value of the diffusion coefficient D by IR microscopy was equal to 6.9×10^{-16} m^2/s. This difference has been explained by the sensitivity of D-value to both temperature and grain matrix.

References

Benhammada A, Trache D (2019) Thermal decomposition of energetic materials using TG-FTIR and TG-MS: a state-of-the-art review. Appl Spectroscopy Rev. https://doi.org/10.1080/05704928.2019.1679825

Boulkadid MK, Lefebvre MH, Jeunieau L, Dejeaifve A (2015) Influence of firing temperature on properties of gun propellants. J Chem Chem Eng 6:415–427

Boulkadid MK, Lefebvre MH, Jeunieau L, Dejeaifve A (2016) Temperature sensitivity of propellant combustion and temperature coefficients of gun performance. Central Euro J Energ Mater 13:1005–1022

Boulkadid MK, Lefebvre MH, Jeunieau L, Dejeaifve A (2017) Local temperature sensitivity coefficients of a deterred spherical single base gun propellant. Central Euro J Energ Mater 14:952–965

Boulkadid MK, Lefebvre MH, Jeunieau L, Dejeaifve A (2019) Burning rate of artificially aged solid double-base gun propellants. J Energ Mater. https://doi.org/10.1080/07370652.2019.1657204

Boulkadid MK, Trache D, Krai S, Lefebvre MH, Jeunieau L, Dejeaifve A (2020) Estimation of the ballistic parameters of double base gun propellants. Propellants Explos Pyrotech. https://doi.org/10.1002/prep.201900341

Brodman BW, Devine MP (1978) Chemical interactions and their effect on the small arms propellant deterring process. J Ballistics 2:51

Brodman BW, Devine MP, Finch RW, MacClaren MS (1974) Autoradiographic determination of the di-n -butyl-phthalate concentration profile in a nitrocellulose matrix. J Appl Polym Sci 18:3739

Fong CW, Cooke C (1982) Diffusion of deterrents into a nitrocellulose-based small arms propellants. The effect of deterrent structure and solvent. J Appl Polym Sci 27:2827

Hanafi S, Trache D, Abdous S, Bensalem Z, Mezroua A (2019) 5-Nitro-1,2,4-triazole-3-one: a review of recent advances. Chin J Energ Mater (Hanneng Cailiao). 27(4):326–347

Jeunieau L, Lefebvre MH, Guillaume P (2007a) Quantitative infra-red microscopy method for the determination of the deterrent diffusion activation energy in a spherical propellant. Cent Eur J Energ Mater 4:109–124

Jeunieau L, Lefebvre MH, Guillaume P (2007b) Ballistic stability of a spherical propellant: comparison with a flattened spherical propellant. Central Euro J Energ Mater 4(1–2):31–44

Louden JD, Kelly J, Phillipson J (1985) Raman Microspectroscopic determination of the methyl centralite (N, N'-Dimethyl-N, N'-Diphenylurea) concentration profile in a nitrocellulose extruded mono-perforated small arms propellant. J Raman Spectrosc 18:137

Louden JD, Kelly J, Phillipson J (1989) Methylcentralite concentration profiles in Monoperforated extruded nitrocellulose and nitrocellulose/nitroglycerine propellant grains by Raman microspectroscopy. ibid 37:3237

Louden JD, Kelly J (1991) Infrared mapping of deterrents (morderents) in nitrocellulose based propellant by Fourier-transform infrared microscopy. Anal Appl Spectrosc 2:90

Pesce-Rodriguez RA, Miser CS, McNesley KL, Fifer RA, Kessel RA, Strauss BS (1992) Characterization of solid propellant and its connection to aging phenomena. Appl Spectrosc 46:1143

Rat M, Lacroix G (1994) A new method to determine migration profiles of low molecular weight constituents in propellants: FTIR microscopy equipped with a motorized stage. In: 25th international annual conference on ICT, Karlsruhe

Stiefel L, Devine MP (1979) Depth of penetration of deterrent in representative U.S. Small Arms Propellants 3:515

Trache D, Tarchoun AF (2018) Stabilizers for nitrate ester-based energetic materials and their mechanism of action: a state-of-the-art review. J Mater Sci 53:100–123

Trache D, Tarchoun AF (2019) Analytical methods for stability assessment of nitrate esters-based propellants. Crit Rev Anal Chem 49(5):415–438

Varriano-Marston E (1987) An infrared microspectroscopy method for determining deterrent penetration in nitrocellulose-based propellant grains. J Appl Polym Sci 33:107

Vogelsanger B, Ossala B, Brönnimann E (1996) The diffusion of deterrents and blasting oils into propellants observed by FTIR microspectroscopy. In: 10th symposium on chemical problems connected with the stability of explosives, p 305

Chapter 4
Study of the Chemical Stability of a Pseudo-Brookite B-Synroc by the MCC1, MCC2, and Microwaves Tests

Fairouz Aouchiche, Nour el Hayet Kamel, Soumia Kamariz, Dalila Moudir, Yasmina Mouheb, and Azzedine Arabi

Abstract Synroc B is a polyphasic ceramic dedicated to the confinement of certain lanthanides and actinides. A Synroc B with 40% pseudo-brookite, 36% zirconolite, and 24% perovskite was synthesized by sintering at 1200 °C. Several leaching tests according to the MCC1 and MCC2 (IAEA) standard tests were conducted on the material to evaluate its chemical stability. The leachates were analyzed by X-ray fluorescence. The elements of the matrix: Ca, Ce, Ti, Zr, Al, and Fe were below the detection limit (≈1 ppm). For both MCC1 and MCC2 tests, Ca was the most dissolved element in the leachates, with about 0.5 and 2.8 wt.% of the total Ca content in the matrix, for the two tests, respectively. The other elements were of very weak dissolutions rates in the leachates, especially those of pseudo-brookite, such as Fe and Ti. Finally, a third test of microwave mineralization resulted 26% of the initial Ca mass in the mineral. One can conclude that the addition of pseudo-brookite to Synroc B significantly increases its chemical durability.

Keywords Synroc B · MCC1 · MCC2 · XRF · Pseudo-brookite · Chemical durability

1 Introduction

The «SYNROC» is the abbreviation of «Synthetic Rock». It is constituted by several mineral phases, namely zirconolite, pérovskite and hollandite, which are able to confine for long term, many radioactive elements. The Synroc was discovered by Ted Ringwood, researcher at ANSTO (Australian Nuclear Science and Technology

F. Aouchiche (✉)
Division of Nuclear Techniques, Algiers Nuclear Research Center,, 2. Bd Frantz Fanon, P.O.Box: 399, Algiers, Algeria
e-mail: Aouchiche.fairouz@hotmail.fr

N. H. Kamel · S. Kamariz · D. Moudir · Y. Mouheb · A. Arabi
Division of Environment, Algiers Nuclear Research Center, Safety of Radioactive Waste, 2 Bd Frantz Fanon, P.O.Box: 399, Alger-RP, Algiers, Algeria

D. Trache et al. (eds.), *Materials Research and Applications*,
Materials Horizons: From Nature to Nanomaterials,
https://doi.org/10.1007/978-981-15-9223-2_4

Organization) at the end of the sixties (Lumpkin et al. 1995; Ringwood et al. 1979; Grote et al. 2018). It is often used for the confinement of both cesium (Cs) and actinides elements (Tripp and Maio 2006). The earlier research results on Synroc showed a good material stability against water leaching, which make this material interesting for the final disposal of radioactive waste (Konynenburg and Guinan 1983; Meng et al. 2016).

New chemical compositions of Synroc were the subject of many studies aiming to improve both the chemical and radiation damage resistance, without altering the constituting minerals material (Tang et al. 2014; Brinkman et al. 2017; Zhang et al. 2016). In Synroc materials, tetravalent actinides are embedded in the zirconolite phase, while trivalent actinides, rare earth elements (REE) and strontium (Sr) are all preferably confined in the perovskite phase (Sabathier 2003). Hence, depending on their composition, there are six types of Synroc materials (Tripp and Maio 2006; Ringwood et al. 1979; Ryerson et al. 1981; Ringwood 1985; Weber 2009; Matzke et al. 1990; Vance 1994; Ewing 1999), namely.

- The Synroc A, which is composed by hollandite, perovskite, zirconolite, Ba-feldspath, kalsilite and leucite (Ringwood et al. 1979).
- The Synroc B which is composed by titanates phases: perovskite ($CaTiO_3$), zirconolite ($CaZrTi_2O_7$), hollandite ($BaAl_2Ti_6O_{16}$), and rutile (TiO_2) (Ringwood et al. 1979; Zhang et al. 2019a and Zhang et al. 2019b).
- The Synroc C is constituted by hollandite ($BaAl_2Ti_6O_{16}$), zirconolite ($CaZrTi_2O_7$), and perovskite ($CaTiO_3$). In this last structure, hollandite mineral accommodates Cs, Rb, and Ba elements; and both zirconolite and perovskite minerals are the main long-lived actinides and REE isotopes host phases (perovskite being also hosting Sr in its crystalline structure). That is why Synroc C was proposed for liquid high-level waste resulting from pressurized water reactors' depleted fuel treatment and can load amounts reaching 30 wt.% of waste (Tripp and Maio 2006; Ghosh et al. 2019; Grote et al. 2018; Ringwood et al.1979; Amoroso et al. 2014).
- The Synroc D is a titaneous-silicated polyphasic ceramic, namely composed by zirconolite ($CaZrTi_2O_7$), perovskite ($CaTiO_3$), spinelle (MN_2O_4-M_2TO_4, M: divalent cation, N: trivalent cation, T: tetravalent cation) and nepheline ($NaAlSiO_4$). Minor phases of sulfur, pollucite ($CsAlSi_2O_6$), and phosphate can also exist (Ryerson et al. 1981).
- The Synroc E is constituted by hollandite, zirconolite, perovskite, pyrochlore, and metallic alloys ($\approx$20 wt.%). It can accommodate less than 7 wt.% of high-level waste (Ringwood 1985).
- The Synroc F contains typically 95 wt.% of pyrochlore $(Ca,Gd,U,Pu,Hf)_2Ti_2O_7$ and 5 wt.% of rutile TiO_2. Its composition can vary depending on the radioactive waste composition and impurities. Initially, it was developed to embed the CANDU reactors depleted fuel, starting from light hydrocarbons (Ringwood et al. 1988).

Synroc B is the purpose of this work, and it has both good resistance to aqueous corrosion and confinement properties toward most elements present in the wastes.

Its composition has been the subject of many studies (Ringwood et al. 1988). It is constantly modified in order to improve the material properties. Many authors report that Synroc B containing pseudo-brookite, and thus iron (Fe), presents satisfactory mechanical and durability properties (Wen et al. 2015).

In the synthesis point of view, Synroc materials, as ceramic materials, can be obtained by both wet and dry routes of synthesis, and the wet route generally consists on a sol–gel reaction in presence of a dispersive material and is employed in order to enhance the isotropic properties of the material (Zhao et al. 2020). The dry route of synthesis which is employed in the present study consists of sintering or calcination of an activated oxides mixture powder, constituting the final material (Donchev et al. 2002). Recent studies dealing with an iron-rich Synroc B report interesting properties of this ceramic, particularly regarding to the material durability (Wen et al. 2015).

The purpose of this work is to study the chemical durability of Synroc B containing pseudo-brookite, by the mean of several aggressive tests, namely MCC1, MCC2, and the microwave test. The kinetics of leaching is followed by *XRF* analysis, after establishing the elemental calibration curves. The evolution of the concentration, the normalized mass loss, and the leaching rate of elements (Ca, Ti, Zr, Fe, Al, and Ce) make the conclusion possible on the leaching behavior of this Synroc B, regarding to the three leaching tests.

2 Experimental

In the MCC1 test, the elements are solubilized in a small volume of water over a limited period of time. The mineral pellet is immersed in a volume of ultrapure water (18 MΩ/cm) at room temperature (≅25 °C) in a dark glass flasks, opened to the atmosphere. The volume of water in which the pellet is immersed is constant and satisfies the ratio $S_0/V_0 = 10$. The latter is an important parameter of the test. S_0 is the surface which is effectively in contact with water and V_0 the liquid volume. A drop of 1 N nitric acid (HNO_3) is added to each sample taken to prevent any deposit on the walls of the test flasks. Regular samplings of 1 mL are taken at defined time periods: 1, 3, 7, 14, 28, 36, and 42 days. (Ojovan 2010). An equivalent volume of ultrapure water is added in the bottles in order to maintain a constant surface/volume (S_0/V_0) ratio. The MCC2 test is performed as for the MCC1 except that the flask is closed to the atmosphere and is set at a temperature of 90 °C (Ojovan 2010). For the microwave test, 0.3 g of synroc matrix B, in a powder form, is put in a teflon container, filled with 4 mL of 65% nitric acid (HNO_3) and 4 mL of 40% hydrofluoric acid (HF). The mineralization is carried out using a Berghof brand speedwave ENTRY microwave oven (application report). For the three tests, the leachates are analyzed by wavelength dispersive X-ray fluorescence (WDXRF) technique using a Panalytical Magix PRO equipment after establishing the calibration curves for the elements of interest: Ca, Ce, Ti, Zr, Al, and Fe. Standard solutions of the elements are prepared from 1000 mg/l ± 4 mg/l ICP/DCP Fluka Analytical Standard solutions, with concentrations of 10, 20, 30, 40, and 50 ppm. The leachates samplings are spotted on Mylar paper supports and dried at 60 °C before XRF measurements. The XRF calibration curves, of the measured intensity (I) as a function of the i element concentration (C_i), $I = f(C_i)$, are established using SuperQ-DS version 3 software program. The latter gives directly the elemental concentration, after each leachate analysis.

The elemental normalized mass loss, *Mi*, considers the effective dissolution of a given *i* element (Strachan 2001). It is expressed in kg/m^2. It is given by the mathematical relation number (1).

$$M_i = \frac{C_i V_0}{S_0 F_i} \tag{1}$$

where *Ci* is the total concentration of the leached "*i*" element (kg/m^3), V_0: the total volume of the leaching solution (m^3), and F_i the "*i*" element weight fraction in the solid phase (wt.%).

The elemental normalized leaching rate (RL_i) expresses the normalized dissolution rate of the material. It is given by the mathematical relation number (2) and is expressed in kg/m^2 (Strachan 2001).

$$RL_i = \frac{dM_i}{dt} = \frac{1}{F_i S_0}\frac{dm_i}{dt} \tag{2}$$

The *i* element average leaching rate is determined by calculating the elemental normalized mass loss at a mean time interval (Δt) taken from the starting time of leaching. It is given by the mathematical relation number (3).

$$RL_{imoy} = \frac{\Delta M_i}{\Delta t} \tag{3}$$

3 Results and Discussions

For both of the MCC1 and MCC2 tests, the results obtained give undetectable concentration values for the following elements: Ce, Ti, Zr, Al, and Fe. These concentrations are below the WRF detection limit (<1 ppm). For Ca, the evolution of concentrations according to time increases slowly in MCC1 test and reaches a maximum of 27.8 10^{-3} kg/m^3 at the 36th day. In MCC2 test, the Ca concentrations seem to decrease and suddenly show a maximum value after the 36th day, highlighting the effect of high temperature that makes water extracting the Ca in the inner layers of the material. The maximum C_{Ca} value is 131.80 10^{-3} kg/m^3, at the 42nd day is four times greater than that obtained at 36th days for both MCC1 and MCC2 tests, as shown in Fig. 1. The evolution of normalized Ca mass losses (M_{Ca}) follows that of concentrations, as shown in Fig. 2.

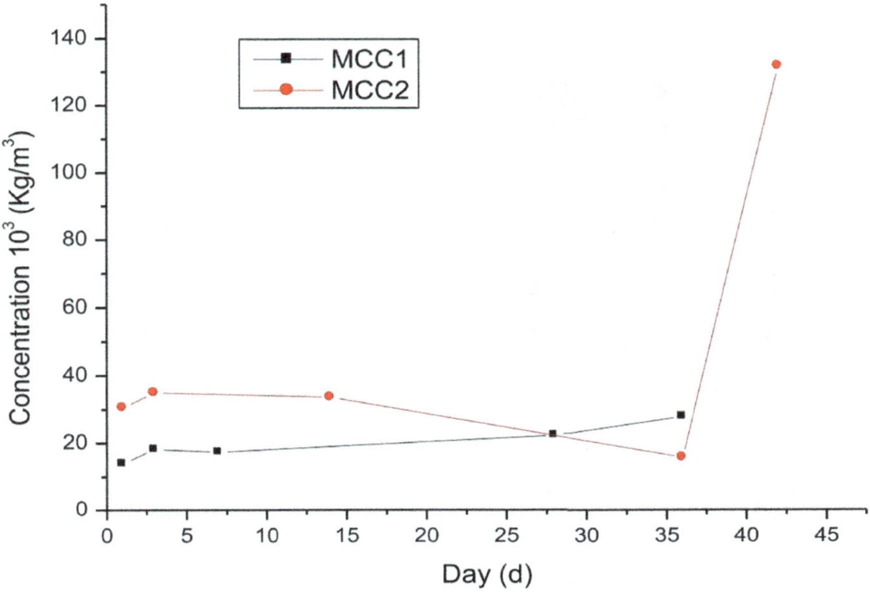

Fig. 1 Evolution of the Ca concentration as a function of time for MCC1 and MCC2 tests

For the MCC1 test, the values grow slowly until the 7th day, then more intensively to reach a maximum of 3.755 10^{-4} kg/m^2 at the end of the 36th day of leaching. Compared to MCC2 test, at this time M_{Ca} is a lower value, about 2.107 10^{-4} kg/m^2. The speed of leaching seems to increase significantly at the 42th day of leaching, with a M_{Ca} value of 17.8 10^{-4} kg/m^2. Expressed in percentage of elemental weight loss, the maximum extracted Ca is about 0.5 wt.% in MCC1 test and 2.8 wt.% in MCC2 test as shown in Fig. 3.

The evolution of the leaching rate in Ca as a function of time expresses the rate of dissolution of the mineral. The corresponding velocity curve is shown in Fig. 4. The maximum leachate is on the first day and reaches: 1.88 10^{-4} kg/m^2.d and 4.1310^{-4} kg/m^2 d, for both of the MCC1 and MCC2 tests, respectively. On the 28th day of the MCC1 test, the leaching rate decreases to a final value of 1.42 10^{-5} kg/m^2.d, after 28 days, showing the establishment of the steady state. Recent studies dealing with the chemical durability of similar ceramics show that MCC1 leaching rates of a Ce-bearing zirconolite are around 6.85 10^{-9} kg/m^2.d after 42 days in an acidic medium (Zhang et al. 2019a). This value is a 10^{-4} times lower than our results showing an excellent chemical stability of zirconolite compared to the Synroc presently studied.

These authors also reported a low leaching values for a Hf-Gd codoped zirconolite for Ca, Cu, Gd, and Hf of about 1.57, 0.13, 4.72 10^{-10}, and 1.59 10^{-11} kg/m^2.d medium (Zhang et al. 2019b). However, Yang et al. (2019) have leached a $(Cs_{0.8-x}Ba_{0.4+x})(Al^{3+}{}_{1.6+x}Ti^{4+}{}_{6.4-x})O_{16}$ hollandite ceramic synthesized by a sol-spray pyrolysis route. These authors report normalized Cs and Ba leaching rates lower than 10^{-5} kg/m^2.d. These values are in the same order of those of the present study.

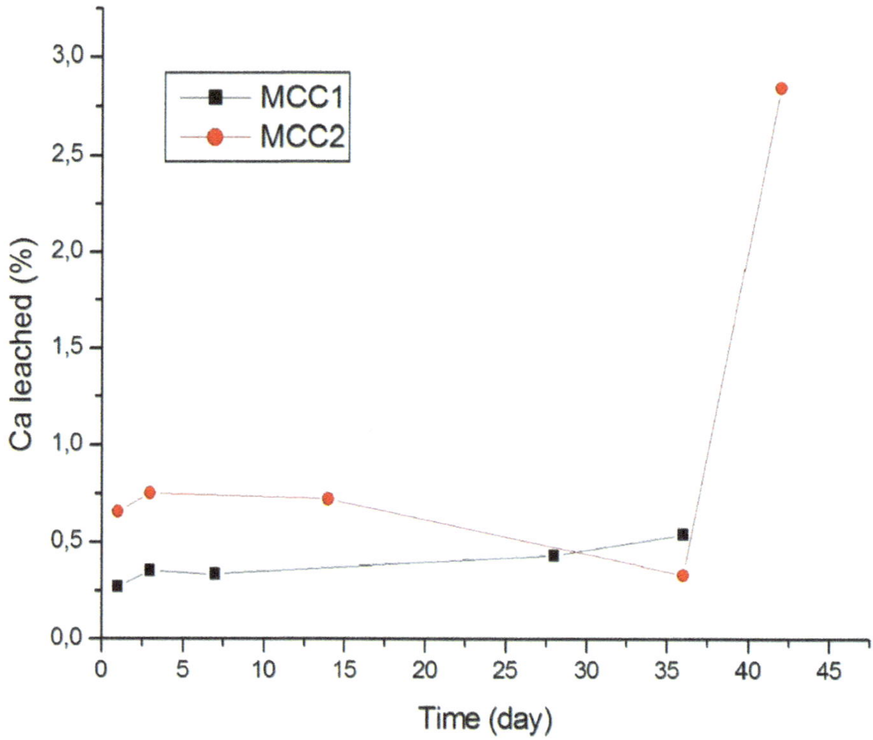

Fig. 2 Evolution of the percentage of leached Ca from the matrices as a function of time for MCC1 and MCC2 tests

For the MCC2 test, it is also toward the 36th day of leaching that the speed reached a minimum of 9.57 10^{-6} kg/m^2.d, after this the high temperature of leaching causes an increase of the leaching speed, and the Synroc structure seems to be altered. The steady state is not reached, and the amounts of dissolved Ca are high (131.80 10^{-3} kg/m^3) compared to the MCC1 test (33.62 10^{-3} kg/m^3 at the 36th day). This phenomenon is characterized by a leaching braking and a stabilization of the normalized mass loss values at the end of the tests (Leturcq et al. 2001). The MCC2 is a leaching test more destructive than the MCC1. For the microwave mineralization test, the leached Ca value is of 150.717 10^{-3} kg/m^3. This represents a content of 26 wt.% of the total Ca in the mineral. As a result, microwave leaching is the most destructive test against the material, compared to MCC1 and MCC2 tests. However, Fe and Ti, which are the building blocks of pseudo-brookite, are not detectable by XRF analysis in the leach solution. This result confirms that the pseudo-brookite is a phase which confers a good chemical stability to Synroc B with respect to aqueous corrosion compared to perovskite which is mainly composed of Ca, element strongly found in leachates.

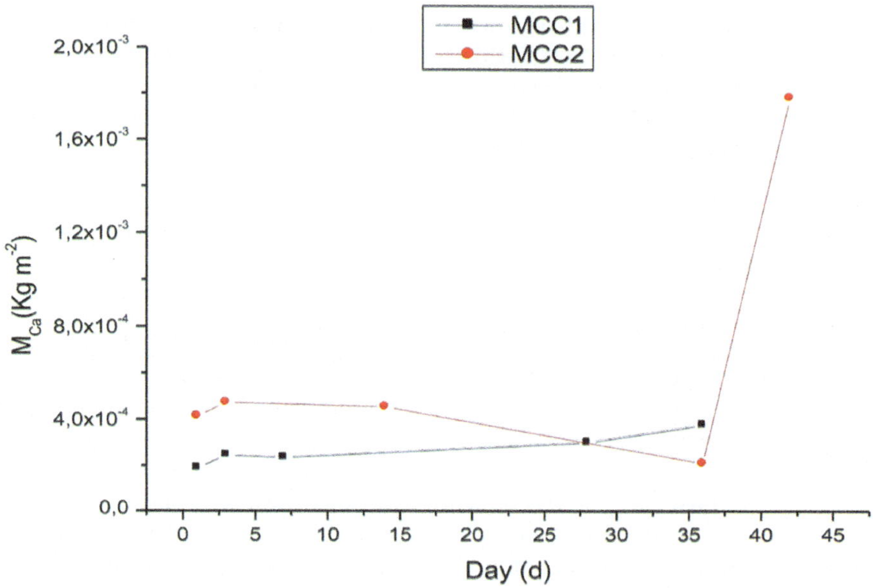

Fig. 3 Evolution of the normalized Ca mass loss as a function of time for Synroc B leached by MCC1 test

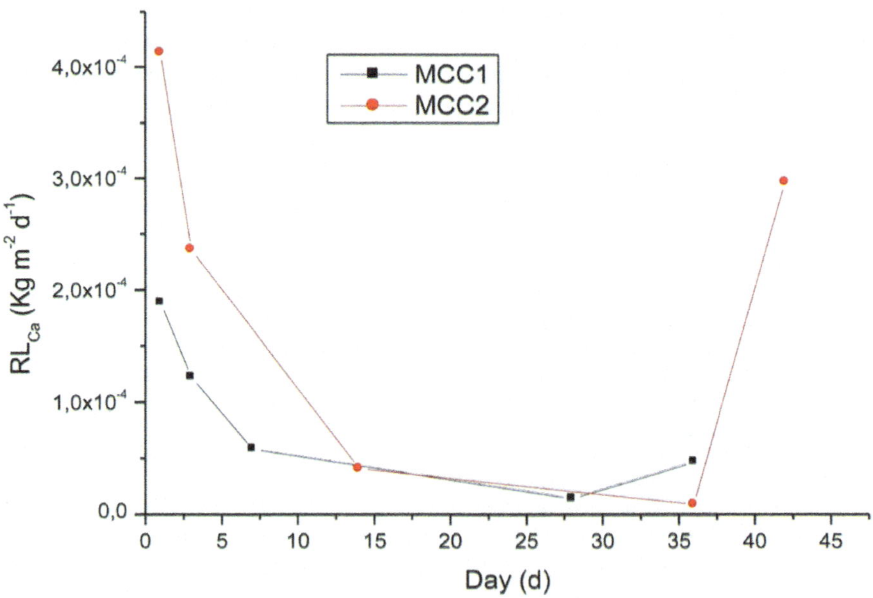

Fig. 4 Evolution of the Ca leaching rate as a function of time for Synroc B leached by MCC1 and MCC2 tests

For Solomah (1980), ICP analysis and atomic absorption of leachates from a Synroc-FA by MCC2 test also yielded undetectable levels of titanium, a constituent element of perovskite, uranite, and hollandite of Synroc-FA.

According to Teng (2014), MCC2 leaching of a cerium-doped titaniferous ceramic gave a leaching rate of Ca below 10^{-2} g m^{-2} d^{-1}.

4 Conclusion

In this study, we synthesized a Fe-rich Synroc B by calcination at 900 °C overnight, followed by sintering at 1200 °C for 6 h. The XRD analysis identified the main crystalline phases of synroc B, namely zirconolite and perovskite. A pseudo-brookite has also been identified. The chemical durability of Synroc B was achieved through three tests: the MCC1 and MCC2 IAEA standard tests and a microwave test. On the basis of the results, the Synroc B presently studied is chemically stable against the dissolution of its major elements as follows: Ce, Ti, Zr, Al, and Fe. The only leached element, Ca, shows percentages of leached amounts in the matrices about 0.5 wt.% of the total Ca for MCC1 test and 2.8 wt.% of the total Ca in MCC2 test. The steady state is reached only at ambient temperature conditions (MCC1) test.

Finally, in the microwave test, the Ca dissolution reaches 26 wt.% of the initial Ca contained in the mineral, and the other elements, namely Ce, Ti, Zr, Al, and Fe, are very resistant to aqueous corrosion, since they are released at levels below the detection limits of XRF analysis. It is raised up from this study that both of the pseudo-brookite and zirconolite phases have good chemical durabilities.

References

Amoroso J, Marra JC, Tang M, Lin Y, Chen F, Su D (2014) Melt processed multiphase ceramic waste forms for nuclear waste immobilization. J Nucl Mater 454:12–21

Application report, Speedwave entry V.9.0, microwave digestion of glas, Quartz

Brinkman K, Bordia R, Reifsnider K, Chiu W, Amoroso J (2017) Final project report CFA-14-6357: a new paradigm for understanding multiphase ceramic waste form performance (No. 14-6357 DOE-Clemson-08260). clemson univ, SC (United States)

Donchev I, Dipchikov F, Petrov O, Tarassov M (2002) Synthesis of artificial rocks (SYNROC TYPE). C R De Acad Bulgare Des Sci 55:15–53

Ewing RC (1999) Nuclear waste forms for actinides. Proc Natl Acad Sci 96:3432–3439

Ghosh DB, Karki BB, Wang J (2019) Utilization of artificial neural network to explore the compositional space of hollandite-structured materials for radionuclide Cs incorporation. J Nucl Mater 530:151957

Grote R, Zhao M, Shuller-Nickles L, Amoroso J, Gong W, Lilova K, Navrotsky A, Tang M, Brinkman KS (2018) Compositional control of tunnel features in hollandite based ceramics: structure and stability of (Ba, Cs)1.33(Zn, Ti)8O16. J Mater Sci 54:112–1125

Konynenburg V, Guinan MW (1983) Radiation effects in SYNROC-D. Nucl Technol 60:206–217

Leturcq G, Advocat T, Hart K, Berger G, Lacombe J, Bonnetier A (2001) Solubility study of Ti, Zr-based ceramics designed to immobilize long-lived radionuclides. Am Miner 86:871–880

Lumpkin GR, Smith KL, Blackord MG (1995) Partitioning of uranium and rare earth elements in Synroc: effect of impurities, metal additive, and waste loading. J Nucl Mater 224:31–42

Matzke HJ, Ray ILF, Seatonberry BW, Thiele H, Trisoglio C, Walker CT, White T (1990) Incorporation of transuranic elements in titanate nuclear waste ceramics. J Am Ceram Soc 73:370–378

Meng C, Ding X, Li W, Zhao J, Yang H (2016) Phase structure evolution and chemical durability studies of Ce-doped zirconolite–pyrochlore synroc for radioactive waste storage. J Mater Sci 51:5207–5215

Ojovan MI (2010) Glassy wasteforms for nuclear waste immobilization. Metall Mater Trans A 42:837–851

Ringwood AE, Kesson SE, Ware NG, Hibberson WO, Major A (1979) The synroc process: a geochemical approach to nuclear waste immobilization. Geochem J 13:141–165

Ringwood AE (1985) Disposal of high-level nuclear wastes: a geological perspective. Mineral Mag 49:159–176

Ringwood AE, Kesson SE, Reeve KD, Levins DM, Ramm EJ (1988) Synroc. Chapter 4. In: Lutze W, Ewing RC (eds) Radioactive waste forms for the future. North-Holland, Amsterdam. 233–334

Ryerson FJ, Burr K, Rozsa R (1981) Formulation of SYNROC-D additives for Savannah river plant high-level radioactive waste. [ADSYN code] (No. UCRL-53237). Lawrence livermore national lab, University of California Report (USA)

Sabathier C (2003) Effets de l'irradiation dans une perovskite: utilisation de cette matrice pour un conditionnement des actinides (Doctoral dissertation N°00004182, Université Paris Sud-Paris XI)

Solomah AG (1980) Demonstration of the feasibility of subsolidus sintering of radwaste-containing synroc B composition. Nucl Technol 49:183–185

Strachan DM (2001) Glass dissolution: testing and modelling for long-term behavior. J Mater Sci 298:69–77

Tang M, Kossoy A, Jarvinen G, Crum J, Turo L, Riley B (2014) Radiation stability test on multiphase glass ceramic and crystalline ceramic waste forms. Nucl Instrum Meth Phys Res Sect B Beam Interact Mater Atoms 326:293–297

Teng Y (2014) Low temperature reactive hot-pressing of cerium-doped titanate composite ceramics and their aqueous stability. J Eur Ceram Soc 34:985–990

Tripp J, Maio V (2006) Evaluation of the use of synroc to solidify the cesium and strontium separations product from advanced aqueous reprocessing of spent nuclear fuel, Idaho national laboratory N°INL/EXT-06-01377

Vance ER (1994) Synroc: A suitable waste form for actinides. MRS Bull 19:28–32

Weber WJ (2009) Materials science of high-level nuclear waste immobilization. MRS Bull 34:46–53

Wen G, Zhang K, Yin D, Zhang H (2015) Solid-state reaction synthesis and aqueous durability of Ce-doped zirconolite-rich ceramics. J Nucl Mater 466:113–119

Yang Y, Xie X, Yang X, Lu X, Yu K, Huang L (2019) Synthesis and phase structure of (Cs0.8-xBa0.4+x)(Al3+1.6+x Ti4+)O16 ceramics using sol-spray pyrolysis route for immobilizing radioactive cesium. J Radioanal Nucl Chem 320:733–739

Zhang K, He S, Yin D, Peng L, Wu J (2016) Self-propagating synthesis and aqueous durability of Nd-bearing zirconolite-rich composites using Ca(NO3)2 as the oxidant. J Nucl Mater 478:315–321

Zhang K, Luo B, Zhang H (2019a) Immobilization of CeO2 using single-phase zirconolite and the chemical stability analysis. Mater Res Expr 6:115526

Zhang K, Yin D, Xu K, Zhang H (2019b) Self-propagating synthesis and characterization studies of Gd-Bearing Hf-Zirconolite. Ceram Waste Forms Mater 12:178

Zhao J, Li J, Liu H, Zhang X, Zheng K, Yu H, Lian Q, Wang H, Zhu Y, Huo J (2020) Cesium perovskite-type Ba1-x(La, Cs)xZrO3 ceramics by sol-gel method. Ceram Int 46:9968–9971

Chapter 5
Behavior Study of New Formulations Based on Plasticized Poly Vinyl Chloride Stabilized with Epoxidized Sunflower Oil

Nadia Lardjane, Farid Hamitouche, Hassiba Habchi Laribi, and Naima Belhaneche Bensemra

Abstract This work focuses on the behavior study of new formulations based on poly vinyl chloride (PVC) and epoxidized sunflower oil (ESO). For that purpose semi-rigid and plasticized sheets were realized. Three plasticizers were considered: the dioctylphtalate (DOP), the di-isodécylphtalate (DIDP) and the di-isononyladipate (DINA). Tests of lixiviation were carried out in two simulators mediums, the acetic acid solution and distilled water at 50 °C while being based on the phenomenon of variation of mass. A soil burial test was realized in a garden (Tizi Ouzou, Algeria)) for six months. The different samples were analyzed by scanning electron microscopy (SEM). The lixiviation test confirmed that the presence of plasticizers and stabilizers affect the phenomenon of migration. The microscopic analysis of bacterial cells in the soil reveals the presence of the *Pseudomonas aeruginosa*.

Keywords Behavior · PVC · Sunflower oil · Epoxidized sunflower oil · Plasticizer · Ageing · Migration · Biodegradation · *Pseudomonas aeruginosa*

1 Introduction

Poly (vinyl chloride) (PVC) is the third most consumed plastic in the world (Visser et al. 2011), and it requires the incorporation of additives (plasticizers and heat stabilizers) to make it flexible (Mersiowsky 2002). Pphthalates are the most used plasticizers; they can be classified into two groups. Low

N. Lardjane (✉) · N. B. Bensemra
Laboratory of Sciences and Techniques of the Environment, Department of Environmental Engineering, Ecole Nationale Polytechnique, BP 182 El-Harrach, Algiers, Algeria
e-mail: nadiaalar@yahoo.fr

N. Lardjane · F. Hamitouche
Production and Safeguarding Threatened Species and Crops Laboratory, Mouloud Mammeri University, Hasnaoua II, 15000 Tizi Ouzou, Algeria

H. H. Laribi
Laboratory of Chemical Engineering, Department of Process Engineering, Faculty of Technology, University of Blida 1, Blida, Algeria

D. Trache et al. (eds.), *Materials Research and Applications*,
Materials Horizons: From Nature to Nanomaterials,
https://doi.org/10.1007/978-981-15-9223-2_5

molecular weight (LMW) phthalates (diethylphthalate, di-n-butylphthalate, di-n-octylphthalate, di-n-isobutylphthalate) are frequently added to shampoos, cosmetics, lotions (Sathyanarayana 2008); whereas high molecular weight (HMW) phthalates (di-2-ethylhexylphthalate DEHP, di-n-octylphthalate, butyl- benzylphthalate) are used to produce vinyl plastics for diverse applications (Schettler 2006). The implementation of PVC requires the use of thermal stabilizers; the main classes currently used are lead salts, metallic soaps, and organotin compounds (Lin et al. 2006). The great PVC production and consumption level generates an important quantity of waste, conducting to serious problems (Brebu et al. 2000). In addition to visual pollution, this waste can interact with the environment and release some of the additives present (Haider and Karlsson 1999). Phthalates are toxicants for jeopardizing immune function, endocrine and reproductive systems, neural and physical development and increase risk of cardiovascular disease (Lind and Lind 2012; Ejaredar et al. 2015; Trasande et al. 2013). DEHP affects males and females fertility and causes severe testicular alterations (testicular atrophy with the possibility of complete absence of spermatogenesis (Gray et al. 2000; Calafat et al. 2006). Lead compounds are classified as toxic, and they can affect fertilization by acting directly on the gonads or by disrupting the hormonal regulation of the gonads by hypothalamic-pituitary factors. It also accumulates in testicles of adult animals resulting in atrophy of leydig cells, inhibition of spermatogenesis and, beyond, testicular degeneration, reduction of intracellular ascorbate content. In the adult female, lead also affects the function of the ovary and the uterus, disrupting the implantation and development of the embryo (Groupe scientifique sur l'eau 2003; Rzymski et al. 2015). Prolonged low-level occupational exposure has caused central nervous system (CNS) damage (Zayed et al. 2003).

In the present work, new formulations of PVC are realized and studied, based on the substitution of lead stabilizers by epoxidized sunflower oil (ESO) as a heat stabilizer. Epoxidized sunflower oil was obtained by epoxidation of commercial sunflower oil; this epoxidation reaction was developed as part of a research project included in national research programs (Benaniba et al. 2001). An acute toxicity study of ESO was performed. It has shown the non-toxicity of the substance in the short term (Atek et al. 2010). The proposed alternative plasticizer is the di-isononyladipate (DINA) known for its simple structure which facilitates the biodegradation process. The objective of this work is to compare the behavior of the formulations based on the proposed substitution additives.

2 Experimental

2.1 Products

- **PVC resin**
  Shintech SE 1200 grade 1 is a PVC resin produced by the American company INC-USA, polymerized in suspension.
- **The plasticizers**
  The plasticizers (DOP, DIDP, and DINA) used in the preparation of the different formulations are clear oily liquids. These plasticizers are provided by the general society of plasticizers (Tunisia).
- **Stearic acid**
  The stearic acid provided by SO.G.I.S SPA (Italy).
- **Epoxidized sunflower oil (ESO)**
  Epoxidized sunflower oil was synthesized at 50 °C, using the peracetic acid prepared in situ by reacting the hydrogen peroxide (30% V/V) with an excess of glacial acetic acid (Atek 2010).

2.2 Preparation of Formulations

Resin and additives were mixed: 7% of ternary stabilizer, 1% of stearic acid, 15%, 30%, 40% or 45% of plasticizer. The obtained mixture was introduced into a rotary two-roll mill heated at 140 °C for 15 min. Finally, the mixture was placed between two trays of a Fontigine tabletop press in order to obtain the desired thickness.

2.3 Lixiviation Tests

Rectangular specimens were placed in test tubes containing 20 ml of acetic acid solution (5000 ppm) stamped to pH 5 with sodium hydroxide (NaOH). This simulator medium has been proposed as simulant for domestic waste (Haider and Karlsson 1999). The same number of samples is introduced in test tubes containing 20 ml of distilled water, for comparison and as simulating medium of rainwater in contact with the waste of a landfill. The test is carried out in an oven at temperature of 50 ± 2 °C for 15 days. The specimens were weighed before and after immersion to follow the mass variation:

$$\zeta(\%) = \frac{m_t - m_0}{m_0} \times 100$$

where m_0 represents the initial mass of the sample and m_t is the mass of the sample weighed after leaching test.

2.4 In Situ Biodegradation of Polymer in Soil

The specimens with dimensions of 40 mm × 20 mm were buried in three plots (1.0 × 0.7 m) for 6 months. The soil had a pH of 7.7, a temperature of 12–27 °C, and water content around 13%. PVC samples were buried at least 20 cm below the soil surface (Woolnough et al. 2010), and they were analyzed by scanning electron microscopy (SEM).

2.4.1 Bacterial Identification

Macroscopic Examination

It consists of the study of bacterial colonies at the macroscopic scale. It is done with the naked eye and the binocular loupe. The study consists in determining the criteria characterizing the colony (Singelton 1999):

- **The appearance**: It corresponds to the diameter of the colony, which can be punctiform if the diameter is less than 1 mm, or not punctiform if the diameter is greater than 1 mm;
- **Form**: The round, irregular, wavy, rhizoid or filamentous colony;
- **Chromogenesis**: It corresponds to the color of the colony and the emission of pigment diffusing into the agar;
- **Opacity**: A colony may be opaque, translucent or transparent;
- **Surface**: Smooth, dry, rough, shiny or blunt.

Microscopic Examination

The microscopic study follows three main stages (Perry et al. 2004):

- **Microscopic observation in the fresh state**
  The observation is useful for estimating the mobility of pure strains.
- **Methylene blue staining**
  This coloration makes it possible to detect the cellular morphology of the pure strains.
- **Gram stain**
  The principle of this staining is based on a double staining which makes it possible to know the shape, the arrangement, the purity as well as the biochemical nature of the wall of the pure cells. It consists of treating a bacterial smear fixed by heat

with a solution of Gentian violet, an iodine-iodide solution (lugol), an organic solvent; ethanol and Fuschine. The cells react in two ways and form two groups:

- The gram negative: The lipids of the bacterial wall dissolve and make it porous, thus causing discoloration of the cytoplasm;
- The others retain their violet color and are called gram positive: Ethanol does not discolor the cell because the wall is predominantly proteinaceous (Marchal and Bourdon 1987).

2.4.2 Morphology Analysis

The PHILIPS ESEMXL30 scanning electron microscope with tungsten filament was used. It is coupled with a complete system of microanalysis energy dispersive of X-rays (EDS-X). The applied voltage is 20 kV.

3 Results and Discussion

3.1 Lixiviation Test

3.1.1 Influence of the Nature of Simulator Medium

Tables 1 and 2 show the evolution of the mass variation rate as a time function of contact with acetic acid and the distilled water at 50 °C in the case of DOP, DIDP, and DINA. The mass variation is important in acetic acid in comparison with distilled water, which is due to the higher polarity of the first simulator medium. Indeed, the most important value was pointed in the case of DINA 45% in acetic acid; it reached the value of 0.943. For distilled water, the recorded value is equal to 0.560. On the other hand, the reported results in the literature show that the extent of possible transfer depends on the liquid nature; in which the solubility of the additive is considered as an important factor (Zeddam and Belhaneche Bensemra 2011).

3.1.2 Influence of the Nature and Rate of Plasticizer

Globally, the results show that the mass variation increases with rate of plasticizer, indeed, by inserting between the macromolecular chains; the plasticizer partially destroys cohesion and increases the free space, therefore, the penetration of the lixiviates (Tables 1 and 2). On the other hand, for the same rate of plasticizer, the mass variation is higher for the samples containing DINA 45% (0.943 in acetic acid and 0.560 in distilled water), which can be explained by the lower steric hindrance of DINA (Atek and Belhaneche Bensemra 2005; Belhaneche Bensemra et al. 2002; Boussoum et al. 2006).

Table 1 Rates of mass variation at 50 °C in acetic acid for the samples containing DOP, DIDP, and DINA at concentrations of 15, 30, and 45%, respectively

Time (Days)	m_0	m_1	$\frac{\Delta m}{m_0} \times 100$
DOP 15% $t = 5$	0.789	0.789	0.088
DOP 15% $t = 10$	0.708	0.709	0.169
DOP 15% $t = 15$	0.663	0.664	0.211
DOP 30% $t = 5$	0.607	0.608	0.197
DOP 30% $t = 10$	0.619	0.621	0.322
DOP 30% $t = 15$	0.637	0.639	0.345
DOP 45% $t = 5$	0.540	0.542	0.295
DOP 45% $t = 10$	0.618	0.620	0.385
DOP 45% $t = 15$	0.596	0.599	0.402
DIDP 15% $t = 5$	0.667	0.668	0.104
DIDP 15% $t = 10$	0.705	0.707	0.240
DIDP 15% $t = 15$	0.690	0.692	0.246
DIDP 30% $t = 5$	0.666	0.667	0.180
DIDP 30% $t = 10$	0.626	0.628	0.223
DIDP 30% $t = 15$	0.654	0.656	0.274
DIDP 45% $t = 5$	0.586	0.588	0.290
DIDP 45% $t = 10$	0.613	0.615	0.293
DIDP 45% $t = 15$	0.591	0.592	0.321
DINA 15% $t = 5$	0.871	0.873	0.206
DINA 15% $t = 10$	0.812	0.815	0.369
DINA 15% $t = 15$	0.782	0.786	0.434
DINA 30% $t = 5$	0.733	0.735	0.313
DINA 30% $t = 10$	0.676	0.679	0.458
DINA 30% $t = 15$	0.676	0.680	0.546
DINA 45% $t = 5$	0.531	0.534	0.500
DINA 45% $t = 10$	0.548	0.552	0.728
DINA 45% $t = 15$	0.529	0.534	0.943

3.2 In Situ Biodegradation of Polymer in Soil

3.2.1 Bacterial Identification

Macroscopic Examination

After 48 h of incubation at 30 °C, bacterial growth was observed by the appearance of colonies on the nutrient agar. The results of the macroscopic study of bacterial colonies of the strain are represented in Table 3.

Microscopic Examination

The results of the microscopic study of bacterial cells are represented in Table 4. From the macro-microscopic observation (Fig. 1), *Pseudomonas aeruginosa* was

Table 2 Rates of mass variation at 50 °C in distilled water for the samples containing DOP, DIDP, and DINA at concentrations of 15, 30, and 45%, respectively

Time (Days)	m_0	m_1	$\frac{\Delta m}{m_0} \times 100$
DOP 15% $t = 5$	0.718	0.718	0.097
DOP 15% $t = 10$	0.689	0.691	0.260
DOP 15% $t = 15$	0.636	0.638	0.298
DOP 30% $t = 5$	0.614	0.615	0.195
DOP 30% $t = 10$	0.640	0.642	0.281
DOP 30% $t = 15$	0.632	0.634	0.300
DOP 45% $t = 5$	0.559	0.560	0.21
DOP 45% $t = 10$	0.619	0.621	0.326
DOP 45% $t = 15$	0.594	0.596	0.420
DIDP 15% $t = 5$	0.756	0.757	0.145
DIDP 15% $t = 10$	0.773	0.774	0.181
DIDP 15% $t = 15$	0.664	0.665	0.255
DIDP 30% $t = 5$	0.635	0.636	0.0629
DIDP 30% $t = 10$	0.594	0.594	0.0841
DIDP 30% $t = 15$	0.643	0.644	0.186
DIDP 45% $t = 5$	0.614	0.615	0.195
DIDP 45% $t = 10$	0.606	0.608	0.240
DIDP 45% $t = 15$	0.581	0.583	0.270
DINA 15% $t = 5$	0.824	0.826	0.254
DINA 15% $t = 10$	0.864	0.866	0.300
DINA 15% $t = 15$	0.865	0.868	0.381
DINA 30% $t = 5$	0.650	0.652	0.353
DINA 30% $t = 10$	0.654	0.656	0.428
DINA 30% $t = 15$	0.636	0.639	0.487
DINA 45% $t = 5$	0.550	0.553	0.435
DINA 45% $t = 10$	0.581	0.584	0.498
DINA 45% $t = 15$	0.588	0.591	0.560

Table 3 Macroscopic characters of bacterial colonies after 48 h of incubation at 30 °C

Strain	S1
Size	<1 mm
Shape	Irregular/mucosa
Chromogenesis	Pink (Pyorubrine)
Opacity	Opaque
Elevation	Plate
Consistency	Dried
Surface	Smooth
Odor	+

Table 4 Results of microscopic observations of bacterial colonies

Characters	Fresh state	Methyl blue staining		Gram staining
Strain		Form	Arrangement	
S_1	Mobile	Bacille	Isolated chain	$Gram^-$

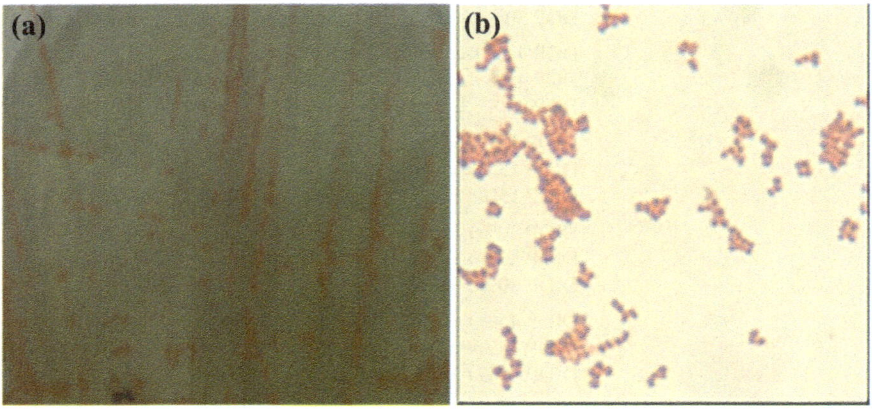

Fig. 1 Macroscopic (**a**) and microscopic (**b**) aspects after Gram staining with a magnification of X100 (*Pseudomonas aeruginosa*: form bacillus, gram negative, catalase positive, aerobic strict, green pigmentation and characteristic odor)

identified by the following characteristics: bacillus form, Gram negative, catalase positive, strict aerobic, green pigmentation and characteristic odor.

3.2.2 Scanning Electron Microscopy

The analysis of the different samples by the scanning electron microscopy (Figs. 2 and 3) reveals the appearance of pores which confirms that biodegradation and/or migration of the additives occurred, in particular for the formulation containing DINA and the ESO. The DINA because of its linear structure is favored by bacterial attack (Lardjane et al. 2013).

4 Conclusion and Future Perspective

The study of the influence of plasticizer according to their content and nature showed that the mass variation increases with the plasticizer content. Indeed, inserting between macromolecular chains, the plasticizer partially destroys cohesion of the material and increases the free space, which facilitates the penetration of simulator medium. The mass variation is higher for the samples containing 45% of DINA.

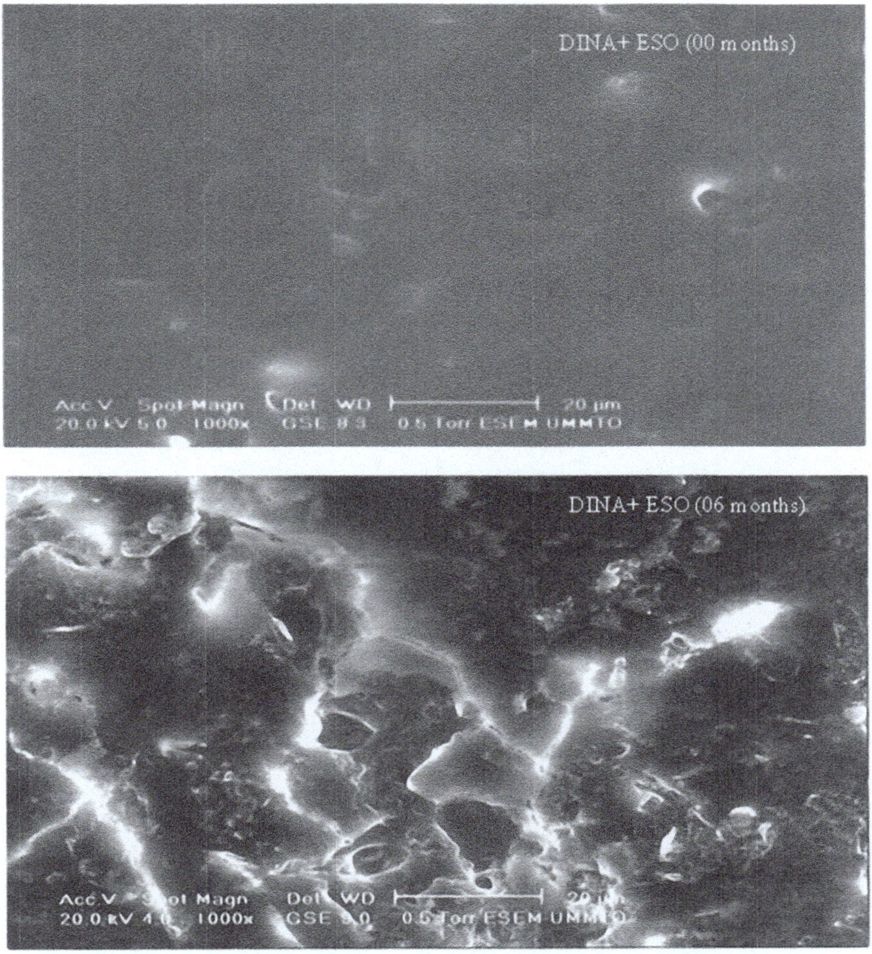

Fig. 2 Scanning electron microscopy analysis of the PVC samples containing 40% of DINA and 5% of ESO after six months of soil burial

The study of the effect of the nature of simulator medium showed that the mass variation is important in acetic acid, comparatively to that of with distilled water, which is principally due to the higher polarity of the first simulator medium. The important value was recorded in the case of DINA 45% in acetic acid; it reached the value of 0.943. The results of the microscopic study of bacterial cells reveal the presence of the *Pseudomonas aeruginosa*. The analysis of the different samples by the scanning electron microscopy confirmed the migration and biodegradation of additives. Thus, compared to the behavior of the new formulation based on DINA and ESO in the soil and the simulator medium, we can say that this formulation is interesting for the environment. We propose to make an in-depth study on the toxicity of DINA in combination with ESO in order to allow its use on an industrial scale.

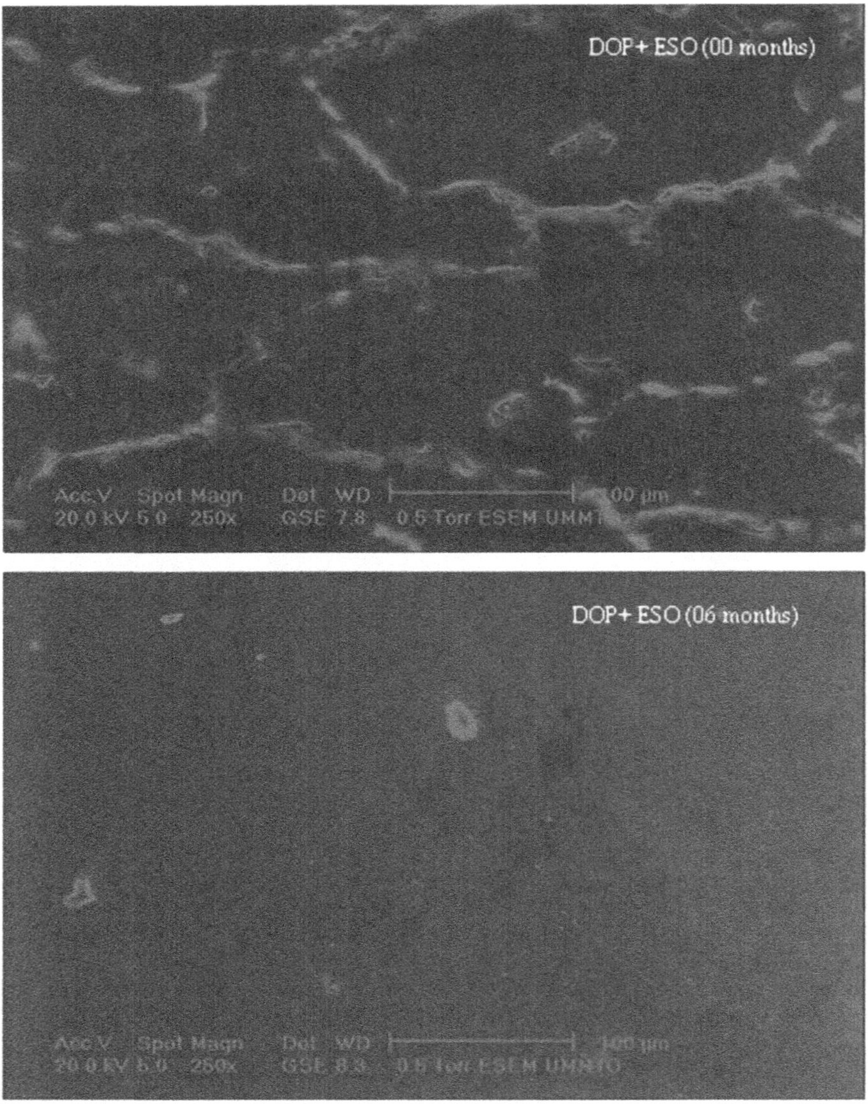

Fig. 3 Scanning electron microscopy analysis of the PVC samples containing 40% of DOP and 5% of ESO after six months of soil burial

Acknowledgements The authors would like to express their gratitude to *Mrs. Noura Arezki* from Professional Formation Center of *Beni Douala* (TiziOuzou, Algeria) for her help.

References

Atek D, Belhaneche Bensemra N (2005) FTIR investigation of the specific migration of additives from rigid poly (vinyl chloride). Eur Polym J 41:707–714

Atek D, Belhaneche Bensemra N, Turki M (2010) Migration of epoxidized sunflower oil and dioctyl phthalate from rigid and plasticized poly (vinyl chloride). Int J Polym Mater 59:342–352

Belhaneche Bensemra N, Zeddam C, Ouahmed S (2002) Study of the migration of additives from plasticized PVC. Macromol Symp 180:191–202

Benaniba MT, Belhaneche Bensemra N, Gelbard G (2001) Stabilizing effect of epoxydized sunflower oil on the thermal degradation of poly (vinyl chloride). Polym Degrad Stab 74:501–505

Boussoum MO, Atek D, Belhaneche Bensemra N (2006) Interactions between poly (vinyl chloride) stabilised with epoxidized sunflower oil and food stimulants. Polym Degrad Stab 91:579–584

Brebu M, Vasile C, Antonie SR, Chiriac M, Precup M, Yang J, Roy Ch (2000) Study of the natural ageing of PVC insulation for electrical cables. Polym Degrad Stab 67:209–221

Calafat AM, Brock JW, Silva MJ, Gray LE, Reidy JA, Barr DB, Needham L (2006) Urinary and amniotic fluid levels of phthalate monoesters in rats after the oral administration of di (2-ethylhexyl) phthalate and di-n-butyl phthalate. Toxicology 217:22–30

Ejaredar M, Nyanza EC, Ten Eycke K, Dewey D (2015) Phthalate exposure and childrens neurodevelopment: a systematic review. Environ Res 142:51–60

Gray I, Rowland I, Foster P, Gangolli S (2000) Species differences in the testicular toxicity of phthalate esters. Toxicol Lett 11:141–147

Groupe scientifique sur l'eau (2003) Fiche Plomb. Institut National de Santé Publique du Québec

Haider N, Karlsson S (1999) Migration and release profile of chimassorb 944 from low-density polyethyléne film (LDPE) in simulated landfills. Polym Degrad Stab 64:321–328

Lardjane N, Belhaneche Bensemra N, Massardier V (2013) Migration of new bio-based additives from rigid and plasticized PVC stabilized with epoxidized sunflower oil in the soil. J Polym Res 20:209

Lind L, Lind PM (2012) Can persistent organic pollutants and plastic-associated chemicals cause cardiovascular disease. J Intern Med 271:537–553

Lin Y, Wang J, Evans DG, Li D (2006) Layered and intercalated hydrotalcite-like materials as thermal stabilizers in PVC resin. J Phys Chem Solids 67:998–1001

Marchal N, Bourdon JL (1987) les milieux de culture pour l'isolement et l'identification biochimique edition. Doin éditeurs, Paris

Mersiowsky I (2002) Long-term fate of PVC products and their additives in landfills. Prog Polym Sci 27:2227–2277

Perry J, Staley J, Lory S (2004) Microbiologie. Cours et questions de revision. Edition Dunod,France

Rzymski P, Tomczyk K, Rzymski P, Poniedziałek B, Opala T, Wilczak M (2015) Impact of heavy metals on the female reproductive system. Ann Agric Environ Med 22:259–264

Sathyanarayana S (2008) Phthalates and children's health. Current problems in pediatric and adolescent health care. Health Care 38:34–49

Schettler T (2006) Human exposure to phthalates via consumer products. Int J Androl 29:134–139

Singelton A (1999) Bactériologie, 4ème edition. Dunod, Paris

Trasande L, Sathyanarayana S, Spanier AJ, Trachtman H, Attina TM, Urbina EM (2013) Urinary phthalates are associated with higher blood pressure in childhood. J Pediatr 163:747–753

Visser HA, Bor TC, Wolters M, Warnet LL, Govaert LE (2011) Influence of physical aging on impact embrittlement of uPVC pipes. Plast Rubber Compos 40:201–212

Woolnough CA, Yee LH, Charlton T, Foster LJR (2010) Environmental degradation and biofouling of 'green' plastics including short and medium chain length poly-hydroxyalkanoates. Polym Int 59:658–667

Zayed J, Panisset M, Mergler D (2003) Système nerveux. Environnement et santé publique-Fondements et pratiques

Zeddam C, Belhaneche Bensemra N (2011) Application of FT-IR spectroscopy to the specific migration study of an organotin heat stabilizer from rigid poly(vinyl chloride) into food simulants. Polimery 56:657–661

Chapter 6
Bi/α-Bi_2O_3/TiO_2 Nanotubes Arrays Heterojunction for Highly Efficient Photocatalytic Applications

Leyla Rekeb, Lynda Ainouche, Salem Boudinar, and Lamia Hamadou

Abstract α-Bi_2O_3/TiO_2 nanotube arrays, (α-Bi_2O_3/TiNT), and Bi/α-Bi_2O_3/TiNT photocatalysts have been successfully constructed by anodization combined with electrodeposition method. The structural and morphological characterizations of the as-prepared heterostructures are carried out by X-ray diffraction (XRD), scanning electron microscopy (SEM), and energy dispersive spectrometry (EDS). UV-Vis diffuse reflectance spectra revealed that the combined effect of both Bi with nanodendritic morphology and α-Bi_2O_3 in Bi/α-Bi_2O_3/TiNT composite considerably improved the optical properties, with important red shift. This improvement is ascribed to the good separation and transportation of charges and to both high surface area and strong localized surface plasmon resonance (LSPR) modes of nanodendritic Bi. Moreover, the photocatalytic (PC) performances of the composites were evaluated by the photocatalytic degradation of methylene bleu (MB) under visible-light illumination. It is demonstrated that Bi/α-Bi_2O_3/TiNT photocatalyst leads to higher degradation of MB under visible light.

Keywords TiO_2 nanotubes · Bi/α-Bi_2O_3/TiNT · LSPR · Photocatalysis

1 Introduction

Heterogeneous photocatalysis using solar energy has been regarded as one of the most prospective approaches to improve and resolve problems of the environmental pollution and the global energy shortage. In recent years, photocatalysis using semiconductor is considered as one of the highly developed physicochemical routes applied in the environmental purification. Among various oxide semiconductor photocatalysts, TiO_2 has received huge attention. Particularly, TiO_2 nanotube arrays (TiNT) have raised a lot of importance due to their highly ordered array structure, large specific surface area, and unidirectional charge transportation properties, which make them

L. Rekeb (✉) · L. Ainouche · S. Boudinar · L. Hamadou
Laboratoire de Physique et Chimie Des Matériaux, LPCM, Université Mouloud Mammeri de Tizi-Ouzou, UMMTO, BP 17, Tizi-Ouzou 15000, Algeria
e-mail: leylarekeb@hotmail.com

D. Trache et al. (eds.), *Materials Research and Applications*,
Materials Horizons: From Nature to Nanomaterials,
https://doi.org/10.1007/978-981-15-9223-2_6

very promising photocatalysts (Ait Ali Yahia et al. 2012, 2016; Ainouche et al. 2014, 2016). Although, low exploitation of visible light due to its large band gap (3.2 eV for anatase) and important recombination rate of photogenerated electron–hole pairs considerably limit its performances in photocatalytic field (Zaho et al. 2015; Fan et al. 2014). So, to overcome these limitations, various strategies were adopted to modify TiNT. One can cite (i) the semiconductor combination (Rekeb et al. 2019; Ma et al. 2015) for which the development of the p–n heterojunction can create the internal electric field in a semiconductor photocatalyst that significantly approves the migration rate of the photogenerated electrons and holes (Huang et al. 2015) and (ii) the incorporation of nanoparticles of noble metal that exhibits LSPR by visible-light absorption (Zhao et al. 2015; Wu et al. 2013). Very recently, LSPR has been employed to improve the photocatalysts mechanism, producing plasmonic photocatalysis, which has been acknowledged by both the photocatalysis and plasmonic communities (An et al. 2019; Bardey et al. 2019; Swearer et al. 2019; Lemos et al. 2018). LSPR highly promotes the utilization of the solar spectrum, which comprises a significant portion of the solar spectrum ($\approx$43%) (Kah Hon et al. 2018). It can be described in terms of a collective oscillation of conduction electrons on the surface of the nanoparticles when they are in resonance with the incident light, contributing to the strong absorption of visible light and to the excitation of charge carriers. Besides, the Schottky junction created at the semiconductor/noble metal interface generates an internal electric field, which forces the holes and electrons to move in different directions. This is of advantage to charge transfer and separation (Zhang et al. 2013). Among plasmonic non-noble metals, low-cost bismuth (Bi) is particularly attractive due to its LSPR effects; it has been reported to be a substitute to noble metals (Rekeb et al. 2017; Li et al. 2018; Li et al. 2016; Tian and Toudert 2018; Boudinar et al. 2019). Besides, Bi has a long carrier mean-free path (Shim et al. 2008), quantum confinement effects (Vandaele et al. 2018; Chang et al. 2019), and an impressive catalytic memory capability (Dong et al. 2014). All these features make Bi a very promising candidate to improve the efficiency of photocatalysts.

As a p-type semiconductor, Bi_2O_3 is a fascinating material for the photooxidation of pollutants. It has been proven to be one of interesting photocatalysts due to its dielectric permittivity, thermal stability (Yang et al. 2014), low band gap energy of 2.8 eV (Chang et al. 2017), and appropriate band edge potentials ($E_{cb} = 0.33$eV, $E_{vb} = 3.13$eV) for transferring and separating the photogenerated charge carriers (Prasad and Karthikeyan 2018).

Despite the fact that plasmonic Bi has proven to be very useful for photocatalysis applications, it is still challenging to develop these materials with optimized morphology. Among the various Bi morphological nanostructures, Bi nanodendrites with multiple branches offer much more hotspots in photocatalysis applications due to their large surface area and to their high proportion of edge and corner atoms. Thus, Bi nanodendrites synthesis with optimized branch size to improve their photocatalytic performance is highly required. It is so very important to develop simple and inexpensive strategies to achieve Bi nanodendrites with improved performance.

Although there are much reports on the synthesis and photocatalytic properties of Bi_2O_3/TiNT heterostructure (Guan et al. 2018; Zou et al. 2019; Peng et al.

2018), to the best of our knowledge, very limited works have been devoted to the synthesis and photocatalytic characteristics of Bi/α-Bi_2O_3/TiNT system. Moreover, although dendritic Bi nanostructures were prepared by various methods, the photocatalytic performance is still to be improved. Motivated by the multiple effects of both Bi (LSPR modes) and α-Bi_2O_3/TiO_2 (p–n heterojunction structure), the present work focuses on the investigation of two types of systems for photocatalytic applications: (i) α-Bi_2O_3/TiNT and (ii) Bi/α-Bi_2O_3/TiNT. We intend synthesizing dendritic Bi nanostructures in order to concurrently attain high surface area and LSPR modes effect. Herein, for the first time, we report the successful synthesize of Bi/α-Bi_2O_3/TiNT photocatalyst, with multi-branched Bi nanodendrites using anodization and electrodeposition methods. We have analyzed the photocatalytic oxidation of methylene blue under visible-light irradiation. The obtained results confirmed that Bi/α-Bi_2O_3/TiNT system has the best photocatalytic performances, which have been attributed to both efficient charge separation and transportation of photogenerated carriers and the presence of strong SPR effect due to the coupled dendritic Bi nanoparticles. The concept of collective excitation of coupled Bi nanoparticles is demonstrated as a valuable approach to improve the exploitation of plasmonic energy.

2 Experimental Methods

2.1 Electrodes Preparation

2.1.1 Synthesis of Pure TiNT

Electrochemical anodization was used to synthesize pure TiNT with vertical alignment. Prior to anodization, titanium sheets (99.9% purity) of 1 cm^2 were first mechanically polished with 600, 800, 1000, and 2000 SiC abrasive papers, respectively, then washed in an ultrasonic bath with acetone, and finally rinsed with distilled water and dried. The electrochemical anodization process was carried out in an electrochemical cell using a conventional two-electrode system connected to the power supply at a constant voltage of 20 V for 2 h. A titanium sheet is used as the anode and a platinum foil as a counter electrode. The electrolyte solution in which the two electrodes are immersed consists of a mixture of organic solvent, glycerol ($C_3H_8O_3$), 8 wt% H_2O, and 0.14 M NH_4F. The obtained TiNT were amorphous, so they were crystallized into anatase crystallographic structure by thermal annealing at 450 °C during 3 h in a furnace and cooled in the open air.

2.1.2 Bi/α-Bi_2O_3/TiNT Heterostructure Synthesis

Bi/α-Bi_2O_3/TiNT electrode was synthesized through the electrochemical deposition. The electrodeposition was achieved via the chronoamperometry technique, using an Autolab potentiostat/galvanostat PGSTAT 30 driven by GPES Software, in a three electrode cell with the annealed TiNT as a working electrode, a saturated Ag/AgCl as a reference electrode, and a platinum sheet as a counter electrode. The electrodeposition was carried out by applying a constant potential of −0.9 V during 600 s in a solution containing a perchloric acid (1 M $HClO_4$), bismuth oxide (10^{-3} M Bi_2O_3), and distilled water. After electrodeposition, the samples were rinsed in deionized water and dried in hot air.

2.1.3 α-Bi_2O_3/TiNT Heterostructure Synthesis

To prepare α-Bi_2O_3/TiNT electrodes, amorphous TiNT were first subjected to electrodeposition using similar operating conditions as mentioned above. After the deposition process, the electrodes were annealed in a furnace at 450 °C during 3 h to acquire both pure α-Bi_2O_3 and pure anatase phases. Figure 1 summarizes the experimental procedures for the preparation of both electrodes.

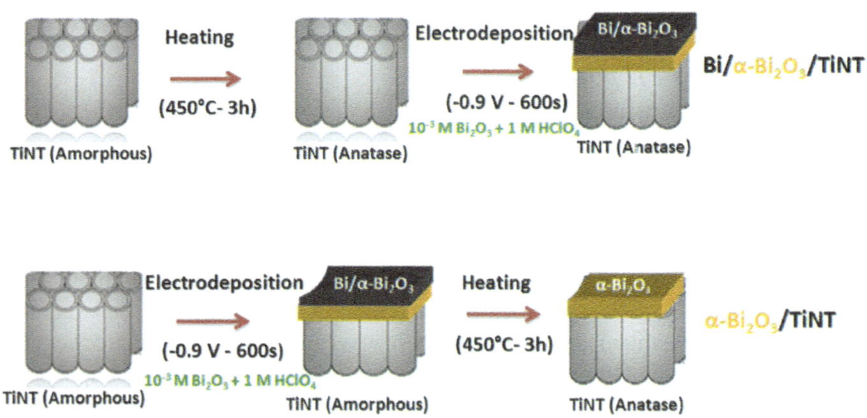

Fig. 1 Schematic representation of the procedures for the preparation of Bi/α-Bi_2O_3/TiNT and α-Bi_2O_3/TiNT electrodes

2.2 Characterization

2.2.1 Morphological Characterization

Morphological examination of the different prepared samples was investigated under low vacuum using scanning electron microscopy (high resolution (~ 1 nm)), with field emission gun, SEM-FEG (Zeiss Ultra 55). The SEM-FEG is coupled to the energy dispersive X-ray spectroscopy (EDS) microanalysis, which allows the determination of the percent elemental chemical composition of the different elaborated heterostructures.

2.2.2 Structural Characterization

Crystalline phase identification was performed by X-ray diffractometer (Phillips PW 3040) operating at 25 kV, using copper anticathode ($Cu - K\alpha$, $\lambda = 1,5405$ Å). X-ray diffractograms were recorded in the 2θ angular range of 20–80° with a scan interval of 0.02°. For phase identification, the resulting DRX spectra were analyzed by means of HighScore Plus software.

2.2.3 Optical Characterization

Optical characterization of the prepared systems was investigated using a diffuse reflectance UV-Vis spectrophotometer (ShimadzuUV-2501) with an integrating sphere module. UV-Vis absorption spectra were registered in the wavelength range 200–800 nm.

2.2.4 Degradation of MB Dye

The degradation of the aqueous MB dye under visible-light irradiation was performed to evaluate the photocatalytic performances of the different systems. A 450 W xenon lamp was used as the visible-light source irradiation of the reaction dye solution, and UV-Vis spectrophotometer was used to evaluate the photocatalytic activity of the synthesized systems. The absorbance of MB at 663 nm was measured at 5 min intervals. In order to obtain a more reliable result, photocatalytic experiments were carried out several times.

3 Results and Discussion

3.1 Electrochemical Analysis (Chronoamperometry)

Amperometric current-time method was used to prepare the deposits, according to the potentials obtained from cyclic voltammetry (CV) experiments presented in our previous work (Rekeb et al. 2017). The current density-time transients conducted at different applied potentials (−0.35 V, −0.85 V and −0.9 V) are illustrated in Fig. 2. As can be seen, during the early stage of the deposition process, the current densities show rapid rising, attaining a maximum and then slight decline for longer deposition time. It is interesting to notice the dependency of the current density on applied potential, which decreases as the deposition potential becomes more anodic. Highest current of 1.15 mA/cm^2 was reached at −0.9 V, whereas lowest value of 0.16 mA/cm^2 was obtained at the potential of −0.35 V.

Electrodeposition of Bi at −0.35 and −0.85 V resulted in a morphology, which is non-uniform and non-dendritic (not shown here), can be ascribed to the different nucleation and growth rates at different active sites. A low overpotential of −0.9 V was necessary to achieve homogeneously dispersed and well-defined nanodendrites.

According to the literature, dendrites are tending to appear at high overpotentials and low electrolyte concentrations. Such conditions usually lead to the "diffusion controlled" nucleation and growth (Lopez and Choi 2006; Nikolic et al. 2005).

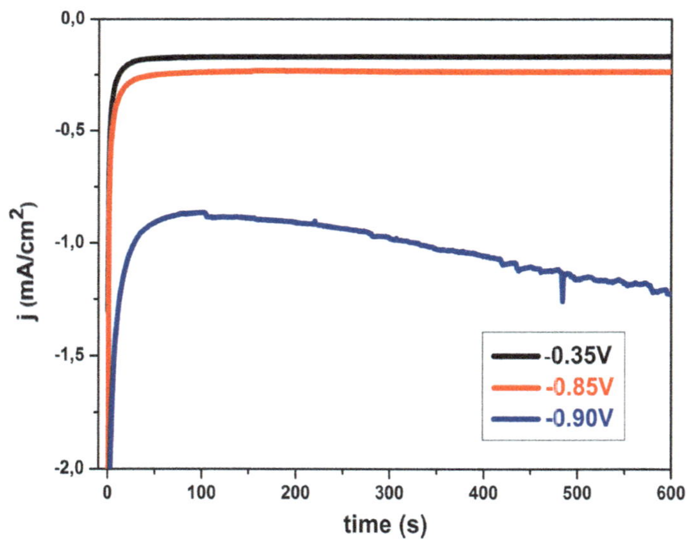

Fig. 2 Current-time transient curves onto TiNT at different growth potentials from the solution containing 0.001 M Bi_2O_3 and 1 M $HClO_4$ at -0.9 V during 600 s

3.2 Analysis of Structural and Morphological Properties of the Systems

3.2.1 XRD Analysis

The structure of the synthesized systems was determined by XRD. The XRD patterns of pure TiNT and the resultant deposits are displayed in Fig. 3.

Figure 3a shows the XRD pattern of annealed TiNT for which all the characteristic peaks of crystalline anatase appear at $2\theta = 25.3^\circ, 48.0^\circ$, and 55° that perfectly match with the (011), (200), and (211) planes of anatase, respectively. These results coincide well with the values of ICDD PDF card #71-1166 of anatase TiO_2. Figure 3b illustrates the X-ray diffraction patterns of Bi/α-Bi_2O_3/TiNT composite, which shows peaks of Bi (003) at 22.59°, (104) at 38.13°, and (211) at 63.10° in agreement with ICDD PDF card No. 44-1246. In addition, intense peak at 2θ values of 27.47° is shown in the pattern. This peak corresponds to the main peak of monoclinic α-bismuth oxide. The elevated intensity of the diffraction peak indicates a high degree of crystallinity. In addition to α-Bi_2O_3 diffraction peak, we notice the presence of Ti and TiO_2 peaks. Figure 3c shows the XRD pattern of the α-Bi_2O_3/TiNT system after annealing, which exhibits a predominant monoclinic α-Bi_2O_3 with preferential orientation along the [120] direction, associated with the ICDD PDF 72-0398. These results revealed that after annealing the deposited crystalline Bi is completely converted to α-Bi_2O_3.

The mass fraction of Bi in the Bi/Bi_2O_3 composite can be determined by means of the relative ratio (Ping et al. 2016):

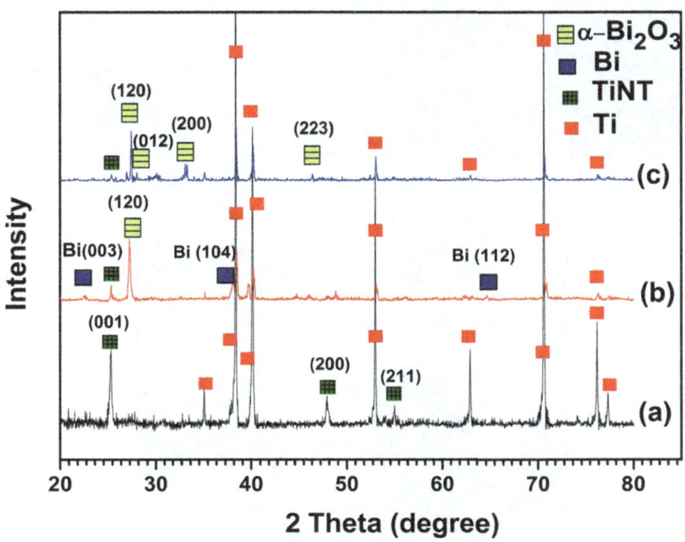

Fig. 3 XRD patterns of **a** TiNT, **b** Bi/α-Bi_2O_3/TiNT, and **c** α-Bi_2O_3/TiNT

$$\frac{I_{\mathrm{Bi}(104)}}{I_{\mathrm{Bi_2O_3}(120)} + I_{\mathrm{Bi}(104)}} \tag{1}$$

where $I_{\mathrm{Bi_2O_3}(120)}$ and $I_{\mathrm{Bi}(104)}$ are the height of the characteristic diffraction peaks of Bi_2O_3(120) and Bi(104) planes, respectively. From the XRD spectrum of Bi/Bi_2O_3/TiNT, the mass ratio between Bi and Bi_2O_3 is about 1:3.

3.2.2 SEM Analysis

Figure 4a–c shows SEM images of TiNT, α-Bi_2O_3/TiNT and Bi/α-Bi_2O_3/TiNT composites, respectively. Highly ordered TiNT arrays with an average wall thickness of 20 nm and inner pore diameter of 70 nm are noticed (Fig. 4a). A cross section of the nanotubular film is presented in the insert. The tube length reached 3.5 μm. SEM micrograph in Fig. 4b shows the top surface morphology of Bi/α-Bi_2O_3/TiNT. Bi/α-Bi_2O_3 deposit consists of dense nanodendrites that grow on the surface. The single nanodendrite is composed of a solid body with length of 3–4 μm and width

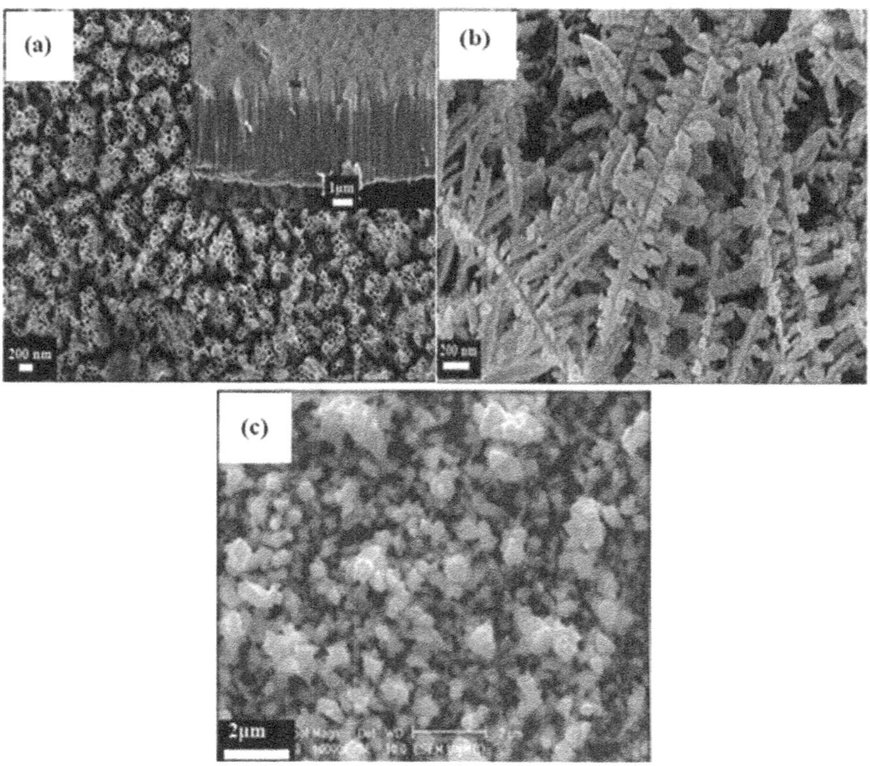

Fig. 4 SEM images of **a** TiO_2 nanotubes layers, **b** Bi/α-Bi_2O_3/TiNT, and **c** α-Bi_2O_3/TiNT

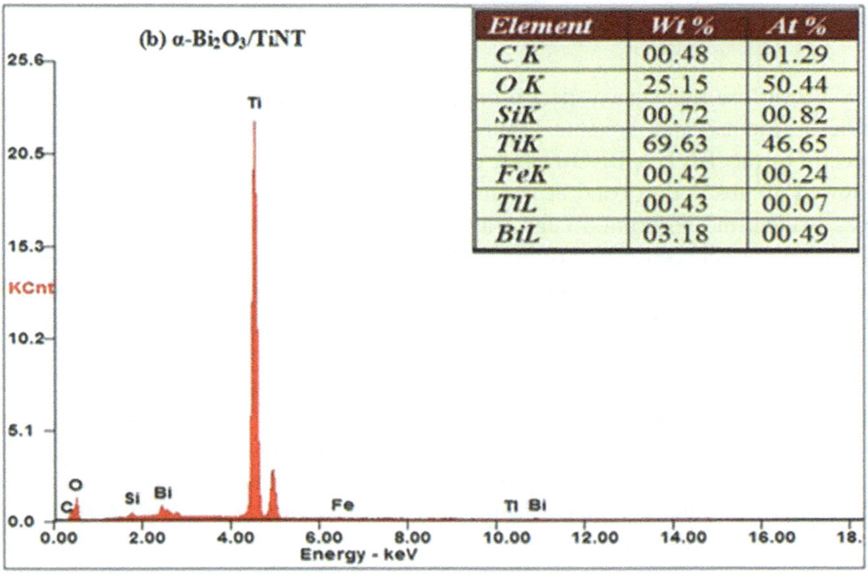

Element	Wt %	At %
CK	00.48	01.29
OK	25.15	50.44
SiK	00.72	00.82
TiK	69.63	46.65
FeK	00.42	00.24
TlL	00.43	00.07
BiL	03.18	00.49

Fig. 5 Representative EDS patterns of the Bi/α-Bi_2O_3/TiNT and α-Bi_2O_3/TiNT heterostructures

of ~ 100 nm and a multiple small branches, which consist of interconnected Bi nanoparticles. It is interesting to notice a black coloration of the electrodeposited film on annealed TiNT, which points out the presence of Bi. Figure 4c illustrates the top surface morphology of α-Bi_2O_3/TiNT. One can notice that a compact deposit covered the TiNT. The deposit is composed of agglomerated and dispersed particles with a size of 200–300 nm. In addition, the color of the deposit was changed to yellow, which proves the predominance of Bi_2O_3.

To determine the chemical composition of the as-prepared deposits, the EDS was used. The representative EDS patterns of the Bi/α-Bi_2O_3/TiNT and α-Bi_2O_3/TiNT heterostructures are shown in Fig. 5. Titanium signals at about 4.5 keV and bismuth peaks at about 2.5 keV and 10.8 keV were observed in both heterostructures. The chemical composition analysis results of the Bi/α-Bi_2O_3/TiNT composite (Fig. 5a) revealed a composition of 59.79 atom % Bi, 26.06 atom % Ti, and 09.56 atom % O. Certain amount of contaminants are also present. EDS analysis of α-Bi_2O_3/TiNT heterostructure (Fig. 5b) shows the presence of Ti, Bi, and O with compositions of 69.63, 03.18, and 25.15 atom %, respectively, implying most probably the stoichiometry of Ti and Bi oxides of the synthesized heterostructure.

3.3 Optical Properties

UV-Vis diffuse reflectance spectroscopy was used to characterize the optical properties of the prepared systems. Figure 6a shows the UV-Visible diffuse reflectance spectra of the TiNT, α-Bi_2O_3/TiNT, and Bi/α-Bi_2O_3/TiNT. The spectrum of TiNT showed an absorption band at around 390 nm in the UV region. When α-Bi_2O_3 was loaded, the absorption edge was extended into visible region, which proves the coupling between α-Bi_2O_3 and TiNT. Concerning Bi/α-Bi_2O_3/TiNT sample, a broader and stronger absorption in both UV and visible-light regions are observed. The absorption edge manifests at a wavelength longer than 650 nm, which is a clear evidence for the efficacy synthesis of Bi nanodendrites with LSPR effect. The nanodendritic Bi film consists of interconnected nanoparticles, and thus, the LSPR impact of each nanoparticle gives rise to the absorption in the broader range of wavelengths.

From the optical absorption measurements, optical band gap energy (E_g) of the different composites can be determined by means of Tauc plot (Singh et al. 2018), based on the formula:

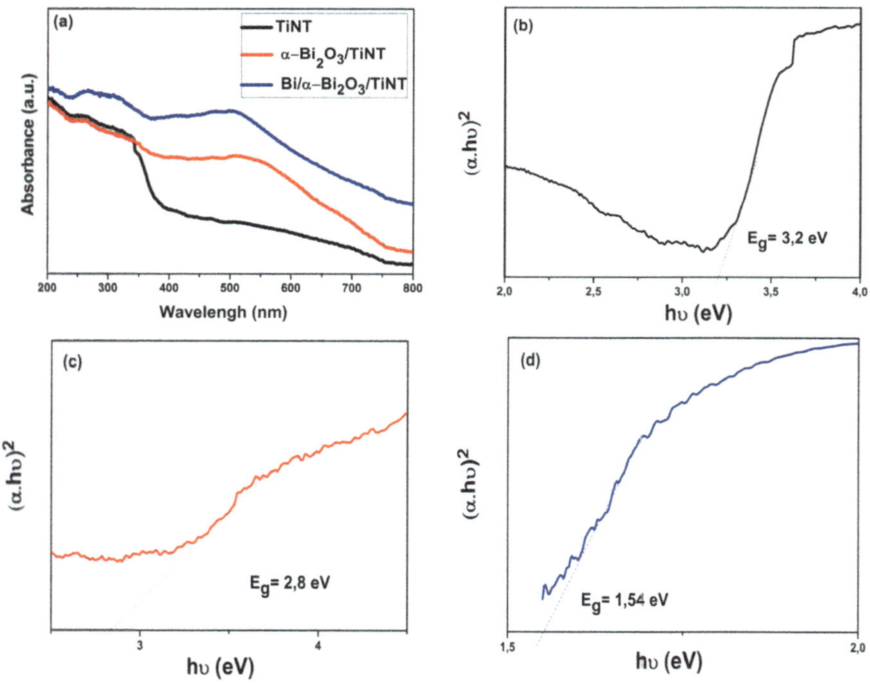

Fig. 6 **a** UV-Vis spectra of the photocatalysts, **b–d** estimated band gaps by Tauc plot of TiNT, α-Bi_2O_3/TiNT, and Bi/α-Bi_2O_3/TiNT, respectively

$$A = \frac{K\left(h\nu - E_g\right)^{\frac{n}{2}}}{h\nu} \tag{2}$$

where A is absorbance, K is a constant of proportionality, ν is frequency, h is Planck's constant and n an integer. n equals 1 for direct transitions.

Taus plots are presented in Fig. 6b–d. As can be noticed, linear portions were found in a large interval of optical energies, indicating that a direct allowed transition governs the excitation process of the films. Direct band gap was already reported for Bi_2O_3 (Leontie et al. 2001; Leontie et al. 2006). The intrinsic band gaps of TiNT and the heterojunctions were determined by extrapolating the linear portions of the plots to the photon energy axis. The energy band gap for unmodified TiNT equals 3.2 eV. Coupling α-Bi_2O_3 with TiNT produces a band gap reduction of the α-Bi_2O_3/TiNT system. The band gap is about 2.8 eV, close to the previously reported value (Jabeen et al. 2015; Lim and Rawal (2017). In the case of Bi/α-Bi_2O_3/TiNT system, the value of E_g is about 1.54 eV only. The band gap narrowing of about 1.6 eV through loading of both Bi and α-Bi_2O_3 would contribute to the enhancement of photocatalysis efficiency.

3.4 Photocatalytic Degradation of MB Solution

The photocatalytic efficiencies of the different systems were examined by photocatalytic degradation of MB under visible-light irradiation. Figure 7a–c depicts the UV-Vis absorption spectra of MB solution irradiated by the visible light for different time durations in the presence of TiNT, α-Bi_2O_3/TiNT, and Bi/α-Bi_2O_3/TiNT electrodes. One can notice that with extension of the irradiation, the characteristic absorption of MB at 663 nm decreases for all electrodes, demonstrating good photodegradation activity of the as-obtained electrodes. Compared to TiNT and α-Bi_2O_3/TiNT, Bi/α-Bi_2O_3/TiNT exhibits improved photocatalytic activity for MB degradation. It can be clearly seen that with increasing irradiation time, the absorbance of the spectra quickly decreased, and the absorption peak of spectra disappeared after only 30 min of irradiation, indicating the almost complete degradation of MB in the solution. The efficiency of degradation of the electrodes (C/C_0) is defined as the ratio between the residual concentration (C) and the initial concentration (C_0) of MB. Figure 7d illustrates the results of photocatalytic MB degradation over the different photocatalysts. Compared with pure TiNT, α-Bi_2O_3/TiNT displays better photocatalytic performance. Enhanced photodegradation activity of MB in the presence of α-Bi_2O_3/TiNT is most probably due to the formation of the p–n junction between the p-type α-Bi_2O_3 and n-type TiNT, leading to the very efficient separation of photogenerated carriers in the electrode. In fact, when α-Bi_2O_3 is in contact with TiNT to form a heterojunction, the Fermi levels of α-Bi_2O_3 and TiNT tend to descend and ascend, respectively. The band gap of α-Bi_2O_3 (2.8 eV) is smaller than that of TiNT (3.2 eV), so under visible-light irradiation, the photogenerated electron–hole pairs were produced in α-Bi_2O_3.

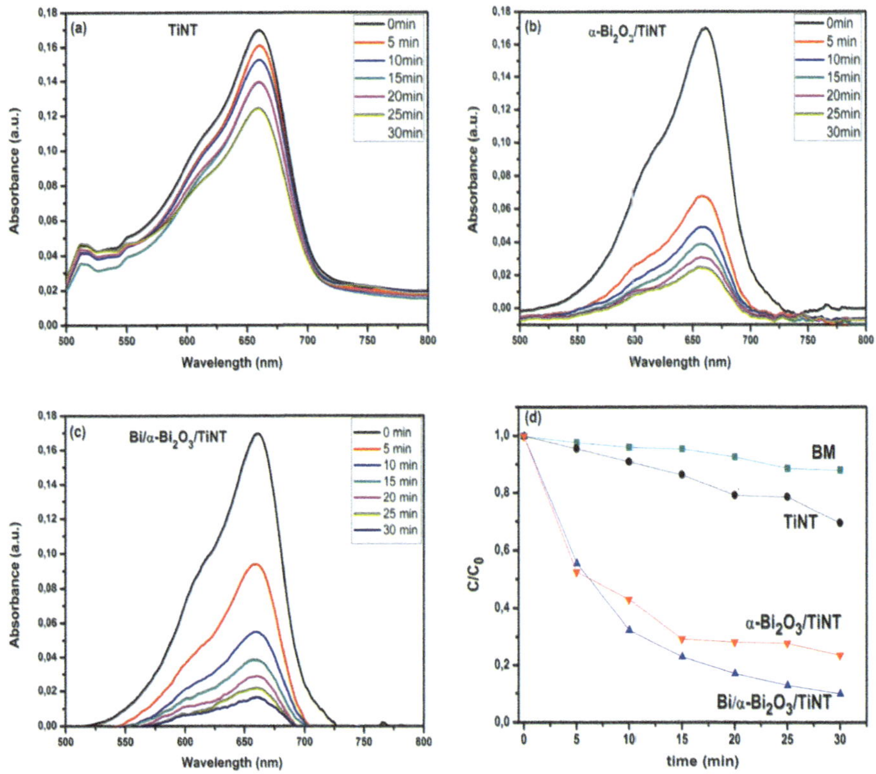

Fig. 7 UV-Vis adsorption spectra of degradation process of MB by **a** TiNT, **b** Bi/α-Bi_2O_3/TiNT, **c** α-Bi_2O_3/TiNT, **d** Time profiles of the photocatalytic degradation of MB for different photocatalysts under UV/Vis irradiation

The generated electrons in the α-Bi_2O_3 migrated to the CB of TiNT, leaving holes in the VB of α-Bi_2O_3. As a result, the higher charge separation rate increased the lifetime of the charge carriers, reduced electron–hole recombination and enhanced the efficiency of the interfacial charges.

Pronounced photocatalytic MB degradation can be achieved by the Bi/α-Bi_2O_3/TiNT system due to the synergistic effect between plasmonic Bi and α-Bi_2O_3/TiNT heterojunction, further confirming the importance of Bi and proves that combining Bi and α-Bi_2O_3 is an efficient way to improve the photocatalytic activity of TiNT.

Bi is more expected to generate electrons through the LSPR effect. In fact, dendritic Bi nanoparticles with high surface area offer much more surface active sites, which facilitate the adsorption of organic molecules, making possible the photodegradation of MB. Furthermore, dendritic structures have also the particularity of possessing several edges, which act as very active sites for the degradation of the pollutant. According to Guo et al. (2017), edge and corner atoms do not only make the entire surfaces of the nanodendrites effectively accessible to the reagents, but also allow them participate in catalytic reactions through the interconnected channels between the branches.

The photocatalytic degradation efficiency is illustrated in Fig. 8a. It revealed that Bi/α-Bi_2O_3/TiNT exhibited the highest BM photodegradation efficiency of about 90%, whereas it was only 30% for TiNT and 76% for α-Bi_2O_3/TiNT for 30 min of irradiation, indicating that Bi/α-Bi_2O_3/TiNT as a photocatalyst was much superior.

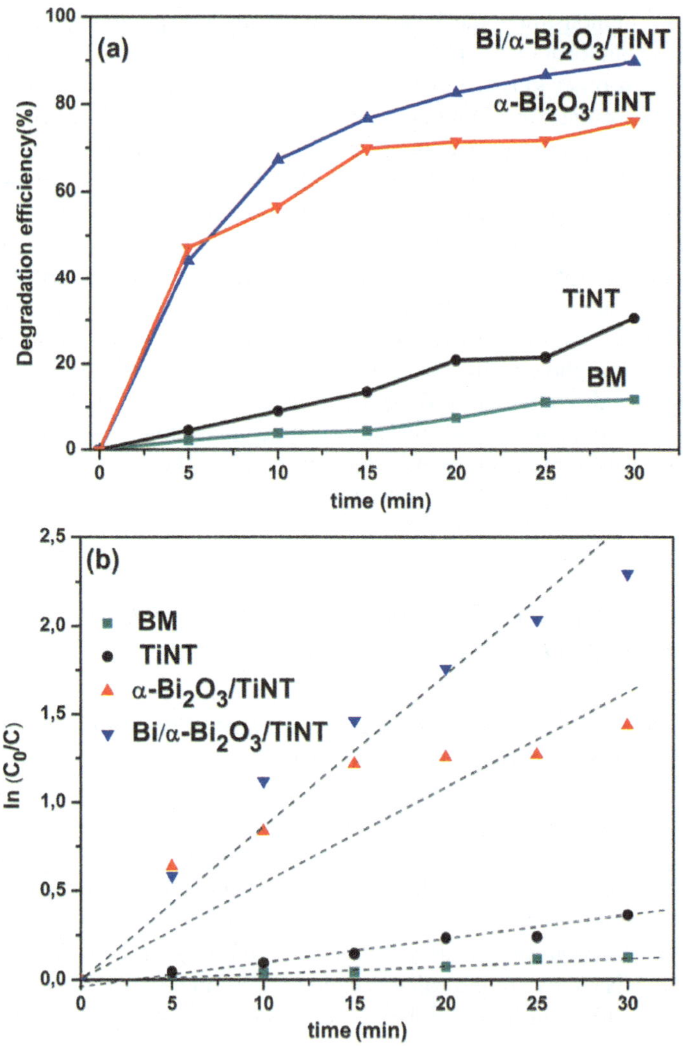

Fig. 8 **a** Photocatalytic degradation efficiency of MB under simulated solar irradiation over TiNT photocatalysts and **b** plots of $\ln\left(\frac{C_0}{C}\right)$ versus irradiation time used to calculate the reaction rate constants

Moreover, to estimate the MB degradation kinetic rate of the photocatalysts, the following pseudo-first-order kinetic equation was used:

$$\ln\left(\frac{C_0}{C}\right) = \mathrm{kt} \tag{3}$$

where k is the rate constant (min^{-1}).

Figure 8b demonstrates the kinetic behaviors of MB photodegradation on the surface of the photocatalysts. The plots show a linear relationship, which reveal that the pseudo-first-order kinetic governs the reactions of photodegradation. From the slope of linear fitting lines, rate constants are measured for all systems. The rate constant of MB degradation over the Bi/α-Bi_2O_3/TiNT reached 0.074 min^{-1}, which is higher than that of TiNT (0.011 min^{-1}) and that of α-Bi_2O_3/TiNT (0.042 min^{-1}).

Strongly coupled plasmonic dendritic Bi nanoparticles give rise to highly intensive electromagnetic field in the metal surface. This improved electromagnetic field increases the absorption of polar molecules and enhances the polarization of the absorbed molecules, which further facilitates the decomposition of these molecules under the photocatalytic process. In addition, LSPR accelerates the movement of electron from the photoexcited plasmonic metal to the semiconductor (Sun et al. 2009; Wang et al. 2019). According to Yu et al. (Yu et al. 2019), the enhanced strength of electromagnetic field caused by coupled plasmonic nanostructure, essentially boosts the formation rate of electron–hole pair in semiconductor and ultimately improves photocatalytic evolution activity of semiconductor photocatalysts.

3.5 *Charge Transfer Mechanism for the MB Degradation by Bi/α-Bi_2O_3/TiNT Composite*

One of the possible mechanisms for the improvement of the photocatalytic activity of Bi/α-Bi_2O_3/TiNT composite is depicted in Fig. 9.

Bi NPs were first excited under visible-light irradiation. Due to the work function difference of Bi and α-Bi_2O_3, the photogenerated hot electrons on Bi via SPR possessed sufficient energy to surmount the Schottky barrier formed at the Bi/α-Bi_2O_3 interface and transfer into the conduction band of the neighboring semiconductor (α-Bi_2O_3). The photogenerated holes in Bi can be utilized for the oxidation reaction. The produced electrons in the α-Bi_2O_3 transfer to the conduction band (CB) of TiNT and holes of TiNT move to the valence band (VB) of the α-Bi_2O_3, leading to more efficient separation. Furthermore, more free electrons on the surface of the TiNT could capture the dissolved O_2 in the solution and generated OH radicals, which can further oxidize MB into simpler molecules.

The photocatalytic reaction paths are shown by the following equations:

$$\mathrm{TiNT}\left(e^-_{\mathrm{BC}}\right) + O_2 \rightarrow \mathrm{TiNT} + O_2^{\cdot -} \tag{4}$$

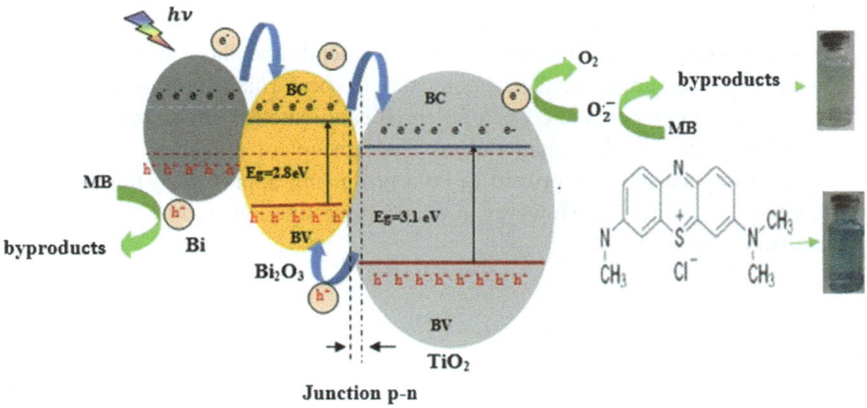

Fig. 9 Schematic illustration of MB degradation mechanism of Bi/α- Bi_2O_3/TiNT heterojunction under visible-light irradiation

$$H^+ + O_2^{\cdot-} \rightarrow HO_2^{\cdot} \tag{5}$$

$$2HO_2^{\cdot} \rightarrow H_2O_2 + O_2 \tag{6}$$

$$H_2O_2 + O_2^{\cdot-} \rightarrow OH^- + OH^{\cdot} \tag{7}$$

$$OH^{\cdot} + BM \rightarrow \text{Degradation products} \tag{8}$$

BM can also be degraded directly by the attack of photogenerated holes.

$$\text{Bi}\ (h_{Bv}^+) + BM \rightarrow \text{Bi} + \text{Degradation products} \tag{9}$$

4 Conclusion and Future Perspectives

In the present work, Bi/α-Bi_2O_3/TiNT heterojunction is successfully prepared using anodization and electrodeposition method. The fabricated Bi/α-Bi_2O_3/TiNT heterojunction has been proved as a better plasmonic photocatalyst comparing to α-Bi_2O_3/TiNT and TiNT due to (i) electron–hole separation at p–n heterojunction (α-Bi_2O_3/TiNT heterojunction) and (ii) high specific surface area of multi-branched Bi nanodendrites, which offer and support LSPR modes in different wavelength values.

Coupling Bi and α-Bi_2O_3 with TiNT produces a band gap narrowing of about 1.6 eV and 90% MB photodegradation after 30 min only of exposure. Therefore, this work represents a novel strategy to prepare plasmonic Bi modified composite. Bi can be considered as a perfect SPR alternate for noble metals to enlarge the light response region of TiNT and improve their photocatalytic performance. Bi/α-Bi_2O_3/TiNT heterojunction can be considered as a potential candidate for environmental photocatalysis/photocatalytic decontamination. The concept of collective excitation of dendritic Bi nanoparticles is proved to be a promising approach to improve plasmonic energy exploitation. However, a deeper investigation on the photocatalytic mechanism is still required.

References

Ainouche L, Hamadou L, Kadri A, Benbrahim N, Bradai D (2014) Interfacial barrier layer properties of three generations of TiO_2 nanotube arrays. J Electrochim Acta 133:597–609

Ainouche L, Hamadou L, Kadri A, Benbrahim N, Bradai D (2016) Ti^{3+} states induced band gap reduction and enhanced visible light absorption of TiO_2 nanotube arrays: Effect of the surface solid fraction factor. Sol Energy Mater Sol Cells 151:179–190

Ait Ali Yahia S, Hamadou L, Kadri A, Benbrahim N, Sutter EEM (2012) TiO_2 nanotubes as alternative cathode in microbial fuel cells: effect of annealing treatment on its performance. J Electrochem Soc 159:K1–K11

Ait Ali Yahia S, Hamadou L, Salar-García MJ, Kadri A, Ortiz-Martínez VM, Hernández-Fernández FJ, Pérez de los Rios A, Benbrahim N (2016) TiO_2 nanotubes as alternative cathode in microbial fuel cells: effect of annealing treatment on its performance. Appl Surf Sci 387:1037–1045

An X, Stelter D, Keyes T, Reinhard BM (2019) Plasmonic photocatalysis of urea oxidation and visible-light fuel cells. Chem 5:1–15

Bardey S, Bonduelle-Skrzypczak A, Fécant A, Cui Z, Colbeau-Justin C, Caps V, Keller V (2019) Plasmonic photocatalysis applied to solar fuels. Faraday Discuss 214 417–439

Boudinar S, Benbrahim N, Benfedda B, Kadri A, Hamadou L (2019) Electrochemical nucleation and optical characterization of highly oriented Bi clusters on Cu substrate. Thin Solid Films 684:68–77

Chang M, Hu H, Zhang Y, Chen D, Wu L, Xinjun Li (2017) Improving visible light-absorptivity and photoelectric conversion efficiency of a TiO2 nanotube anode film by sensitization with bi_2o_3 nanoparticles. Nanomaterials 7:104

Chang T-R, Lu Q, Wang X, Lin H, Miller T, Chiang T-C, Bian G (2019) Band topology of bismuth quantum films. Crystals 9:510

Dong F, Xiong T, Sun Y, Zhao Z, Zhou Y, Feng X, Wu Z (2014) Semimetal bismuth element as a direct plasmonic photocatalyst. Chem Commun 50:10386–10389

Fan X, Fan J, Hu X, Liu E, Kang L, Tang C, Ma Y, Wu H, Li Y (2014) Preparation and characterization of Ag deposited and Fe doped TiO_2 nanotube arrays for photocatalytic hydrogen production by water splitting. Ceram Int 40:15907–15917

Guan Z-C, Wang H-P, Wang X, Hu J, Du R-G (2018) Fabrication of heterostructured β-Bi_2O_3-TiO_2 nanotube array composite film for photoelectrochemical cathodic protection applications. Corros Sci 136:60–69

Guo L, Huang LB, Jiang WJ, Wei ZD, Wan LG, Hu JS (2017) Tuning the branches and composition of PtCu nanodendrites through underpotential deposition of Cu towards advanced electrocatalytic. Act J Mater Chem A 5:9014–9021

Huang H, He Y, Du X, Chu P, Zhang P (2015) A general and facile approach to heterostructured Core/Shell $BiVO_4$/BiOI *p–n* Junction: room-temperature in situ assembly and highly boosted visible-light photocatalysis. ACS Sustain Chem Eng 3:3262–3273

Jabeen Fatima MJ, Niveditha CV, Sindhu S (2015) α-Bi_2O_3 photoanode in DSSC and study of the electrode electrolyte interface. RSC Adv 5:78299–78305
Kah Hon L, Azrina A, Lan Ching S, Pichiah S, Min J, Detlef B (2018) Mechanistic insights into plasmonic photocatalysts in utilizing visible light. Beilstein J Nanotechnol 9:628–648
Lemos de Souza M, Pereira dos Santosbc D, Corio P (2018) Localized surface plasmon resonance enhanced photocatalysis: an experimental and theoretical mechanistic investigation. RSC Adv 8:28753–28762
Leontie L, Caraman M, Delibas M, Rusu GI (2001) Optical properties of bismuth trioxide thin films. Mater Res Bull 36:1629–1637
Leontie L, Caraman M, Visinoiu A, Rusu GI (2006) On the optical properties of bismuth oxide thin films prepared by pulsed laser deposition. Thin Solid Films 473:230–235
Lopez CM, Choi KS (2006) Electrochemical synthesis of dendritic zinc films composed of systematically varying motif crystals. Langmuir 22:10625–10629
Li X, Zhang W, Cui W, Sun Y, Jiang G, Zhang Y, Huang H, Dong F (2018) Bismuth spheres assembled on graphene oxide: Directional charge transfer enhances plasmonic photocatalysis and *in situ* DRIFTS studies. Appl Catal B : Environ 221:482–489
Li K, Chen P, Li J, Sun Y, Chu Y, Dong F (2016) Enhanced plasmonic photocatalytic disinfection on noble-metal free bismuth nanospheres/graphene nanocomposites. Catal Sci Technol 00:1–3
Lim H, Rawal SB (2017) Integrated Bi_2O_3 nanostructure modified with Au nanoparticles for enhanced photocatalytic activity under visible light irradiation. Prog Nat Sci: Mater Int 27:289–296
Ma B, Kim J, Wang T, Li J, Lin K, Liu W, Woo S (2015) Improvement of photocatalytic oxidation activity on WO_3/TiO_2 heterojunction composite photocatalyst with broad spectral response. RSC Adv 5:79815–79819
Nikolic ND, Popov KI, Pavlovic J, Pavlovic MG (2005) The effect of hydrogen codeposition on the morphology of copper electrodeposits.The concept of effective overpotential. J Electroanal Chem 588:88–98
Prasad N, Karthikeyan B (2018) Broad band and enhanced photocatalytic behaviour of Ho^{3+}doped Bi_2O_3 micro-rods. Appl Phys A 124:421
Peng Q, Peng G, Wu L, Wang X, Yang X, Li X (2018) Entrapment of Bi_2O_3 nanoparticles in TiO_2 nanotubes for visible light-driven photocatalysis. Res Chem Intermed 44:6753–6763
Ping B, Mihua S, Chengwen S, Shuaihua W, Murong C (2016) Enhanced photocatalytic activity of Cu_2O/Cu heterogeneous nanoparticles synthesized in aqueous colloidal solutions on degradation of methyl orange. Rare Metal Mater Eng 45:2214–2218
Rekeb L, Hamadou L, Kadri A, Benbrahim N (2017) Bi/a-Bi_2O_3/TiO_2 nanotubes heterojunction with enhanced UV and visible light activity: role of Bismuth. Electrochim Acta 256:162–171
Rekeb L, Hamadou L, Kadri Benbrahim N, Chainet E (2019) Highly broadband plasmonic Cu film modified Cu_2O/TiO_2 nanotube arrays for efficient photocatalytic performance. Int J Hydrogen Energy 44:10541–10553
Shim W, Ham J, Lee K, Jeung WJ, Johnson M, Lee W (2008) On-film formation of Bi nanowires with extraordinary electron mobility. Nano Lett 9:18–22
Singh M, Jampaiah D, Kandjani AE, Sabri YM (2018) Oxygen-deficient photostable Cu_2O for enhanced visible light photocatalytic activity. Nanoscale 10:6039–6050
Sun S, Wang W, Zhang L, Shang M, Wang L (2009) Ag@C core/shell nanocomposite as a highly efficient plasmonic photocatalyst. Catal Commun 11:290–293
Swearer DF, Robatjazi H, Martirez JMP, Zhang M, Zhou L, Carter EA, Nordlander P, Halas NJ (2019) Plasmonic photocatalysis of nitrous oxide into N2 and O2 using aluminum-iridium antenna–reactor nanoparticles. ACS Nano 13:8076–8086
Tian Y, Toudert J (2018) Nanobismuth: fabrication, optical, and plasmonic properties-emerging applications. Journal of Nanotechnology 2018:1–23
Vandaele K, Otsuka M, HasegawaY Heremans JP (2018) Confinement effects, surface effects, and transport in Bi and Bi_{1-X} Sb_X semiconducting and semimetallic nanowires. J Phys: Condens Matter 30:403001

Wang H, Zhao W, Xu C-H, Chen H-Y, Xu J-J (2019) Electrochemical synthesis of Au@semiconductor core–shell nanocrystals guided by single particle plasmonic imaging. Chem Sci 10:9308–9314

Wu F, Hu X, Fan J, Liu E, Sun T, Kang L, Hou W, Zhu C, Liu H (2013) Potocatalytic activity of Ag/TiO_2 nanotube arrays enhanced by surface plasmon resonance and application in hydrogen evolution by water splitting. Plasmonics 8:501–508

Yang J, Wang X, Dai J, Li J (2014) Efficient visible-light-driven photocatalytic degradation with Bi_2O_3 coupling silica doped TiO_2. Ind Eng Chem Res 53:12575–12586

Yu G, Qian J, Zhang P, Zhang B, Zhang W, Yan W, Liu G (2019) Collective excitation of plasmon-coupled Au-nanochain boosts photocatalytic hydrogen evolution of semiconductor. Nat Commun 10:4912

Zhang X, Chen YL, Liu RS, Tsai DP (2013) Plasmonic Photocatalysis. Rep Prog Phys 76:046401

Zhao W, Guo Y, Wang SM, He H, Sun C, Yang SG (2015) A novel ternary plasmonic photocatalyst: ultrathin g-C_3N_4 nanosheet hybrided by Ag/$AgVO_3$ nanoribbons with enhanced visible-light photocatalytic performance. Appl Catal B Environ 7:335–343

Zou Q, Li H, Yang YP, MiaoYC Huo YN (2019) Bi_2O_3/TiO_2 photocatalytic film coated on floated glass balls for effcient removal of organic pollutant. Appl Surf Sci 467–468:354–360

Chapter 7
Synthesis and Characterization of Microcrystalline Cellulose from Giant Reed Using Different Delignification Processes

Moufida Beroual, Oussama Mehelli, Lokmane Boumaza, Djalal Trache, Ahmed Fouzi Tarchoun, Mehdi Derradji, and Kamel Khimeche

Abstract In this work, cellulose was extracted from giant reed (GR) via alkaline treatment and two different delignification processes (acidified $NaClO_2$ and totally chlorine-free (TCF). After that, microcrystalline cellulose (MCCs) particles were successfully isolated from two prepared cellulose samples using HCl acid hydrolysis. The different properties of each sample were studied with various characterization tools such as infrared spectroscopy (FTIR), X-ray diffraction (XRD), scanning electron microscopy (SEM), thermogravimetry (TGA), and differential scanning calorimeter (DSC). The FTIR spectra confirmed the removal of hemicellulose and lignin of the giant reed through the applied extraction processes. The XRD spectra exhibited that both MCCs belong to cellulose I type and showed higher crystallinity index compared to pure cellulose samples. DSC and TGA/DTG results showed that MCC samples have a higher thermal stability. Moreover, the use of a totally chlorine-free process allowed obtaining MCC with higher crystallinity and thermal stability. Consequently, giant reed can be considered as a potential source to extract MCC for which the green totally chlorine-free is a prominent process to produce MCC from cellulosic substrate.

Keywords Microcrystalline cellulose · Giant reed · Extraction · Delignification · Thermal properties · Crystallinity

M. Beroual · O. Mehelli · L. Boumaza · D. Trache (✉) · A. F. Tarchoun · M. Derradji · K. Khimeche
UER Procédés Energétiques, Ecole Militaire Polytechnique, BP 17, Bordj El-Bahri, 16046 Algiers, Algeria
e-mail: djalaltrache@gmail.com

D. Trache et al. (eds.), *Materials Research and Applications*,
Materials Horizons: From Nature to Nanomaterials,
https://doi.org/10.1007/978-981-15-9223-2_7

1 Introduction

In recent few decades, the employment of micro/nanoparticles of cellulose in several applications has attracted much attention from researchers due to their important physicochemical features such as good thermal stability, great tensile strength, biodegradability, biocompatibility, low density, and low environmental influence (Trache 2017; Trache et al. 2020b; Fodil Cherif et al. 2020). The characteristics of these micro and nanoscale materials open new ways for their utilization in many industrial fields (Capron et al. 2017; Hussin et al. 2019; Trache et al. 2014, 2017; Trache et al. 2020a).

Microcrystalline cellulose (MCC) is considered as an effective micro-sized biomaterial that is largely used in pharmaceutical, food packaging and beverage, cosmetic, biocomposite reinforcements, and other applications. MCC is a purified, odorless, crystalline, and partially disintegrated cellulose (Trache et al. 2016; Tarchoun et al. 2019b). The variation of MCC structure and features can be attributed to the natural cellulosic properties, the exploited source in addition to the different conditions used for its isolation. MCC is a snowy powder that can be extracted from various lignocellulosic biomass using different chemical, physical and biological procedures, and a combination of two or more of these processes (Ilyas et al. 2018; Ren et al. 2019; Tarchoun et al. 2019a; Trache et al. 2016; Zeni et al. 2015). Besides, for the production of MCC from cellulosic fibers, the acid hydrolysis approach stays the most utilized process for its several benefits such as effectiveness, low cost, and short time of reaction. Over the acid hydrolysis method, H_3O^+ ions provoke the rupture of glycosidic linkages, where the cellulose amorphous regions are easily hydrolyzed, while the crystalline parts are more resistant to the attack of acids (Tarchoun et al. 2019).

Many researchers are currently seeking for new non-wood fibers as alternative sources of cotton and wood for the isolation of cellulose and their derivatives (Hussin et al. 2019). MCC extraction from different lignocellulosic sources such as Alfa grass (Trache et al. 2014), jute (Jahan et al. 2011), oil palm (Haafiz et al. 2013; Hussin et al. 2018; Owolabi et al. 2017; Ramli et al. 2015; Xiang et al. 2016; Yiin et al. 2019), corncob (Azubuike and Okhamafe 2012), posidonia oceanic brow algae (Tarchoun et al. 2019a), among others, has been reported.

Giant reed (Arundo Donax) is one of abundant aquatic plants that mainly grows in hot and tropical regions such as canals, rivers around marshes, and other humid lands. Giant reed is an interesting fast growing annual herb owing to its highly cellulose content (31–42%). It can be considered as powerful candidate for many fields (Lamb and Dixon 1997; Somerville et al. 2004). Thus, it is expected to be a potential source to produce MCC. Therefore, the aim of this study is to improve the efficient procedure for MCC isolation from Algerian giant reed through a series of chemical treatments like a delignification with two different methods (acidified $NaClO_2$ and totally chlorine-free TCF), alkaline treatment, and then the acid hydrolysis. The produced samples are deeply characterized using FTIR, XRD, SEM, and TGA to inspect the influence of the used delignification method on the chemical structure, crystallinity, morphology, and thermal stability properties of the obtained MCC.

2 Material and Methods

2.1 *Materials*

Giant reed fibers utilized in this work were gathered from the Oum El Bouaghi area, in Algeria. These fibers were cleaned with distilled water then dried in an oven at 110 °C for 24 h. After that, the giant reed fibers milled and sifted into particle size of 35 mesh, then preserved in a vacuum desiccator. Ethanol (96%), toluene (99.5%), hydrogen peroxide (H_2O_2, 30%), acetic acid glacial (CH_3COOH, 96%), nitric acid (HNO_3, 67%), and hydrochloric acid (HCl, 37%) were purchased from VWR Prolabo. Sodium hydroxide (NaOH) pellet and sodium chlorite ($NaClO_2$) were supplied by Sigma Aldrich. All reactants are utilized without any prior purification.

2.2 *Methods*

2.2.1 Preparation of Giant Reed Cellulose Samples

In order to isolate cellulose, many treatments were applied to giant reed fibers. The natural fibers (15 g) are treated with 220 mL of toluene/ethanol (2/1) mixture using a Soxhlet extractor at a fixed temperature during 6 h to remove organic solvent extracts (Tappi 1997). Then, the obtained fibers are treated with 200 mL of distilled water in an oil bath at 105 °C for 5 h to remove sugars, coloring matter ad starch (Kilic and Niemz 2012). After that, the delignification of fibers has been carried out with two different delignification processes using either acidified $NaClO_2$ process or the totally chlorine-free beaching with hydrogen peroxide, as schematized in Fig. 1.

The acidified sodium chlorite process is employed in accordance with the method described by Ilyas et al. (Ilyas et al. 2018). Briefly, 10 g of the giant reed fibers are placed in a 500 mL round bottom flask containing 325 mL of distilled water. 4 mL of acetic acid and 5 g of sodium chlorite are added to the mixture (once every one hour) for 7 h, and the mixture temperature was adjusted at 70 °C. The obtained holocellulose (holocellulose-giant reed-$NaClO_2$) is washed with distilled water during the filtration and then dried in oven at 110 °C.

The TCF process, however, is employed with three successive steps. The first step is performed following Kuznetsov et al. method with slight modification (Kuznetsov et al. 2017). 10 g of the giant reed fiber with no extractives is soaked in a 250 mL round bottom flask containing 150 mL of hydrogen peroxide (6%)/acetic acid (25%) mixture at 100 °C for 5 h. Then, the mixture is filtered and rinsed with distilled water. In the second stage, this residue is placed in a 250 mL round bottom containing 150 mL of nitric acid (20%)/acetic acid (25%) mixture at 120 °C for 20 min. The obtained product is filtered and washed with distilled water (Rosa et al. 2012). In the final step, the residue is treated with 200 mL of hydrogen peroxide (5%)/sodium hydroxide (4%) mixture at 50 °C for 90 min. The obtained holocellulose, named

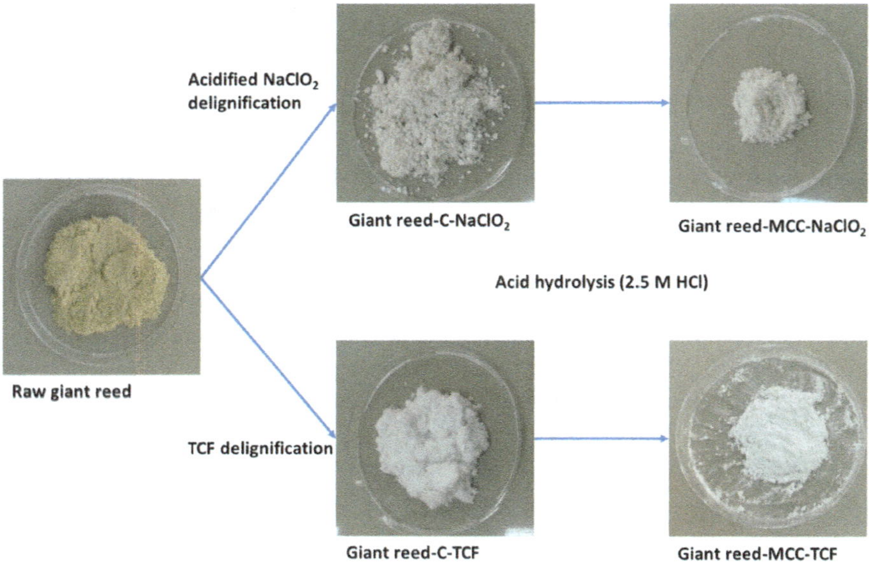

Fig. 1 Scheme of the isolation of microcrystalline cellulose samples from giant reed fibers

holocellulose-giant reed-TCF, is filtered, washed with distilled water, and dried in oven at 105 °C overnight (Sun et al. 2004).

The both samples of holocellulose are treated with sodium chloride in order to remove the residual lignin and hemicellulose to isolate pure samples of cellulose. The process is performed following the method of Jiang et al. (2015). The samples are mixed with NaOH (5%) solution with a ratio of (1/20) during 24 h under magnetic stirring at room temperature, which is then increased to 90 °C for 2 h. The isolated giant reed cellulose samples are filtered, washed with distilled water and diluted acetic acid, and then dried at 90 °C during 48 h.

2.2.2 Preparation of Microcrystalline Cellulose

To isolate giant reed MCCs, the hydrolysis acid of giant reed cellulose samples is performed according to the process reported in the literature with small modification (Trache et al. 2014). Both cellulose samples are hydrolyzed using 2.5 M of hydrochloric HCl acid at 100 °C under reflux for 30 min with the ratio of 1/20 (g/Ml) fiber to liquor. After that, large amount of cold distilled water was added to the mixture to stop the reaction. The resultant cellulose suspensions were then rinsed with distilled water and NaOH (0.5 M) solution to reach a neutral pH. Finally, the generated MCC was dried in oven at 60 °C for 24 h until constant weight was reached. The obtained solids were snowy-white in appearance. The different stages of giant reed MCCs extraction are presented in Fig. 1.

2.3 *Characterization*

2.3.1 Chemical Composition

The chemical composition of giant reed fibers was obtained according to Technical Association of the Pulp and Paper Industry (TAPPI) standard methods and previous investigations (Reddy et al. 2016; Hu et al. 2014). The ash content is evaluated with the standard T 211 om-07, the toluene/ethanol solvent extractives content is evaluated using T 204 cm-07, the hot water extractives are quantified with the standard T-257, the lignin content is quantified using the standard T-222 cm-06, and then, the α-cellulose content is obtained with the standard T 203 cm-99. The holocellulose content was obtained using the method reported in the work of Wise et al. (1946), whereas the hemicellulose content was measured as the difference between holocellulose and α-cellulose contents. The measurements were performed in triplicate.

2.3.2 FTIR Analysis

FTIR spectroscopic analysis was carried out to study the modifications of functional groups and chemical transformation in the structure of the different specimens. The dried samples (0.1–2 mg) were pelletized with KBr (100 mg) and were analyzed with Parkin Elmer FTIR spectrometer. The spectra were recorded in transmittance band mode, in the wavenumber region of 400–4000 cm^{-1} by averaging of 64 scans at resolution of 4 cm^{-1}. Before each measurement, background spectra were obtained at room temperature and then subtracted automatically from the sample spectrum.

2.3.3 X-ray Diffraction Analysis

X-ray diffraction analysis is performed to investigate the structure and to gather the diffractograms of the different cellulose and microcrystalline cellulose samples. The results were collected using a PANalytical X'Pent PRO Multi-purpose diffractometer with Cu Kα radiation at a generator voltage of 45 kV and current of 40 mA. An X'celerator detector was employed to record the data over a 2θ angular range of 5–50° with a step size of 0.017°/2θ and a count time of 50.1650 s at each step. The crystalline indexes of raw giant reed and all cellulose samples and MCCs were obtained from different diffractograms, according to Segal's method (Segal et al. 1959).

$$\mathrm{CrI}(\%) = \frac{I_{200} - I_{\mathrm{amp}}}{I_{200}} \times 100 \tag{1}$$

where CrI is the crystallinity index, I_{200} is the maximum intensity of the 200 peak at $2\theta = 22°$ and I_{am} is the intensity at $2\theta = 18°$.

2.3.4 Scanning Electron Microscopy (SEM)

The morphologies of all cellulose samples and MCCs were studied with FEI Quanta 250 scanning electron microscope tool at a 10 mm working distance and 10 kV accelerating voltage. Dry powders of different cellulosic samples are placed on double-sided conductive adhesive tape. Then, they were analyzed by using the secondary electrons (SE) for morphology. The particle size of the samples was measured using Image J software (Ilyas et al. 2018).

2.3.5 Thermal Analysis

The thermal behavior of cellulose samples and MCCs was examined using a thermogravimetric analysis (TGA). The TGA experiments were conducted using a TGA Q500 V20.13 Build 39. A sample mass of 4 mg was used for the different analyses, which are conducted in the temperature range of 50–500 °C at a heating rate of 10 °C min^{-1}. The different analyses were carried out under inert nitrogen gas atmosphere.

3 Results and Discussion

3.1 Chemical Composition

The chemical composition of raw giant reed fibers is shown in Table 1. It can be seen that the raw fibers are composed of 35.4% cellulose, 21.29% of hemicellulose, and 21.2%, lignin. The hot water extractives content is about 7.2%, that of toluene/ethanol

Table 1 Chemical composition of raw giant reed fibers

Component	Mass fraction (%)
Ash	03.16 ± 0.24
Toluene/ethanol solvent extractives	08.82 ± 0.71
Hot water extractives	07.30 ± 0.79
Lignin	21.20 ± 0.32
Hemicellulose	21.92 ± 0.82
Cellulose	35.40 ± 0.65
Other	02.20 ± 0.27

extractives is 8.82%, whereas the ash content is 3.16%. This composition of the utilized giant reed is in agreement with that reported in the literature (Tarchoun et al. 2019).

3.2 *FTIR Analysis*

FTIR spectra of the untreated and the treated samples are presented in Fig. 2. The spectra showed that practically all fibers have similar patterns, indicating that the different employed processes did not have any unfavorable influence on the chemical functionality of cellulose structure. Moreover, the absorption peaks approximately of 3450 cm^{-1} are assigned to –OH group linkages, for which the intensity decreases after delignification and alkaline treatments owing to the formation of Na-cellulose bonds that minimizes the intra/intermolecular hydrogen bonds in relation to –OH groups (Hussin et al. 2019; Kalita et al. 2013; Zhao et al. 2018). However, the intensity of these peaks raised after acid hydrolysis in both giant reed MCC spectra, which is referred to the highly exposed cellulose area created by the scission of cellulose

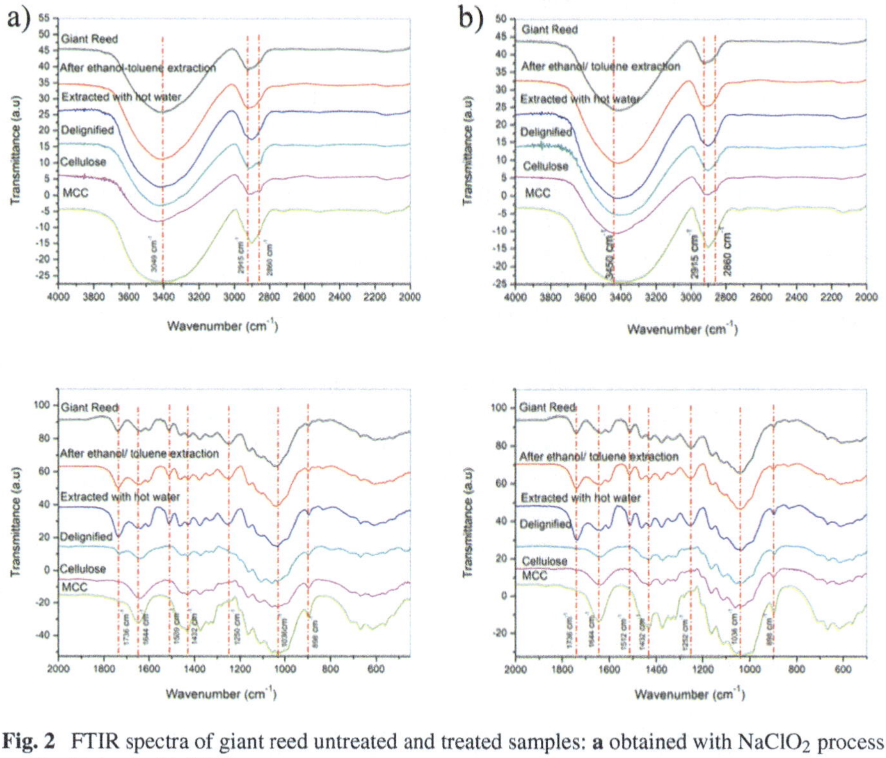

Fig. 2 FTIR spectra of giant reed untreated and treated samples: **a** obtained with $NaClO_2$ process and **b** obtained with TCF process

chains. In addition, the appearance of peaks at 2915 cm^{-1} is attributed to C–H symmetric stretching vibrations of the cellulose structure. The peak at 2860 cm^{-1} corresponded to the asymmetric and symmetric stretching vibration of the CH_2 groups of hemicellulose in giant reed fibers. These latter decreased after the treatment carried out with either $NaClO_2$ or totally chlorine-free. The totally chlorine-free (TCF) with hydrogen peroxide process seems to be more efficient compared to the acidified $NaClO_2$. Besides, the acidified sodium chlorite treatment initially dissolves the lignin without taking into account the removal of hemicellulose (Reddy et al. 2018; Liu et al. 2019), while the totally chlorine-free process yields to a better and effective delignification. The hydroperoxide anions (HOO^-) in the TCH treatment eliminate the lignin chromophore groups by attacking ethylenic and carbonyl groups, whereas the other radical species such as (OH^-) dissolve the hemicellulose, giving rise to purified cellulose (Ditzel et al. 2017). In addition, the bond at about 1736 cm^{-1} is associated to the C–O bonds of uronic acids and acetyl ester groups of hemicellulose or the carboxylic groups of the ferulic and p-coumeric acids of lignin and/or hemicellulose. This bond is present for raw giant reed fibers; however, it is disappeared in spectra of giant reed cellulose and giant reed MCC fibers. This finding proved that the residual amount of non-cellulosic components hemicellulose and/or lignin is dissolved with alkaline treatment and the two different used delignification processes (Bian et al. 2012). After delignification, the absence of characteristic peaks absorption at about 1509 and 1250 cm^{-1} is linked to C = C of aromatic vibration and C–O–C of aryl-alkyl-ether in lignin, respectively, indicating the removal of lignin for both treatments. Meanwhile, the peak at approximately 1646 cm^{-1} is related to the cellulose water interaction in those fibers. The peak is enlarged after hydrochloric acid hydrolysis indicating that the giant reed MCC has higher hygroscopic property. In addition, the peak at around 1420 cm^{-1} is linked to CH_2 symmetric stretching and related to the crystallinity of cellulose. The peak is increased after acid hydrolysis in MCC spectra raveling the hydrolysis of mostly cellulose amorphous parts. Moreover, the peaks at about 898 and 1036 cm^{-1} are corresponding to the β-glycosidic bond stretch, C–O–C pyranose ring skeletal stretch, and C–H asymmetric deformation (Adel and El-Shinnawy 2012; Rosa et al. 2012). These peaks are augmented in giant reed MCC spectra, confirming the breaking of cellulose chains. From the literature and the obtained FTIR spectra, it can be pointed out that the different produced cellulose samples are successfully synthetized using different delignification processes (Trache 2016; Trache et al. 2013, 2014). On the other hand, the prepared MCCs from giant reed using hydrochloric acid hydrolysis present typical chemical composition than that obtained in the literature (Trache et al. 2014).

3.3 XRD Analysis

The crystalline structure and the crystallinity index of the raw giant reed, cellulose samples, and giant reed MCCs are investigated with X-ray diffraction analysis. The XRD diffractograms of raw giant reed fibers, giant reed cellulose, and giant reed

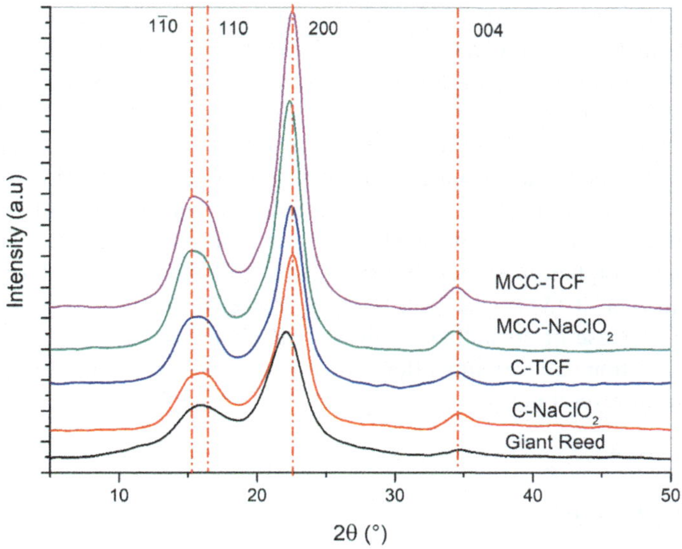

Fig. 3 X-ray diffraction patterns of raw giant reed fibers, giant reed cellulose, and giant reed MCC

MCC are presented in Fig. 3. From the obtained results, all samples exhibited similar peaks of cellulose I at about 15.04°, 16.39°, 22.43°, and 34.53°, which corresponded to 1$\bar{1}$0, 110, 200, and 004 planes, respectively. It is essential to point out that the different obtained patterns showed the non-appearance of cellulose II because there is no doublet in the main peak situated at 22.43° (Hu et al. 2014; Kian et al. 2017). In addition, it is obvious that the peak at about 22.43° becomes tight and sharper, designating the crystallinity augmentation of microcrystalline cellulose compared to the cellulose and raw fibers, respectively. This finding indicates that the different used treatments had no influence on the chemical structure of raw fibers.

The crystallinity indexes of the different samples are measured and given in Table 2. The raw giant reed fibers consist of amorphous non-cellulosic components with lower crystallinity index of 52.36% compared to the pure cellulose samples

Table 2 Crystallinity indexes, diameter size, and the thermal analysis data of different giant reed cellulose samples and MCCs

Material	Crystallinity index (%)	Diameter size (μm)	T peak (°C)	ONSET	OFFSET
Raw Giant reed	55.42	/	/	/	/
Giant reed-C-$NaClO_2$	72.25	11.37 ± 2.4	345.55	317.04	359.61
Giant reed-C-TFC	69.90	10.46 ± 1.6	347.33	306.77	365.52
Giant reed-MCC-$NaClO_2$	76.95	07.82 ± 2.1	336.15	312.65	347.91
Giant reed-MCC-TCF	75.61	07.28 ± 1.7	319.33	289.43	339.41

and MCCs (Kumar et al. 2013). However, the giant reed MCC prepared with TCF exhibited higher crystallinity index compared to that obtained with acidified $NaClO_2$. During acid hydrolysis of cellulose derived from the TCF process, a higher amount of protons (H^+) from HCl opens the fiber structure to simplify the access to cellulose microstructure, distracts the farthest amorphous cellulosic regions, and then breaks the β1,4-glycosidic bonds, conducting to higher crystalline MCC. However, in the case of the acidified $NaClO_2$, the obtained cellulose contains a small amount of hemicelluloses, which allows the acid hydrolysis to selectively dissolve hemicellulose than the amorphous parts of cellulose (Adel and El-Shinnawy 2012; Haafiz et al. 2013; Ramli et al. 2015; Trache et al. 2013; Wyman et al. 2004; Kishani et al. 2018). Hence, the increase of crystallinity value of MCC obtained with totally chlorine-free delignification will certainly offer better physical and mechanical properties for applications in several fields.

3.4 *Morphology*

The morphological structure of giant reed cellulose samples and MCCs extracted with different delignification processes (acidified $NaClO_2$ and totally chlorine-free) is examined with SEM. The obtained SEM micrographs are presented in Fig. 4.

From cellulose images (a and c samples), the cellulose fibers showed clean smooth surface, long, individualized filaments owing to the removal of non-cellulosic components during the used pretreatments, delignification processes, and alkaline treatment (Xiang et al. 2016). These findings are in accordance with FTIR results, which evidencing the elimination of lignin and hemicellulose after the different treatments. The diameter of the analyzed samples was calculated by Image J processing software (IJ 1.46) from the SEM images. More than 30 fibers have been employed to certify the reproducibility of the obtained data. Moreover, the giant reed cellulose produced with TCF process showed lower diameter than that obtained with the acidified $NaClO_2$ process (Table 2). It can be attributed to the effect of nitric and acetic acid mixture used in the procedure of TCF at high temperature and to the impact of oxidation with hydrogen peroxide under alkaline and acidified conditions, leading to more degradation of cellulose amorphous parts (Francis et al. 1998; García Hortal and Vidal Lluciá 1984; Ramos et al. 2008; Sun et al. 2004). It is clear that MCCs present irregular shapes with rough surfaces, which can be assigned to the degradation of the structure of cellulose fibers due to HCl acid hydrolysis. Short length particles have been obtained after hydrolysis because of the cleavage of the amorphous parts linkages of cellulose (Trache et al. 2014). This finding is confirmed by XRD results. Otherwise, from SEM images, it is observable that the MCCs length is lower than their corresponding diameters. It is obvious that the isolated MCCs show some quantity of aggregates owing to the strong cohesion of hydrogen bonding between microfibers.

Fig. 4 SEM micrographs of all cellulose samples: **a** Giant reed-C-$NaClO_2$, **b** Giant reed-MCC-$NaClO_2$, **c** Giant reed-C-TCF, and **d** Giant reed-MCC-TCF

3.5 Thermal Stability

The TGA analyses are performed to study the thermal stability of the produced cellulose samples and to estimate the efficiency of the material and its application in different fields. The thermal properties of cellulose and cellulose microcrystalline samples are displayed in Table 2.

The TGA and DTG curves of the produced giant reed cellulose (a) and MCCs (b) are depicted in Fig. 5. As can be seen, all fibers showed a weight loss within temperature range of 280–370 °C because of the degradation of cellulose caused by the decarboxylation, depolymerization, and the decomposition of glycosidic units followed by the formation of a charred residue (Adel and El-Shinnawy 2012; Trache et al. 2016). It can be also observed that the MCCs illustrated the better thermal stability with higher temperature of decomposition compared to their corresponding pristine cellulose. The giant reed cellulose and MCC samples isolated with acidified $NaClO_2$ treatment displayed lower thermal stability in comparison with giant reed MCC isolated using the TCF delignification treatment. This lower temperature

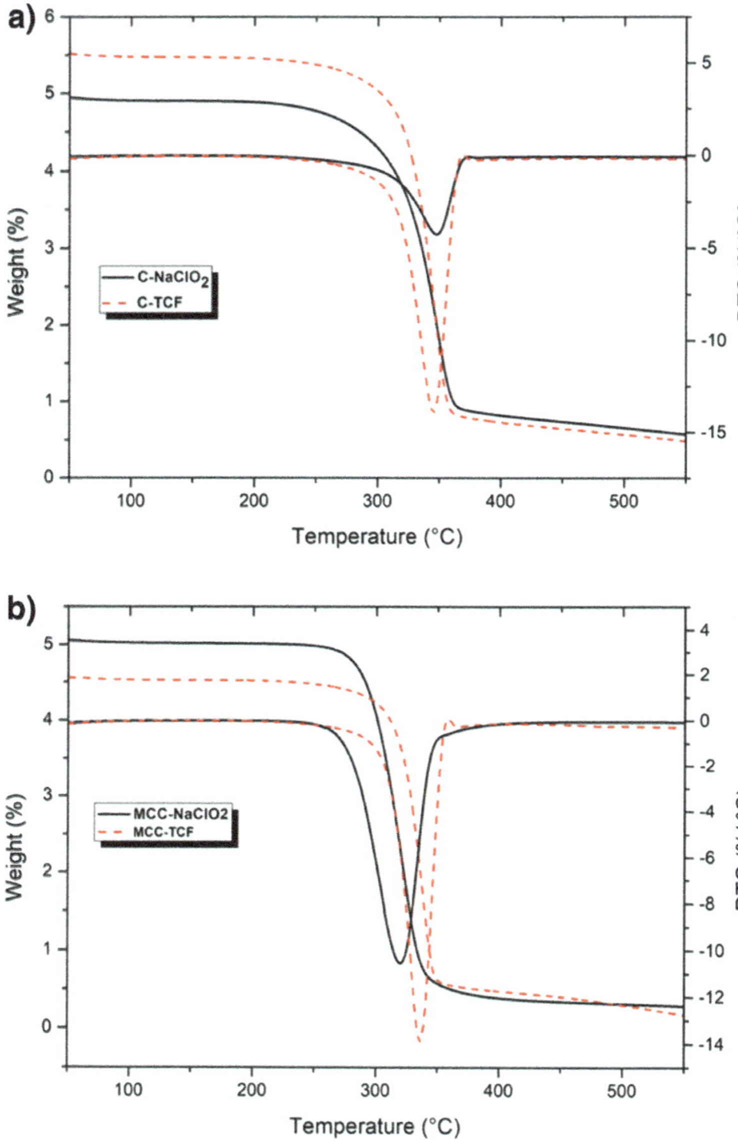

Fig. 5 TGA and DTG curves of **a** giant reed cellulose samples **b** giant reed MCCs

of decomposition is referred to the residual amount of amorphous hemicellulose, which degrades at lower temperature, presenting negative influence on the stability of cellulose fibers. The isolated giant reed MCC produced with totally chlorine-free treatment and acid hydrolysis process presents better thermal stability, which makes it a potential candidate to produce high-value products.

4 Conclusion

The present study is carried out to investigate the effect of different delignification media (acidified $NaClO_2$ and the green media TCF), alkaline treatment, and HCl acid hydrolysis to produce giant reed MCC. The FTIR examination demonstrated that the employed delignification processes have significant impact on the isolation of cellulose without any effect on its chemical structure. The eco-friendly process (TCF) provided higher cellulose purity compared to the acidified sodium chlorite. The XRD results revealed that the isolated MCCs presented typical cellulose type I with high crystallinity index values of 75–77%. The thermal stability of giant reed MCC produced using TCG is better than that of MCC prepared with acidified $NaClO_2$. Consequently, the obtained MCC using TCF can be employed in several promising applications.

References

Adel AM, El-Shinnawy NA (2012) Hypolipidemic applications of microcrystalline cellulose composite synthesized from different agricultural residues. Int J Biol Macromol 51(5):1091–1102

Azubuike CP, Okhamafe AO (2012) Physicochemical, spectroscopic and thermal properties of microcrystalline cellulose derived from corn cobs. Int J Recycl Org Waste Agric 1(1):9

Bian J, Peng F, Peng XP, Peng P, Xu F, Sun RC (2012) Acetic acid enhanced purification of crude cellulose from sugarcane bagasse: structural and morphological characterization. BioResources 7(4):4626–4639

Capron I, Rojas OJ, Bordes R (2017) Behavior of nanocelluloses at interfaces. J Curr Opin Colloid Interface Sci 29:83–95

Ditzel FI, Prestes E, Carvalho BM, Demiate IM, Pinheiro LA (2017) Nanocrystalline cellulose extracted from pine wood and corncob. Carbohyd Polym 157:1577–1585

Fodil Cherif M, Trache D, Brosse N et al (2020) Comparison of the physicochemical properties and thermal stability of Organosolv and Kraft Lignins from hardwood and softwood biomass for their potential valorization. Waste Biomass Valor. https://doi.org/10.1007/s12649-020-00955-0

Francis RC, Rodriguez SK, Bose SK, Granzow SG, Evans TD (1998) The critical role of transition metals in high-temperature peroxide (PO) bleaching. Annual Meeting-Technical Section Canadian Pulp and Paper Association, vol 84, pp A49–A56

Haafiz MKM, Eichhorn SJ, Hassan A, Jawaid M (2013) Isolation and characterization of microcrystalline cellulose from oil palm biomass residue. Carbohyd Polym 93(2):628–634

Hortal G, Antonio J, Llluciá TV (1984) Blanqueo de pastas en la industria papelera: UPC, ETSII de Terrassa

Hu Y, Tang L, Lu Q, Wang S, Chen X, Huang B (2014) Preparation of cellulose nanocrystals and carboxylated cellulose nanocrystals from borer powder of bamboo. Cellulose 21(3):1611–1618

Hussin MH, Husin NA, Bello I, Othman N, Abu Bakar M, Haafiz MKM (2018) Isolation of Microcrystalline Cellulose (MCC) from oil palm frond as potential natural filler for PVA-$LiClO_4$. Polymer Electrolyte. Int J Electrochem Sci 13:3356–3371

Hussin MH, Trache D, Chuin CTH, Fazita MRN, Haafiz MKM, Hossain MdS (2019) Extraction of cellulose nanofibers and their eco-friendly polymer composites. In: Sustainable polymer composites and nanocomposites. Springer, pp 653–691

Ilyas RA, Sapuan SM, Ishak MR (2018) Isolation and characterization of nanocrystalline cellulose from sugar palm fibres (Arenga Pinnata). Carbohyd Polym 181:1038–1051

Jahan MS, Saeed A, He Z, Ni Y (2011) Jute as raw material for the preparation of microcrystalline cellulose. J Cellulose 18(2):451–459

Jiang F, Hsieh YL (2015) Cellulose nanocrystal isolation from tomato peels and assembled nanofibers. Carbohyd Polym 122:60–68

Kalita RD, Nath Y, Ochubiojo ME, Buragohain AK (2013) Extraction and characterization of microcrystalline cellulose from fodder grass; Setaria glauca (L) P. Beauv, and its potential as a drug delivery vehicle for isoniazid, a first line antituberculosis drug. Colloids Surf, B 108:85–89

Kian LK, Jawaid M, Ariffin H, Alothman OY (2017) Isolation and characterization of microcrystalline cellulose from roselle fibers. Int J Biol Macromol 103:931–940. https://doi.org/10.1016/j.ijbiomac.2017.05.135

Kilic Aand Niemz P (2012) Extractives in some tropical woods. Eur J Wood Wood Prod 70(1–3):79–83

Kishani S, Vilaplana F, Xu W, Xu C, Wågberg L (2018) Solubility of softwood hemicelluloses. Biomacromol 19(4):1245–1255

Kumar R, Hu F, Hubbell CA, Ragauskas AJ, Wyman CE (2013) Comparison of laboratory delignification methods, their selectivity, and impacts on physiochemical characteristics of cellulosic biomass. Biores Technol 130:372–381

Kuznetsov BN, Sudakova IG, Garyntseva NV, Djakovitch L, Pinel C (2017) Kinetic studies and optimization of abies wood fractionation by hydrogen peroxide under mild conditions with TiO_2 catalyst. React Kinet Mech Catal 120(1):81–94

Lamb C, Dixon RA (1997) The oxidative burst in plant disease resistance. Annu Rev Plant Biol 48(1):251–275

Liu Y, Nie Y, Lu X, Zhang X, He H, Pan F, Zhou L, Liu X, Ji X, Zhang S (2019) Cascade utilization of lignocellulosic biomass to high-value products. Green Chem 21(13):3499–3535

Owolabi AF, Haafiz MKM, Hossain MS, Hussin MH, Fazita NMR (2017) Influence of alkaline hydrogen peroxide pre-hydrolysis on the isolation of microcrystalline cellulose from oil palm fronds. Int J Biol Macromol 95:1228–1234

Ramli R, Junadi N, Beg M, DH and Yunus RM (2015) Microcrystalline cellulose (MCC) from oil palm empty fruit bunch (EFB) fiber via simultaneous ultrasonic and alkali treatment. Chem Mol Nuclear Mater Metall Eng 9(1):8–11

Ramos E, Calatrava SF, Jiménez L (2008) Bleaching with hydrogen peroxide. A review. Afinidad 65(537)

Reddy KO, Maheswari CU, Muzenda E, Shukla M, Rajulu AV (2016) Extraction and characterization of cellulose from pretreated ficus (peepal tree) leaf fibers. J Nat Fibers 13(1):54–64

Reddy KO, Maheswari CU, Dhlamini MS, Mothudi BM, Kommula VP, Zhang J, Zhang J, Rajulu AV (2018) Extraction and characterization of cellulose single fibers from native African Napier grass. Carbohyd Polym 188:85–91

Ren H, Shen J, Pei J, Wang Z, Peng Z, Fu S, Zheng Y (2019) Characteristic microcrystalline cellulose extracted by combined acid and enzyme hydrolysis of sweet sorghum. Cellulose 26(15):8367–8381

Rosa SML, Rehman N, de Miranda MIG, Nachtigall SMB, Bica CID (2012) Chlorine-free extraction of cellulose from rice husk and whisker isolation. Carbohyd Polym 87(2):1131–1138

Segal LGJMA, Creely JJ, Martin AE Jr, Conrad CM (1959) An empirical method for estimating the degree of crystallinity of native cellulose using the X-ray diffractometer. Text Res J 29(10):786–794

Somerville C, Bauer S, Brininstool G, Facette M, Hamann T, Milne J, Osborne E, Paredez A, Persson S, Raab T (2004) Toward a systems approach to understanding plant cell walls. Science 306(5705):2206–2211

Sun JX, Sun XF, Zhao H, Sun RC (2004) Isolation and characterization of cellulose from sugarcane bagasse. Polym Degrad Stab 84(2):331–339

Tappi S (1997) Solvent extractives of wood and pulp. TAPPI T 204

Tarchoun AF, Trache D, Klapötke TM, Chelouche S, Derradji M, Bessa W, Mezroua A (2019a) A promising energetic polymer from Posidonia Oceanica brown algae: synthesis, characterization, and kinetic modeling. Macromol Chem Phys 220(22):1900358

Tarchoun AF, Trache D, Klapötke TM, Derradji M, Bessa W (2019b) Ecofriendly isolation and characterization of microcrystalline cellulose from giant reed using various acidic media. Cellulose 26(13–14):7635–7651

Trache D, Khimeche K, Donnot A, Benelmir R (2013) FTIR spectroscopy and X-ray powder diffraction characterization of microcrystalline cellulose obtained from alfa fibers. 39th Edition of the Joint European Days on Equilibrium between Phases

Trache D, Donnot A, Khimeche K, Benelmir R, Brosse N (2014) Physico-chemical properties and thermal stability of microcrystalline cellulose isolated from Alfa fibres. Carbohyd Polym 104:223–230

Trache D (2017) Microcrystalline cellulose and related polymer composites: synthesis, characterization and properties. In: Thakur VK, Thakur MK, Kessle MR (eds) Handbook of composites from renewable materials, structure and chemistry. Wiley, pp 61–92

Trache D, Hussin MH, Chuin CTH, Sabar S, Fazita MRN, Taiwo OFA, Hassan TM, Haafiz MKM (2016a) Microcrystalline cellulose: isolation, characterization and bio-composites application—a review. Int J Biol Macromol 93:789–804

Trache D, Khimeche K, Mezroua A, Benziane M (2016b) Physicochemical properties of microcrystalline nitrocellulose from Alfa grass fibres and its thermal stability. J Therm Anal Calorim 124(3):1485–1496

Trache D, Hussin MH, Haafiz MKM, Thakur VK (2017) Recent progress in cellulose nanocrystals: sources and production. Nanoscale 9(5):1763–1786

Trache D, Tarchoun AF, Derradji M, Hamidon TS, Masruchin N, Brosse N, Hussin MH (2020a) Nanocellulose: from fundamentals to advanced applications. Front Chem 8:392

Trache D, Tarchoun AF, Derradji M, Mehelli O, Hussin MH, Bessa W (2020b) Cellulose fibers and nanocrystals: preparation, characterization and surface modification. In: Kumar V, Guleria P, Dasgupta N, Ranjan S (eds) Functionalized nanomaterials I: fabrication. Taylor & Francis

Wise LE (1946) Chlorite holocellulose, its fractionation and bearing on summative wood analysis and on studies on the hemicelluloses. Paper Trade 122:35–43

Wyman CE, Decker SR, Himmel ME, Brady JW, Skopec CE, Viikari L (2004) Hydrolysis of cellulose and hemicellulose. In: Dumitriu S (ed) Polysaccharides: structural diversity and functional versatility. CRC Press, 1023–1062

Xiang LY, Mohammed MAP, Baharuddin AS (2016) Characterisation of microcrystalline cellulose from oil palm fibres for food applications. Carbohyd Polym 148:11–20

Yiin CL, Yusup SHS, Quitain AT, Chan YH, Loy ACM, Gwee YL (2019) Recovery of cellulose fibers from oil palm empty fruit bunch for pulp and paper using green delignification approach. Biores Technol 290:121797

Zeni M, Favero D, Pacheco K, Grisa A (2015) Preparation of microcellulose (Mcc) and nanocellulose (Ncc) from eucalyptus kraft ssp pulp. J Polym Sci 1:1–5

Zhao T, Chen Z, Lin X, Ren Z, Li B, Zhang Y (2018) Preparation and characterization of microcrystalline cellulose (MCC) from tea waste. Carbohyd Polym 184:164–170

Chapter 8
A Green Synthesis and Polymerization of N-Alkyl Methacrylamide Monomers with New Chemical Approach

Samira Derkaoui, Mohammed Belbachir, and Sara Haoue

Abstract In the present work, we have developed an original method to synthesize and to polymerize N-alkyl methacrylamide monomers with a new chemical approach by using an heterogeneous catalyst "Maghnite Na^{+}" as green and recyclable catalyst in one step with respect to the green chemistry principals. The polymerization of N-alkyl methacrylamide monomers was carried out using an environmentally friendly catalyst, during which the monomers were synthesized at room temperature for 2 h by reacting an amino derivative (heterocyclic and aryl amine) with methacrylic anhydride in presence of "Maghnite H^{+}" a green catalyst in a bulk (without solvent). The undertaken green reaction has yielded 85% of conversion and 100% selectivity of the obtained product, and contributed to substitute former used toxic reactant methacryloyl chloride and triethylamine, frequently used in previous published studies. Infrared spectroscopy (FTIR) and nuclear magnetic resonance (NMR) spectroscopy ^{1}H and ^{13}C subjected the monomers and polymers structures to confirmation.

Keywords Green synthesis · Montmorillonite clay · N-alkyl methacrylamide · Anionic polymerization · Methacrylic anhydride

1 Introduction

In latest years, the employ of solid (heterogeneous) catalysts: zeolites, montmorillonite clays, silica gel, and alumina have been of great interest in various fields of organic synthesis (Dumesic et al. 2008), due to their compatibility with the environment, their use, their high selectivity, their ease of use, their low cost, and the ease of recovery (Sauvik et al. 2017). The products based on acrylic/methacrylic monomers are found everywhere: paints, lacquers, and adhesives, water treatment products and plastics or textile fibers. They are an important class of functional

S. Derkaoui (✉) · M. Belbachir · S. Haoue
Laboratory of Polymer Chemistry, Department of Chemistry, Faculty of Exact and Applied Sciences, University Oran 1 Ahmed Ben Bella, BPN 1524 El'Menouer, 31000 Oran, Algeria
e-mail: derkaoui.samira@gmail.com

D. Trache et al. (eds.), *Materials Research and Applications*,
Materials Horizons: From Nature to Nanomaterials,
https://doi.org/10.1007/978-981-15-9223-2_8

polymers having ionic, non-ionic, flexible, rigid, hydrophilic or hydrophobic properties (Sigma Aldrich 2018). Methacrylamide is employed in the production of (polymethacrylamide) which is used as an additive in paper manufacturing, coating agents, and flocculation of minerals and water clarifying agents (Donald 2005). Many examples are found in applications fields such as poly (N-phenyl methacrylamide) has various applications due to the presence of the aromatic ring, which can undergo easily substitution reactions (Jayasimha et al. 2003). 1,4-bis (methacryloyl) piperazine is widely used in the medical area to prepare cross-linking agents that serve to separate amino acids, peptides, proteins, and viruses (Shundrina et al. 2015; Dubois and Lavignac 2013). Derivatives of poly acryloylmorpholine are of great application in the biological and medical field: the manufacture of blood plasma separating membranes, peptide synthesis, and the immobilization of enzymes, in drug delivery (Hülyaet al. 2013).

In previous published works, one can find the description of synthesis and polymerization of similar monomers; by using acryloyl chloride or methacryloyl chloride with triethylamine as reactants in various solvents: chloroform, hexane, and by using benzoyl peroxide/AIBN or potassium persulfate as initiator. Polymethacrylamide has been generally synthesized by free radical polymerization (Anbarasan et al. 2011; Muserref and Meltem 2006; Sadegui et al. 2013) and methacrylamide monomer is synthesized by Matsuda (1986) by reacting methacrylonitrile with water using raney copper as a catalyst or by Wiley and Walter 1948). The polymerization of 1,4-bis(acryloyl)piperazine is described by Shundrina et al. (2015) and Dubois and Lavignac (2013) and poly N-acryloylmorpholine is synthesized by Boursier et al. (2015) and Gorman et al. (2014) and poly N-phenylmethacrylamide is synthesized by Brooks and McCormick (2016) and Jayasimha et al. (2014) (Fig. 1).

The originality of this work is the study of anionic polymerization of N-alkylmethacrylamide monomers initiated by non-toxic and recyclable catalyst a "Maghnite-Na$^+$". Firstly, the synthesis of the monomers started by reaction of primary amine (ammonia and aniline) or heterocyclic secondary amines such as piperazine and morpholine with methacrylic anhydride in the presence of the green catalyst "Maghnite H$^+$" in bulk medium(absence of any extra solvent), at 0 °C for 2 h as reported by our previous work (Derkaoui and Belbachir 2019). Secondly, obtained monomers were polymerized under suitable conditions initiated by anionic catalyst

Fig. 1 General formula of the N-alkylmethacrylamide

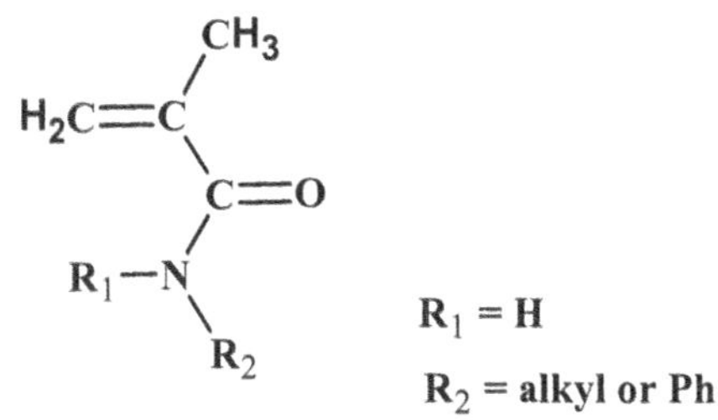

"Maghnite Na^{+}". The monomers and polymers structures were characterized and confirmed by infrared spectroscopy (FTIR) and nuclear magnetic resonance (NMR) spectroscopy ^{1}H and ^{13}C.

2 Experimental Section

2.1 Materials and Methods

Methacrylicanhydride and dichlomethane were purchased from (SIGMA-ALDRICH), piperazine (ACOROS-ORGANICS), morpholine (PROLABO), and ammonia from (RIEDEL-DE-HAEN). The aniline (PRONALYS) was vacuum distilled immediately prior to use, $MgSO_4$ (BIOCHEM). Raw-Maghnite clay was obtained from ENOF (NATIONAL COMPANY OF NON-FERROUS MINING PRODUCTS) Maghnia (Algeria). Fourier-transform infrared (FTIR) spectra was obtained on an Alpha-PATR Brukerfrom 4000 cm^{-1} to 400 cm^{-1}. ^{1}H and ^{13}C nuclear magnetic resonance (NMR) measurements were carried out on a 300 MHz Bruker NMR spectrometer in $CDCl_3$, D_2O and DMSO-d_6. UV–Visible spectra were measured in MeOH solution at 10^{-3} molar concentration using a Shimadzu UV–Visible 2401 recorder spectrophotometer.

2.2 Preparation of Catalyst

Raw-Maghnite (20 g) was placed in an Erlenmeyer flask together with 500 ml of distilled water and then 500 ml of sulfuric acid solution 0.25 M concentration are added. The mixture was stirred for 48 h at room temperature (T = 25 °C). The clay was filtered and then washed with distilled water until neutralization pH = 7. The activate clay was put in the oven at 105° C for 48 h to dry it (to remove a water trapped in the sheets of clay). The "Maghnite-H^{+}" is then crushed and sieved to obtain a powder; it was put back in the oven for further drying, and then it was enclosed an airtight bottle in the shelter of moisture. "Maghnite-Na^{+}" was prepared according to the following method (Belbachir and Bensaoula 2006): Use an Erlenmeyer flask to mix (20 g) of Raw-Maghnite with 500 ml of 1 M NaCl solution. Then thoroughly stirring the maghnite/water mixture until saturation for 24 h at room temperature, and then filtering the mixture and wash it several times with distilled water to remove the chloride, and finally dried at 105 °C. It is worth to emphasize the necessity of clay grounding the clay in order to enhance the good contact of monomer within the clay layers (better catalyst support).

2.3 Syntheses

2.3.1 General Procedure for Synthesis of the Monomers

Synthesis of monomers was carried out by mixing 0.1 mol of amine (ammonia, aniline, piperazine, and morpholine)with various amounts of "Maghnite-H^+" 0.25 M (3, 5, 10 and 15% wt) for 30 min in bulk; then after, we added 0.1 mol (15 mL) of methacrylic anhydride. Except for piperazine, 0.2 mol (30 ml) of methacrylic anhydride was added (with respect to a molar ratio of 1:2 piperazine to methacrylic anhydride) in order to be activated on both sides. After 2 h, we filtered the solution and recover the clay catalyst, which was transferred to a separating funnel, to be washed thoroughly with a 5% sodium hydroxide solution and water, and extracted with dichloromethane (3 × 30 mL). The organic layers were combined and dried over anhydrous magnesium sulfate $MgSO_4$ and filtered. After the evaporation of dichloromethane in vacuum, the products were purified by precipitation and recrystallization in methanol/diethyl ether mixture (MAM) or in an ethanol–water mixture (NPM) or in diethyl ether (NBMP) or purified by column chromatography (NMM). According to our previous study (Derkaoui et al. 2019a, b).

2.3.2 Polymerization of the Monomers

Monomers polymerization was carried out in sealed tubes. Each tube contains a mixture of 1 g of monomer, 5 ml of solvent, and (0.15 g) 15% "maghnite-Na^+". The mixtures were kept in an ice bath at 0 °C and stirred with a magnetic stirrer for 24 h. The resulting polymers were precipitated in methanol, washed several times, and dried at 40 °C and then weighed (Scheme 1).

3 Results and Discussion

3.1 Characterization of Monomers

- **Methacrylamide (MAM)**:

The product is white powder. Yield: 85%. C_4H_7NO; M.P: 109 °C. IR (cm^{-1}): $\upsilon_{NH_2 \text{ primary amine}}$ 3374.45, 3180. 88, 607, υ_{CH_2, CH_3} 2930.75, 2984, $\upsilon_{C=O\text{amide}}$ 1665.34, $\upsilon_{C=C}$ 1600,15, υ_{CN} 1405,22. The ^1H NMR ($^\delta$ ppm in $CDCl_3$): 1.951 (s, 3H of $-CH_3$), 5.39–5.76 (d, 2H of $=CH_2$), 6.027–6.48 (2 s, 2H of $-NH_2$). The ^{13}C NMR ($^\delta$ ppm in $CDCl_3$):18.63($-CH_3$); 120.88 ($=CH_2$), 139.12 (C = C), 170.95(C = O)$_{amide}$. λ_{MAX} = 222.50 nm.

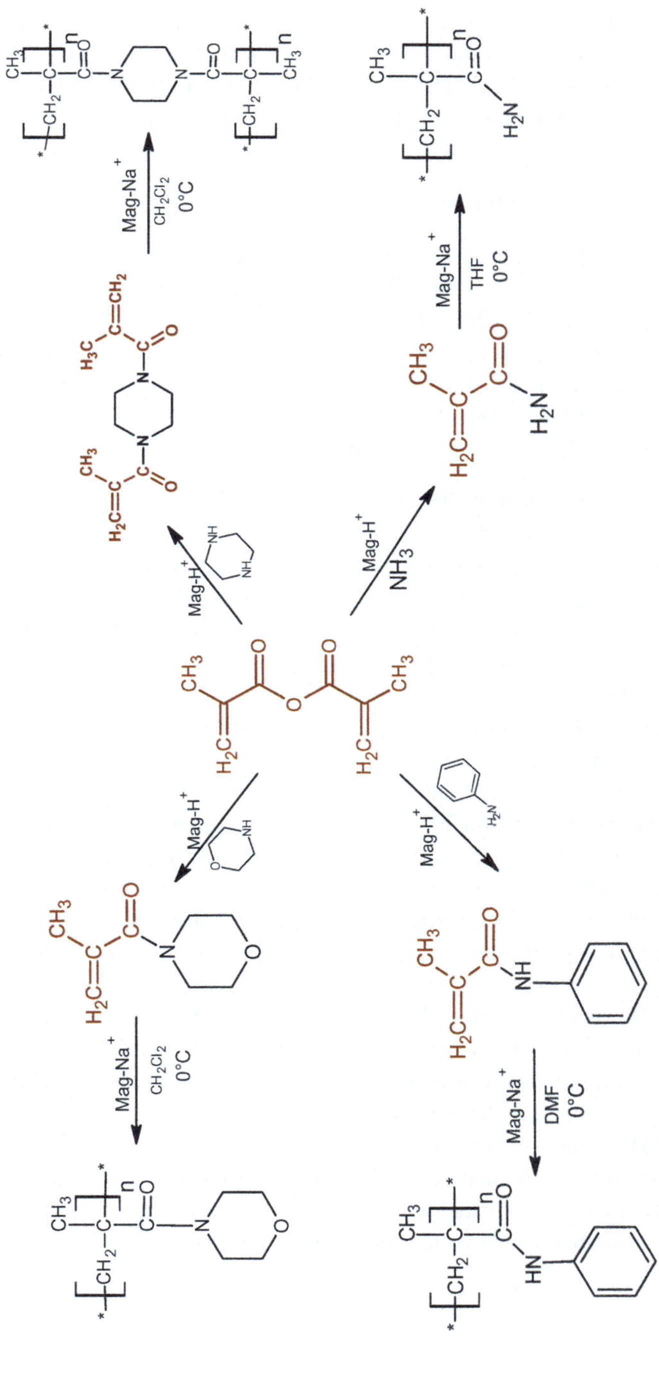

Scheme 1 General procedure for synthesis and polymerization of monomers

- **1,4-bis(Methacryloyl) Piperazine (NBMP)**:

The product is a white crystal solid. Yield: 70%; $C_{12}H_{18}N_2O_2$; M.P. 112°. IR (cm^{-1}): υ_{CH_2, CH_3} 2864, 2974, $\upsilon_{C=Oamide}$ 1716, $\upsilon_{C=C}$ 1614, 522, υ_{CN} 1430; ^1H NMR ($^\delta$ ppm in $CDCl_3$) 1.97 (s, 3H of CH_3), 5.06–5.25 (d, 2H of $=CH_2$), 3.60 (m, 8H of NCH_2CH_2N); ^{13}C NMR ($^\delta$ ppm in $CDCl_3$): 20.53 ($-CH_3$); 31.25–38.15 ($N-(CH_2)_2-N$); 116.17 ($=CH_2$), 139.92 ($C=C$), 171.43 $(C=O)_{amide}$. λ_{MAX} = 211.50 nm.

- **N-Methacryloylmorpholine (NMM)**:

The product obtained is a colorless liquid and is purified by column chromatography using CH_2Cl_2/CH_3OH (3/2) as the eluent Rf = 0.65. Yield = 65%; $C_8H_{13}NO_2$. IR (cm^{-1}): υ_{CH_2, CH_3} 2856, 2961, $\upsilon_{C=Oamide}$ 1721, $\upsilon_{C=C}$ 1609, 558, υ_{C-N} 1433, $\upsilon_{C-O-Cheterocyclic}$ 3492.59, $\upsilon_{C-O-C\ oxide}$ 1112; ^1H NMR ($^\delta$ ppm in $CDCl_3$) 1.89 (s, 3H of CH_3), 4.97–5.15 (d, 2H of $=CH_2$), 3.54–3.60 (m, 8H of NCH_2CH_2O); ^{13}C NMR ($^\delta$ ppm in $CDCl_3$) 20.33 ($-CH_3$); 66.82 ($O-(CH_2)_2-N$); 115.86 ($=CH_2$), 139.83 ($C=C$), 171.27$(C=O)_{amide}$. λ_{MAX} = 239 nm.

- **N-Phenyl Methacrylamide (NPM)**:

The obtained product is a white solid. Yield = 72%; $C_{10}H_{11}NO$; M.P. 85–86 °C. IR (cm^{-1}): $\upsilon_{NHprimary\ amine}$ 3291, 935, 758, υ_{CH_2, CH_3} 2927, 2982, $\upsilon_{C=Oamide}$ 1721.50, $\upsilon_{C=C}$ 1620, 511, υ_{C-N} 1437, $\upsilon_{C-Cphenyl\ ring}$ 757.61, 694.16; ^1H NMR ($^\delta$ ppm in $CDCl_3$) 2.054 (s, 3H of $-CH_3$), 4.45–5.80 (d, 2H of $=CH_2$), 7.12–7.60 (m, 1H of $-CH)_{phenyl\ ring}$, 7.875(s, 1H of $-NH$); ^{13}C NMR ($^\delta$ ppm in $CDCl_3$) 18.78 ($-CH_3$); 124.44 ($=CH_2$), 140.88 ($C=C$), 166.90$(C=O)_{amide}$, 120.26, 128.97, 137.85 $(-CH)_{phenyl\ ring}$. λ_{MAX} = 213.50 and 246 nm.

The FTIR analysis confirms the obtained methacrylic monomers. All of them present bands between 2856 and 2982 cm^{-1} corresponding to C–H symmetrical and asymmetrical stretching of CH_2 and CH_3 groups, respectively, the strong band at 1721 cm^{-1} was attributed to amide carbonyl ($C=O$) stretching vibration, another strongest band at 1632 cm^{-1} corresponds to the ($C=C$) stretching vibration and the second strongest and sharp band at 1454 $cm^{-1}$was attributed to the C–N stretching vibration (Fig. 2).

The ^1H NMR spectra for MAM, NBMP, NMM, and NPM are approximately identical; all spectra confirm the presence of the strongest and sharp peak between 1.87–2.05 ppm corresponds to methyl protons ($-CH_3$) and the two medium broad peaks at 4.45–6.25 ppm are attributed to the vinylic group ($=CH_2$).

The ^{13}C NMR spectrum of all monomers shows: the sharp peak centered at 18.63 ppm corresponds to methyl carbon ($-CH_3$) Sp^3; two peaks at 120.88 and 139.12 ppm characteristic to the vinylic carbon ($C=C$) Sp^2. The last peak at 170.95 ppm is attributed to the amide function ($C=O$) (Fig. 3).

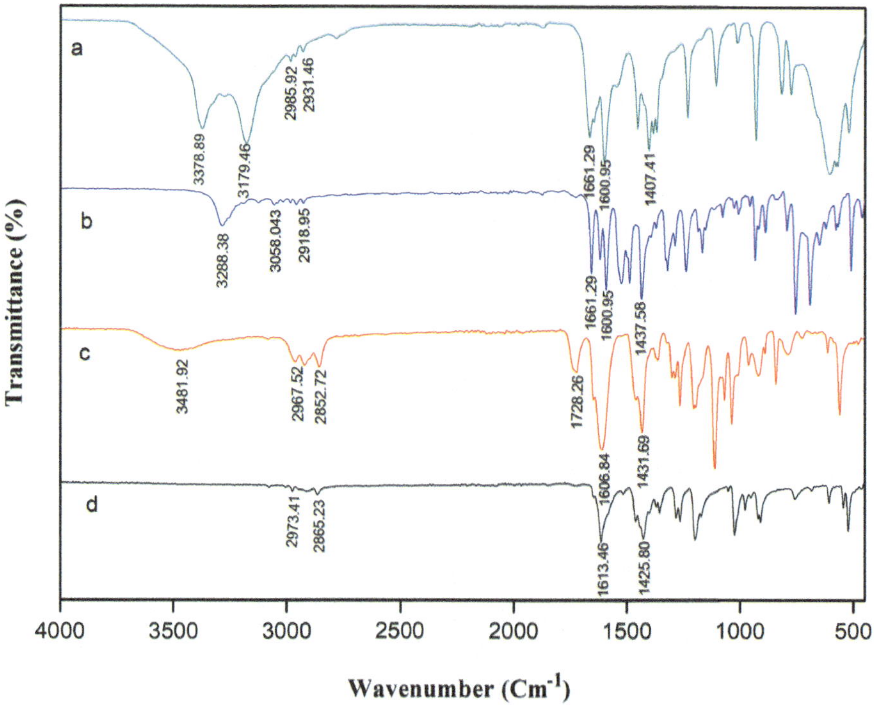

Fig. 2 FTIR spectra for the monomers **a** MAM; **b** NPM; **c** NMM, and **d** NBMP

3.2 *Characterization of the Polymers*

- **Poly (Methacrylamide) PMAM**:

Figure 4 shows the IR spectrum of poly (MAM): we observe the increase of the intensity of the bands. Two large and intense bands at 3315 cm^{-1}correspond to N–H stretching vibration and a large bands between 2836 and 2990 cm^{-1} were attributed to C–H symmetrical and asymmetrical stretching of CH_2 and CH_3 groups, respectively. The band at 1653 cm^{-1} is assignable to amide carbonyl ($C = O$) stretching vibration. The decrease in intensity of the band at 1600 cm^{-1} which corresponds to the stretching vibration $C = C$ due to the opening of double bond allowing the connection between the monomer units.

The ^1H NMR spectrum in Fig. 5 confirms the structure of the polymer obtained. Two broad peaks observed at 1.058 and 1.68 ppm are assigned to methyl (CH_3) and methylene protons (CH_2) of the polymer repeating unit. The second broad peak between 6.80–7.55 ppm corresponds to the protons of amine group (–NH_2). The small peak at 1.82 ppm and at 3.23 ppm is attributed to the methyl protons groups (–CH_3) at the end of the polymer chain; as well as a doublet at 5.39–5.76 ppm

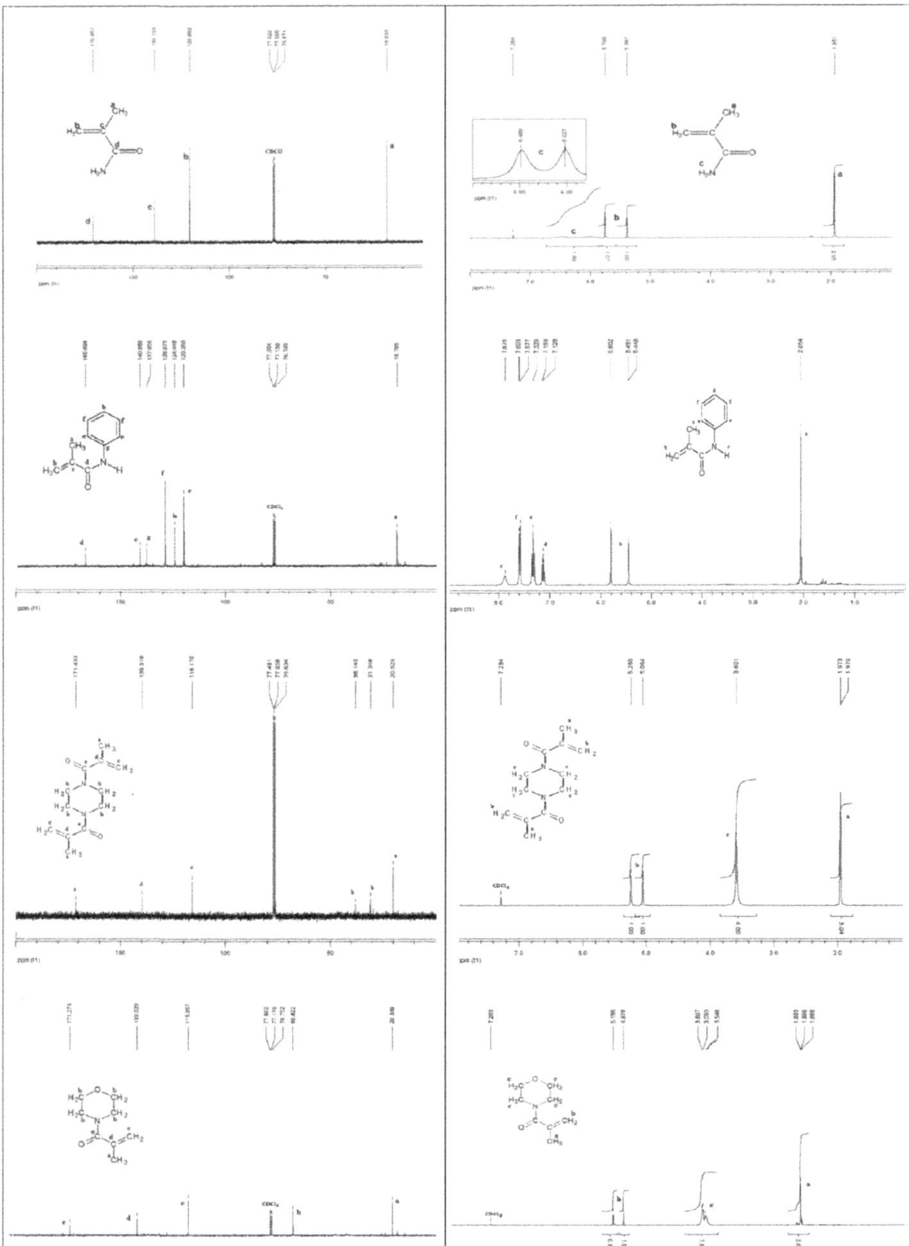

Fig. 3 ^{1}H NMR and ^{13}C spectrum ($CDCl_3$) of monomers

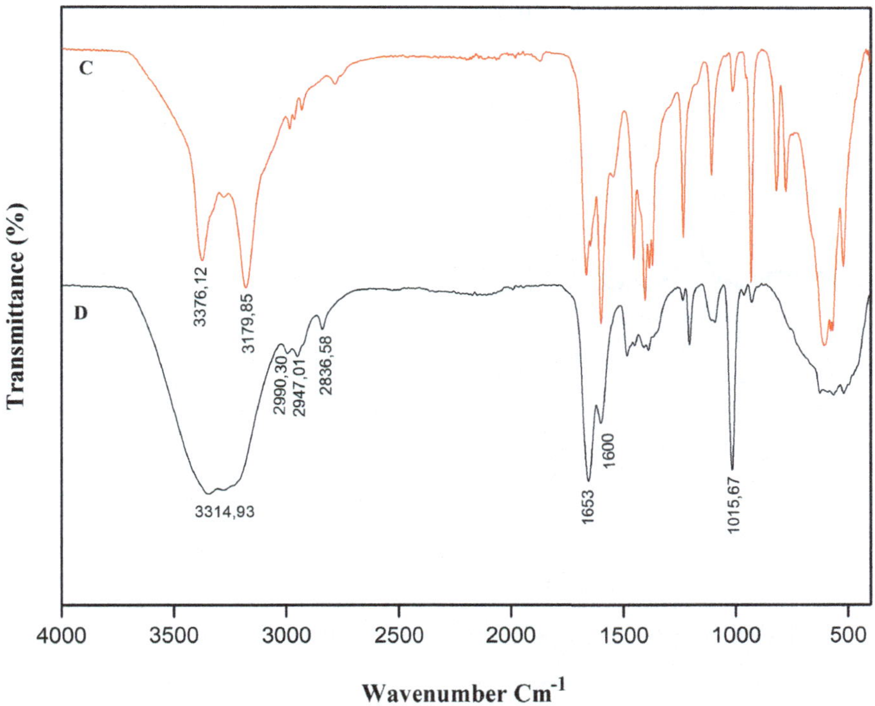

Fig. 4 FTIR spectra for the methacrylamide (MAM) (C) and poly(methacrylamide) (PMAM) (D)

is assigned to the vinylic group (=CH). The strongest and sharp peak centered at 4.70 ppm corresponds to water deuterated solvent (D_2O), respectively.

- **Poly 1,4-bis(Methacryloyl)Piperazine (PNBMP)**:

We observe the increase of the intensity of the bands. A large and intense bands between 2920 and 2952 cm^{-1} were attributed to C–H symmetrical and asymmetrical stretching of CH_2 and CH_3 groups, respectively; and the disappearance of the intense band at 1617 cm^{-1} corresponded to stretching vibration (C = C) (Fig. 6).

The ^1H NMR spectrum of poly (NBMP) shows: The strongest peak at 0.76 ppm corresponds to methyl proton ($-CH_3$) groups, the broad peak between 0.99–1.185 ppm due to methylene protons ($-CH_2$) and the other broad peak at 1.49 ppm characteristic to heterocyclic protons ($N-CH_2CH_2-N$) (Fig. 7).

- **Polymethacryloylmorpholine (PNMMo)**:

A large and intense band at 3324.44 cm^{-1} characteristics to stretching vibration of methylene group (CH_2) to heterocyclic and a broad band between 2852.72 and

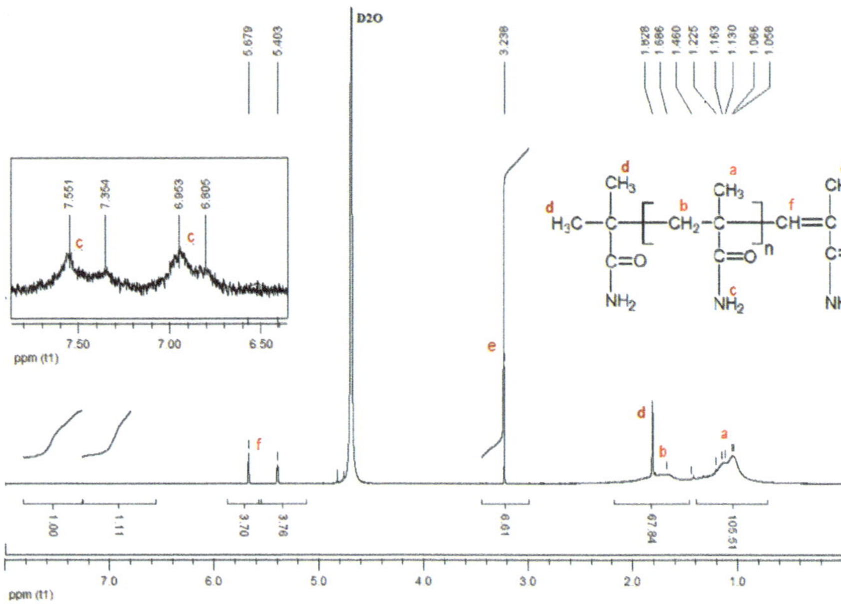

Fig. 5 ^{1}H NMR spectrum (D_2O) of poly(methacrylamide) (PMAM) obtained

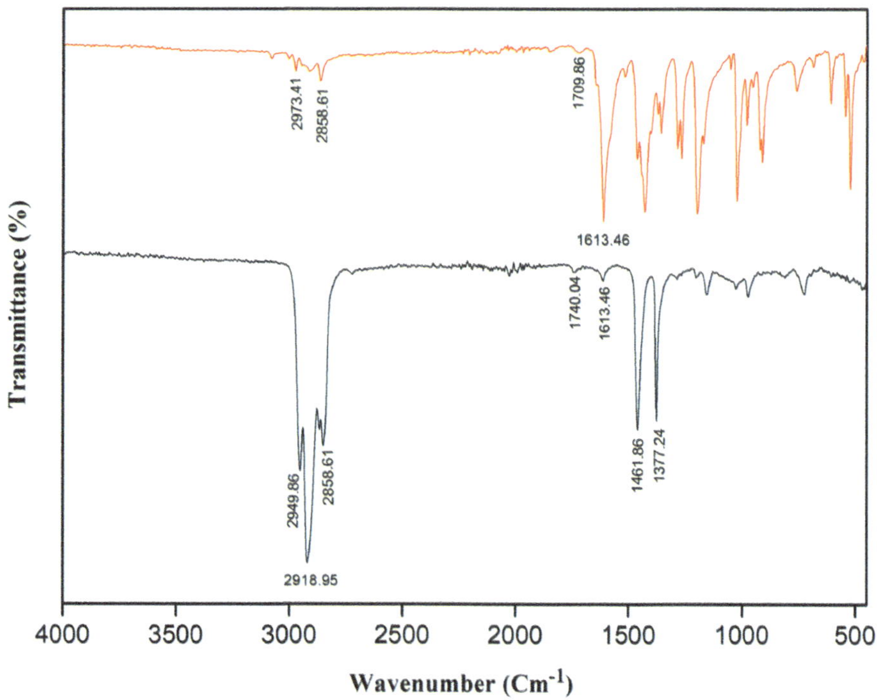

Fig. 6 FTIR spectra for the Poly 1,4-bis(Methacryloyl)Piperazine (PNBMP) and monomer

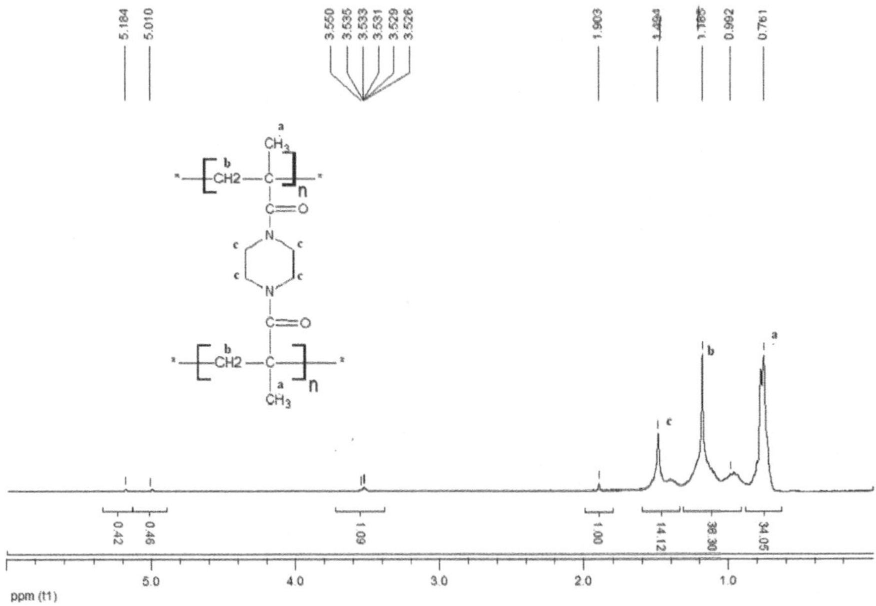

Fig. 7 ^{1}H NMR spectrum ($CDCl_3$) of Poly1,4-bis(Methacryloyl)Piperazine (PNBMP)

2949 cm^{-1} were attributed to C–H symmetrical and asymmetrical stretching vibrationof CH_2 and CH_3 groups, respectively. The disappearance of the intense band at 1606 cm^{-1} corresponds to stretching vibration of the vinylic group (C = C) (Fig. 8).

The ^{1}H NMR spectrum of poly (NMMo) shows: The broad peaks at 1.27–1.68 ppm correspond to methyl proton (–CH_3) groups, the strongest peak between 1.90–2.127 ppm due to methylene protons (–CH_2) and the other broad peak at 3.69 ppm correspond to heterocyclic protons (N–CH_2CH_2–O) (Fig. 9).

- **Poly N-Phenyl Methacrylamide (PNPM)**:

We observe the increase of the intensity of the bands. Two large and intense bands at 2852 and 2919 cm^{-1} were attributed to C–H symmetrical and asymmetrical stretching vibration of CH_2 and CH_3groups, respectively; and a broad band around 3288 cm^{-1} assigned to stretching vibration of amide (N–H) function. The absorptions at 1722 cm^{-1} are characteristic to amide carbonyl (C = O) stretching vibration (Fig. 10).

The ^{1}H NMR spectrum of poly (NPM) shows: The strongest peaks between 0.86–0.59 ppm correspond to methyl proton (–CH_3) groups, the broad peaks at 1.86–2.04 ppm due to methylene protons (–CH_2) and the aromatic protons peaks were observed at 7.04–7.66 ppm. A weak peak at 7.96 ppm was attributed to the (–NH) proton. The strongest and sharp peaks centered at 3.17 ppm, 2.90–2.75 ppm and 2.51 ppm correspond to water, DMF and the deuterated solvent (DMSO-d_6), respectively (Fig. 11).

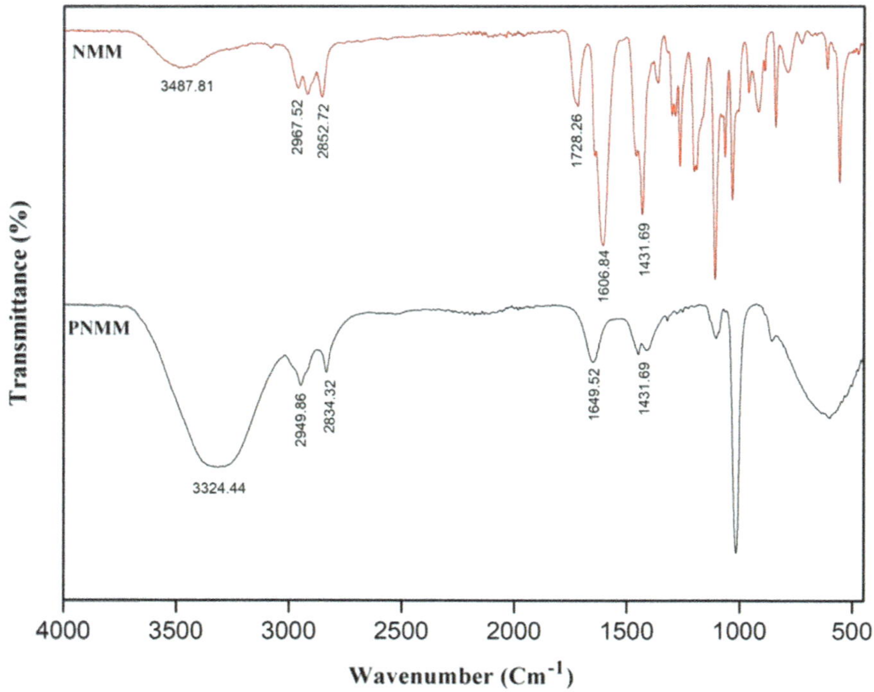

Fig. 8 FTIR spectra for the polymethacryloymorpholine (PNMMo) and monomer

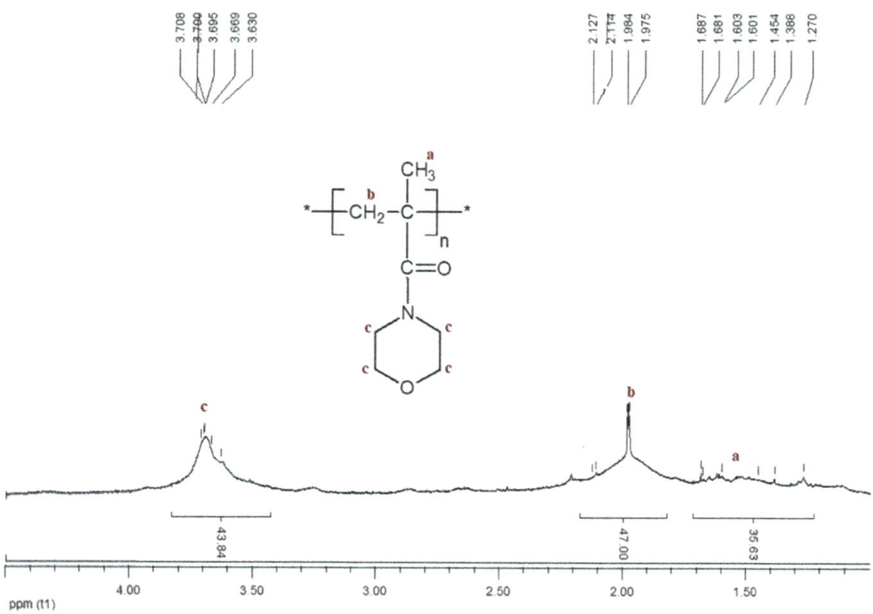

Fig. 9 ^{1}H NMR spectrum ($CDCl_3$) of polymethacryloylmorphline (PNMMo)

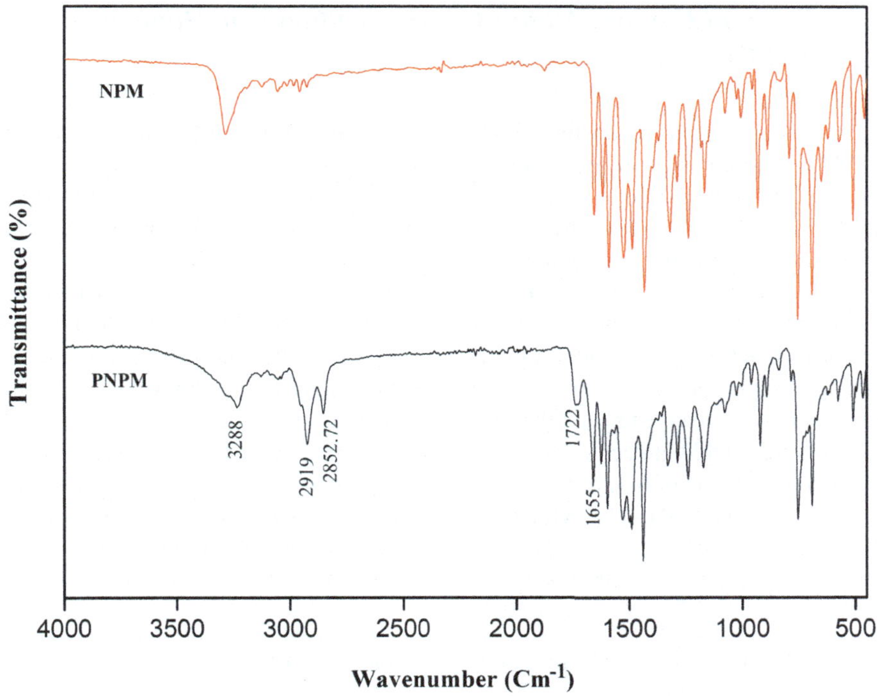

Fig. 10 FTIR spectra for the poly N-phenyl methacrylamide (PNPM) and monomer

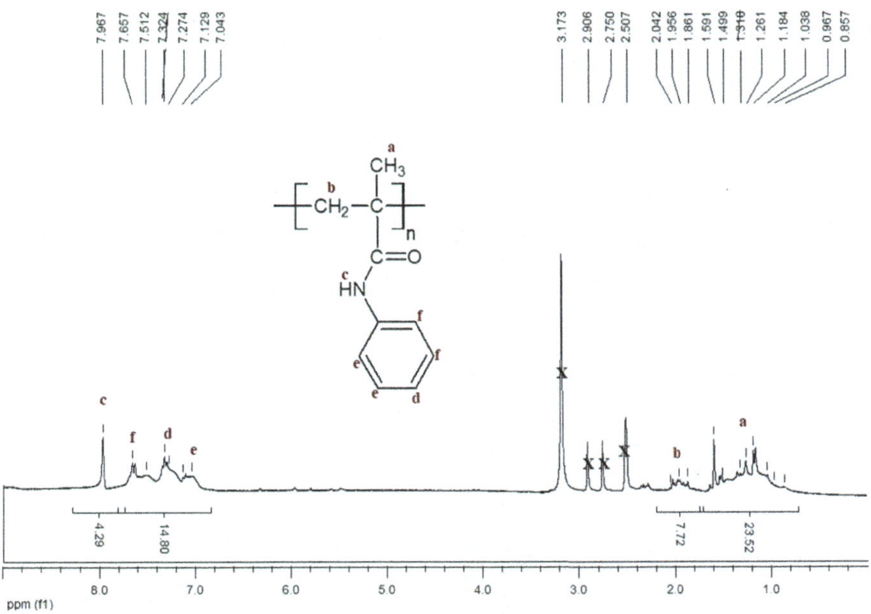

Fig. 11 ^{1}H NMR spectrum (DMSO-d_6) of polyN-phenyl methacrylamide (PNPM)

3.3 Effect of Catalyst Amount on the Synthesis of Monomers and Polymers

These new monomers disubstituted NBMP and monosubstituted NMMo, NPM, and MAM were synthesized by nucleophilic substitution of primary amine (ammonia and aniline) and secondary cyclic amine (piperazine and morpholine) with methacrylic anhydride as indicated in Scheme 1. We have carried out a series of experiments by maintaining the time and temperature constant and varying the quantities of the catalyst Mag-H^+ (3, 5, 10, and 15% wt) (Table 1). After obtaining higher yields around 85%, we selected the amount of catalyst to be around (10%), for the monomers synthesis reaction. We can see from Table 1 and Fig. 12 that the yield of monomers synthesisis proportional to the quantities of the catalyst "Maghnite H^+" 0.25 M and reaches its steady state at 10%. The monomers synthesis reaction was obtained with a better yield between 65–85% and with a selectivity of 100% to pure product without production of by-product, to confirm the effectiveness of catalyst "Mag H^+" in the synthesis of the N-alkylmethacrylamide monomers. The conversion of the polymerization reaction reached a threshold of 62% using 15% Maghnite-Na^+ as an initiator for anionic polymerization is clearly shown. This is mostly owing to the

Table 1 Effect of catalyst amount (Mag H^+) on monomers yield, time of reaction 2 hours; temperature 0–5 °C

Monomers	Mag H^+ (weight %)	3	5	10	15
MAM	Yield %	42	70	85	85
NBMP		40	60	70	70
NMM		38	55	65	65
NPM		40	68	72	72

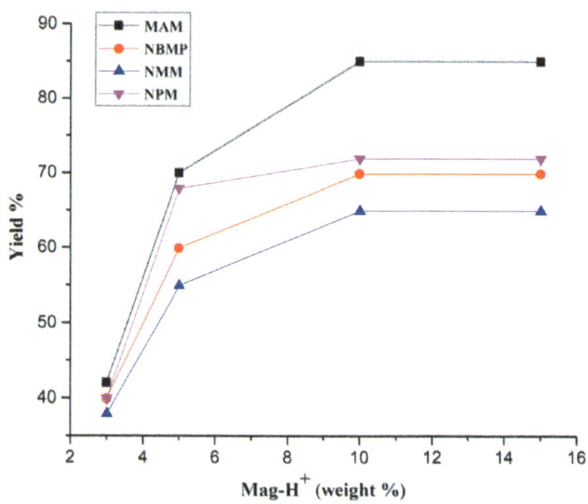

Fig. 12 Effect of catalyst (Mag H^+) on yield of monomers, time of reaction 2 hours; $T = 0–5$ °C

Table 2 Effect of catalyst amount (Mag Na^+) on monomers yield, time of reaction 24 hours; temperature 0–5 °C

Polymers	Mag Na^+ (weight %)	3	5	10	15
Poly(MAM)	Conversion %	10	35	45	20
Poly(NBMP)		5	30	40	59
Poly(NMM)		5	25	38	53
Poly(NPM)		15	38	49	62

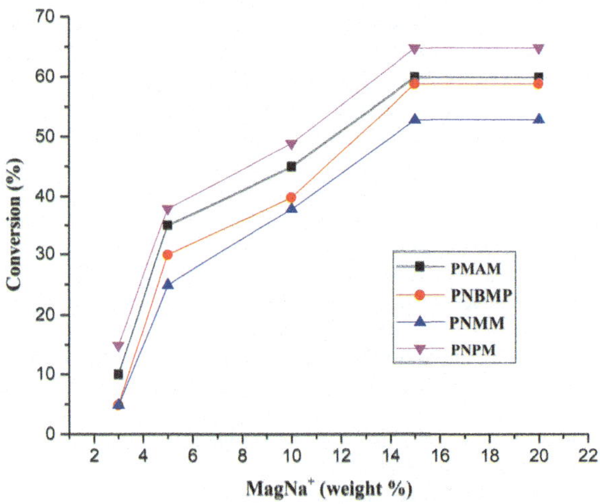

Fig. 13 Effect of catalyst (Mag Na^+) on yield of polymers, time of reaction 24 hours; $T = 0$–5 °C

number of active sites and explained by the fact that the increase for catalyst leads to the growth of the active initiator sites (Table 2 and Fig. 13).

4 Conclusion and Future Perspective

A new method based on an ecological and recyclable catalyst "Maghnite H^+" and "Maghnite Na^+", has been adopted to synthesis and to polymerize the new N-alkylmethacrylamide monomers: Methacrylamide, 1,4-bis(Methacryloyl)Piperazine, N-Methacryloyl morpholine and N-phenyl methacrylamide. All of this has been achieved through the activation of different primary aromatic and heterocyclic secondary amines: ammonia, aniline, piperazine and morpholine, respectively, using methacrylic anhydride, under mild chemo-selective and effective conditions in bulk without the production of toxic by-products. The anionic polymerization of the monomers is carried out for the first time using the

"Maghnite Na^{+}" as green initiator. The operating conditions of synthesis and polymerization of the monomers are carried out under mild conditions which follow the principles of "green chemistry,"; inexpensive, in bulk and at low temperature at 0 °C.

References

Anbarasan R, Arvind P, Dhanalakshmi V (2011) Synthesis and Characterization of Polymethacrylamide Clay Nanocomposites. J Appl Polym Sci 121:563–573

Belbachir M, Bensaoula (2006) A Composition and method for catalysis using bentonite. U.S Patent. 7, 094,823 B2

Boursier TG, Mosquet M, Rinaldi D, D'Agosto F (2015) Synthesis of Poly(N-acryloylmorpholine) macromonomers using RAFT and their copolymerization with Methacrylic acid for the design of graft copolymer additives for concrete. PolymerChemestry. https://doi.org/10.1039/C5PY01730K

Brooks AA, McCormick CL (2016) Mechanistic insights into temperature dependent trithiocarbonate chain-End degradation during the RAFT Polymerization of N-Arylmethacryamide. Macromolecule 49:465–474

Derkaoui S, Belbachir M (2019) A new approach to synthesis methacrylic monomers n-alkylmethacrylamide using an ecologic and friendly catalyst maghnite H^{+}. Journal of Fundamental Appl Sci 11(1):142–153

Derkaoui S, Kherroub DE, and Belbachir M (2019a) Green synthesis, anionic polymerization of 1,4-bis(methacryloyl)piperazine using Algerian clay as catalyst. Green Process Synth 8:611–621

Derkaoui S, Belbachir M, Haoue S, Zeggai FZ, Rahmouni A, Ayat M (2019b) Homopolymerization of methacrylamide by anionic process under effect of Maghnite-Na^{+} (Algerian MMT). J Organomet Chem 893:52–60

Donald EH (2005) Hand book of polymers synthesis, 2nd edn. Marcel Dekker, South Carolina

Dubois JLN, Lavignac N (2013) Poly (amidoamine)s synthesis, characterization and interaction with BSA. Polymer Chem. https://doi.org/10.1039/C3PY01121F

Dumesic JA, Huber GW, Boudart M (2008) Principles of Heterogeneous Catalysis 1–15

Gorman M, Ya Hua C, Hart A, Riehle MO, Urquhart AJ (2014) Poly (N-acryloylmorpholine): A simple hydrogel system for temporal and spatial control over cell adhesion. J Biomed Mater Res Part a 102:1809–1815

https://www.sigmaaldrich.com/materials-science/material-scienceproducts:Acrylicmonomers (2018). Accessed on 21 April 2018

Hülya E, Merve B, Vezir KM, Kayaman AN (2013) Synthesis of 4-Acryloylmorpholine-based Hydrogels and Investigation of their Drug Release Behaviors. J Brazilian Chem Soc 24(5):814–820

Jayasimha RG, Venkata NS, Reddy R (2003) Synthesis and characterization of poly (N-phenylmethacrylamide-co-methylmethacrylate) and reactivity ratios determination. J Appl Polym Sci 90:2179–2186

Jayasimha RG, Reddy M, Venkata NS (2014) Synthesis and characterization of functional acrylic copolymers of Acrylonitrile and N-Phenyl Mehacrylamide. Indian J Adv Chem Sci 2(4):264–270

Matsuda F (1986) Process for the production of Acrylamide and Methacrylamide, U.S Patent 4593123

Muserref O, Meltem C (2006) Polymethacrylamide/Na-Montmorillonite Nanocomposites Synthesized by Free-Radical Polymerization. Matter Lett 60:48–52

Sadegui M, Sarari S, Shahsavari H, Sadeghi H, Soleimani F (2013) Graft copolymerization methacrylamide onto pectin in homogeneous solution. Asian J Chem 8:4615–4618

Sauvik C, Piyali B, Luna P, Mahammad A, Bhaumik A (2017) MnAPO-5 as an efficient heterogeneous catalyst for selective liquid phase partial oxidation reactions. Royal Soc Chem. https://doi.org/10.1039/C7DT03897F

Shundrina IK, Bukhtoyarova AD, Russkikh V, Parkhomenko DA, Shelkovnikov V (2015) Synthesis and properties of novel random copolymers made from N-acryloylpiperazine-based monomers and fluoroalkylmethacrylate. Polym Bull 72:2783–2796

Wiley RH, Walter WE (1948) Preparation of Methacrylamide. J Org Chem 13:421–423

Chapter 9
Synthesis of Hydroxyapatite by Wet Method: Effect of Acid Solution Concentration on Powder Granulometry

Ouardia Zamoume and Rym Mammeri

Abstract Hydroxyapatite was obtained after heat treatment (600 °C) of the powder synthetized by neutralizing a solution of calcium hydroxide $Ca(OH)_2$ with phosphoric acid H_3PO_4 under atmospheric conditions and reduced time aging. The Ca/P molar ratio of the reagents was 2.0. The concentration of the solution of phosphoric acid influences the pH of the mixture and particle size distribution of solid products. When the $Ca(OH)_2$ solution was neutralized with concentrated phosphoric acid, the average particle size obtained is about 9.13 μm, whereas when the $Ca(OH)_2$ solution was neutralized with diluted solution of phosphoric acid, the average particle size is about 79.40 μm. Agglomerated particles were formed when phosphoric acid solution was used in diluted form. FTIR and DRX analyses have shown that the as-dried powder (at 100 °C) was a calcium carbonate–phosphate with low crystallinity, which after sintering at temperature higher than 600 °C leads to the formation of crystalline hydroxyapatite (HA) with CaO second phase in minor quantities. No decomposition of hydroxyapatite single phase was observed even at high temperature treatment of the powder (1100 °C).

Keywords Hydroxyapatite · Neutralization · Rich calcium solution · Thermal stability

1 Introduction

Hydroxyapatite $Ca_{10}(PO_4)_6(OH)_2$ (Ca/P = 1.67) is the most calcium phosphate used in biomedical field as biomaterial, adsorbent and catalytic support. As biomaterial, hydroxyapatite (HA) is bioactive, biocompatible, and osteoconductive but possesses a very low solubility in vivo. To overcome this problem, hydoxyapatite is sometimes associated with tricalcium phosphate (βTCP), which has good solubility. This biphasic material (HA/βTCP) combines the properties of HA and βTCP, and thus,

O. Zamoume (✉) · R. Mammeri
Faculty of Sciences, Department of Chemistry, University M'hammed Bouguara Boumerdes, Boumerdes, Algeria
e-mail: zamouar@yahoo.fr

D. Trache et al. (eds.), *Materials Research and Applications*,
Materials Horizons: From Nature to Nanomaterials,
https://doi.org/10.1007/978-981-15-9223-2_9

its resorption properties in vivo are improved. Hydroxyapatite biomaterial can be obtained by heat treatment of bones, which are composed by carbonated hydroxyapatite or by synthetic routes. Due to its low solubility, hydroxyapatite has a great stability in water, which is an important parameter for adsorbent materials. Hydroxyapatite crystallizes in hexagonal system ($P_{6/3\ m}$ space group), with a lattice parameters: $a = 9.37$ A° and $c = 6.86$ A°. Its structure is composed of PO_4^{3-}, Ca^{2+}, and OH^- ions, which can be substituted by monovalent, divalent, and trivalent ions. This ionic substitution capacity makes these materials very efficient in the field of wastewater treatment, removal of dyes, and chromatography (Maya et al. 2020). Hydroxyapatite possesses very good retention capacities for heavy metal ions such as Pb^{2+}, Cd^{2+}, Cr^{3+}.... (Haifeng et al. 2019; Michele et al. 2019). The retention capacities of heavy metals on surface of hydroxyapatite depend on the surface affinity of this material. Thus, the phosphate and hydroxyl groups in hydroxyapatite surface are hard Lewis bases and have good affinity for hard Lewis acid as lead cations. The ionic radii of the cation are another parameter on which depends the retention capacity. Cations with ionic radii bigger than Ca^{2+} are more likely to be incorporated in HA structure. In addition, hydroxyapatite with low crystallinity that corresponds to high surface area is more beneficial to remove heavy metal ions (Doan et al. 2012a, b; Sangeetha et al. 2018; Maya et al. 2020).

Hydroxyapatite can be prepared by many synthetic routes, which include dry methods, high-temperature processes, wet methods, etc. (Mehdi et al. 2013). Among these techniques, the most widely used one is the wet chemical route. The advantages of this method are that different particle sizes and morphologies can be obtained by varying the preparation conditions (Peipei et al. 2010; Elhammari et al. 2007). The common reagents used to synthesize hydroxyapatite are calcium nitrate and calcium chloride as calcium sources and ammonium orthophosphate salts as orthophosphate sources because of their high solubility in aqueous solutions. However, these reagents result in non-used ions as NO_3^-, Cl^-, K^+, and Na^+, which require several washing in order to eliminate them. Calcium hydroxyde and orthophosphoric acid seem to be good reagents for the synthesis of hydroxyapatite by an acid–base neutralization (Florin et al. 2017). This process results in any non-used ion and the reagents are available at low cost (Doan et al. 2014). Hydroxyapatite used as adsorbent is stoichiometric ($Ca/P = 1.67$). According to previous work, in order to obtain a stoichiometric hydroxyapatite, the molar ratio of the reagents must be equal to 1.67, the temperature of the reaction must be high (90 °C), and the aging time of the obtained precipitate must be long (above 24 h).

The objective of this work is to synthesize hydroxyapatite powder by acid–base neutralization at low temperature, reduced aging time, and different acid solution concentration. In these conditions and in order to obtain stoichiometric hydroxyapatite with a $Ca/P = 1.67$, the reagents were mixed in a molar ratio of 2 ($Ca/P = 2$). Based on the results of the analyses, the effect of phosphoric acid solution concentration on the pH of the suspension, the particle size distribution, and thermal stability of the Ca-HA are studied.

2 Materials and Methods

2.1 Reagents

The reagents were $Ca(OH)_2$ and H_3PO_4 (85%) in water. All of them were of high purity (99%) and purchased from Prolabo.

2.2 Synthesis Methods

2.2.1 Powder n°1 (P1)

A solution of $Ca(OH)_2$ was obtained by dissolution of 0.1 mol of $Ca(OH)_2$ in 500 mL of distilled water. The solution was stirred with a magnetic stirrer at 400 rpm and maintained at room temperature (23 °C). A volume of 3.2 mL of H_3PO_4 (85% wt., in water), at room temperature, was then added drop wise (1 mL/min) into the $Ca(OH)_2$ solution with a graduated burette. After the total addition of H_3PO_4, the mixture was continuously agitated for an additional one hour at the same temperature. The pH of the acid–base solution was monitored with pH meter. The precipitate was separated, washed several time with distilled water, filtered then dried à 100 °C during 24 h.

2.2.2 Powder n°2 (P2)

A solution of $Ca(OH)_2$ was obtained by dissolution of 0.1 mol of $Ca(OH)_2$ in 250 mL of distilled water. The solution was stirred with a magnetic stirrer at 400 rpm and maintained at room temperature (23 °C). 3.2 mL of H_3PO_4 solution was introduced in 250 mL of distilled water and then added drop wise (12.5 ml/min) into the $Ca(OH)_2$solution. The total time addition of H_3PO_4 solution was 20 min. The obtained precipitate was continuously agitated for an additional one hour at the same temperature. The obtained suspension was washed several times with distilled water, filtered then dried à 100 °C during 24 h.

In order to study their thermal stability, the as-dried powders are sintered at 600, 800, 1000, and 1100 °C in a furnace under air atmosphere, with a heating rate of 5 °C/min, kept for 1 h, and then the sample was cooled down to room temperature.

2.3 *Characterization Methods*

2.3.1 Powder X-ray Diffraction (XRD)

The phase composition of the as-synthesized and calcined powders was analyzed by X-ray diffraction with Philips XPERT PRO diffractometer using CuKα radiation ($\lambda = 1.54056$ Å) at the X-ray tube voltage 45 kV and a current of 40 mA. The X-ray data were collected over the 2θ range of 20–70° at a step size of 0.0167°.The average crystallite size of the precipitates was estimated from the Debye–Scherrer equation:

$$D_{hkl} = \frac{K\lambda}{ß \cos\theta}$$

where D_{hkl} is the size (in nm), ß is the full width at half maximum of the peak at half of the maximum intensity, FWHM (in degree), λ is the diffraction wavelength of X-rays (1.5418 Å), θ is the diffraction angle for each reflection, and *K* the shape factor equals to 0.94.

2.3.2 Particles Size Distribution

A particle size analyzer (Malvern Mastersizer 2000) was used to determine the size distribution of the powder samples.

2.3.3 FTIR Spectroscopy

The typical groups of the synthesized and calcined powders were identified by Fourier-transform infrared spectroscopy (FTIR-4100//ATR ESPECAC). The measurements were obtained at 400–4000 cm^{-1} with a resolution of 0.4 cm^{-1}.

2.3.4 Elementary Analysis

The elemental analysis was performed by fluorescence X spectroscopy (Magix PRO, PW 2540 vrc).

3 Results and Discussions

As shown in Fig. 1, the pH of the mixture varies slowly with time when the phosphoric acid solution used is in concentrated form (pH ≈ 12.3), whereas when the phosphoric acid solution used for neutralization is diluted, the pH of the mixture was slowed

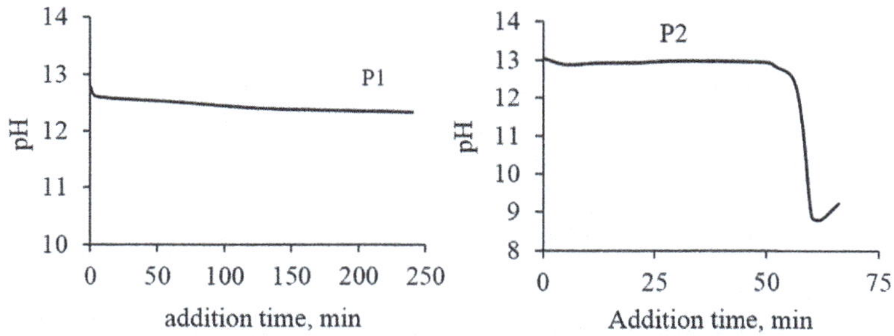

Fig. 1 Evolution of pH with time addition of phosphoric acid to $Ca(OH)_2$ solution

down to 8.8. These results indicate that the formation mechanism of the powders P1 and P2 is different. Indeed, at a pH of 12.3, the predominant species in solution are PO_4^{3-} ions, while at a pH of 8.8, there are both PO_4^{3-} and HPO_4^{2-}.According to the results showed in Table. 1, the Ca/P molar ratio of powder P2 (Ca/P = 1.20) is higher when compared to powder P1 (Ca/P = 1.15), which can be due to a more substitution of PO_4^{3-} groups by CO_3^{2-}.

The X-ray diffraction patterns of the as-dried powders are similar as shown in Fig. 2. All the peaks are identified and correspond to HA phase (ICDD N°96-900-2217). The broad and weak peaks diffraction indicates that the powder is a poorly crystallized apatitic phosphate with small crystallite size. This is confirmed by the

Table 1 Ca/P molar ration of as-dried powders P1 and P2 at 100 °C and 24 h

Powder	Ca (wt %)	P (wt%)	Ca/P	Ca/P (molar ration)
P1	48.93	32.88	1.48	1.15
P2	52.74	33.81	1.55	1.20

Fig. 2 XRD patterns of as-dried powder P1 and P2 at 100 °C for 24 h

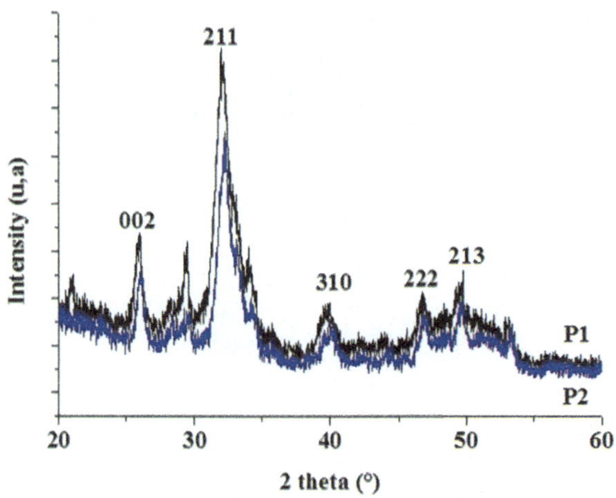

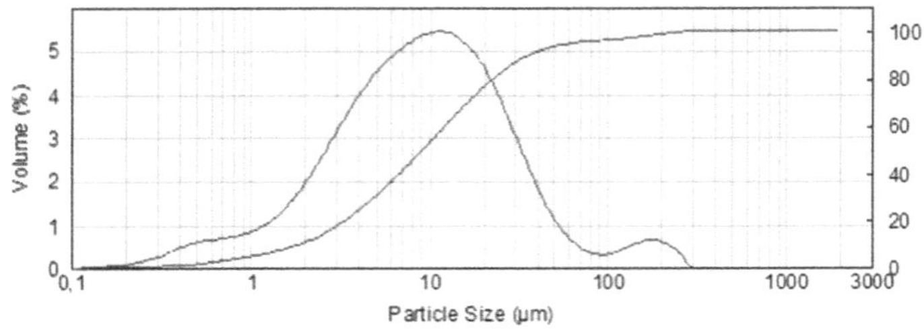

Fig. 3 Particle size distribution of as-dried powder P1 at 100 °C and 24 h

crystallite sizes estimated from the Debye–Scherrer equation, which are about 8.77 and 10.20 nm for powder P1 and P2, respectively. The low crystallinity of the formed apatitic phosphate can be attributed to the low synthesis temperature used (Elhammari et al. 2007; Elena et al. 2004), which lead to fast precipitation that prevented the crystallization. Moreover, the presence of B-carbonate in anapatite lattice causes a decrease in crystallinity (Elena 2004). Souzaa et al. (2019) showed that the amorphous calcium phosphate (ACP) is very reactive, thus adsorbs atmospheric CO_2 and forms a carbonated apatite.

The particle size distribution of the powder P1 is homogeneous (Fig. 3). 90% of particles have smaller size than 34.85 µm (d_{90}), and 50% of them present smaller size than 9.13 µm (d_{50}). Visually and during synthesis process, the precipitate P2 formed has spherical agglomerates compared to the precipitate P1, which has a very fine appearance. The formation of agglomerates in the case of powder P2 was confirmed by particle size analyses (Fig. 4), which showed that 90% of particles have smaller size than 350.23 µm (d_{90}), and 50% (d_{50}) of them present smaller than 79.40 µm. Our results are not really in agreement with those obtained by Azade and Suat (2018), who have shown that slow or rapid addition of H_3PO_4 to $Ca(OH)_2$ solution has no influence on the particles size distribution.

The FTIR spectra of the as-dried powders P1 and P2 are shown in Fig. 5. All the bands detected for powder P1 are characteristics of functional groups of an apatitic phase (Peipei et al. 2004; Heughebaert et al. 1977; Blumenthal et al. 1972). Intense absorption bands of PO_4^{3-} groups have been found at 1045 cm^{-1} (ν_3 asymmetric stretching mode), 604 and 571 cm^{-1} (ν_4 bending mode). The ν_1 asymmetric

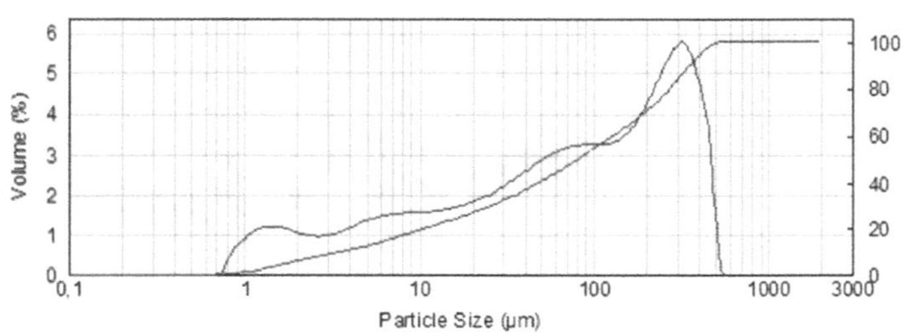

Fig. 4 Particle size distribution of as-dried powder P2 at 100 °C end 24 h

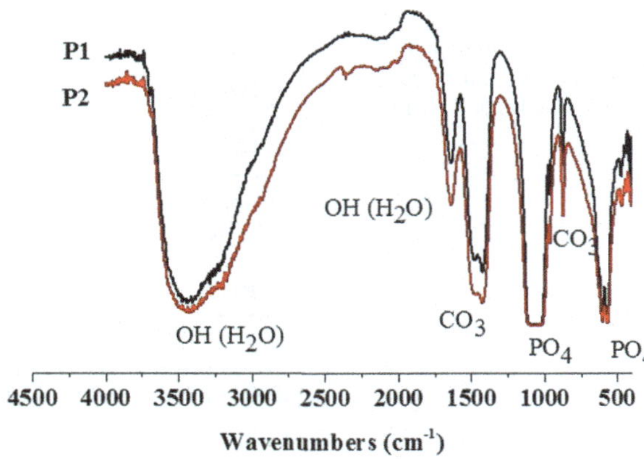

Fig. 5 FTIR spectra of as-dried powder P1 and P2 at 100 °C during 24 h

stretching mode at 960 cm^{-1} and ν_2 at 470 cm^{-1} is not observed in the sample P1. The absorption band at 1422 cm^{-1} is attributed to CO_3^{2-} groups (ν_3 stretching mode) which resulted from partial substitution of PO_4^{3-} groups by carbonate ions (B-type substitution). The presence of CO_3^{2-} groups in the lattice of apatitic phase formed is resulted from atmospheric CO_2 contamination during the precipitation process. The reactor exposed to air, medium alkaline reaction (pH > 8), and low precipitation temperature, which increases the solubility of CO_2 are the main factors that caused this contamination.

The reaction of atmospheric CO_2 dissolution in an alkaline medium is:

$$CO_2(g) + 2OH^-(aq) \rightarrow CO_3^{2-}(aq) + H_2O \quad (1)$$

The weak band at 874 cm^{-1} is attributed to CO_3^{2-} (A-type substitution) or HPO_4^{2-} groups (Moghimian et al. 2012). The high intensity bands at 3441 and 1641 cm^{-1} are attributed to absorbed water molecules. The absence of the bands at 3750 and 630 cm^{-1} characteristic of OH^- groups of hydroxyapatite is assigned to the substitution of these groups by carbonate ions. Although, Gema et al. (2018) have found the same result and indicate that the presence of strong carbonate bands in the range of 1400–1460 cm^{-1} and the band at 875 cm^{-1} implies the presence of AB type carbonated substituted apatites ($Ca_{10-x}(PO_4)_{6-x}(CO_3)_x(OH)_{2-x\text{-}2y}(CO_3)y$).

The FTIR spectrum of the as-dried powder P2 is similar to that of powder P1. However, two supplementary bands have been identified at 960 cm^{-1} and 470 cm^{-1}, which are, respectively, attributed to $\nu_1$symmetric stretching mode and ν_2 bending mode of PO_4^{3-}. The broad band at 1045 cm^{-1} corresponding to PO_4^{3-} groups indicates that the synthetized apatite is poorly crystalline, which confirms the XRD results.

Infrared spectra of the powder P1 calcined at 600, 800, 1000, and 1100 °C are reported in Fig. 6. The absorption bands of OH groups of hydroxyapatite appeared at about 600 °C and all other peaks in the spectra were characteristics of hydroxyapatite.

Fig.6 FTIR spectra of the powder P1: as-dried and calcined at 600, 800, 1000, 1100 °C for 1 h at 5 °C/min

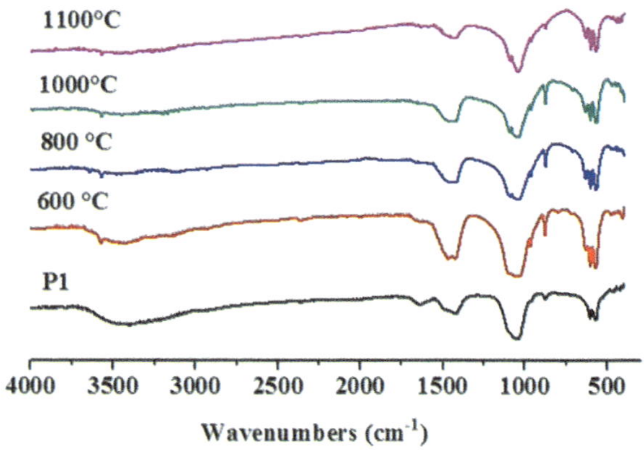

The sharp peaks of PO_4^{3-} groups, when the calcination temperature increases, indicate that the crystallinity of hydroxyapatite increases. No decomposition of hydroxyapatite was observed even at 1100 °C, which confirms their stability (presence of OH groups at 3570 cm^{-1}) (Kang et al. 2019).The same result was obtained with the calcined powder P2 (Fig. 7).The diminution of the band at 1450 cm^{-1} is assigned to the decomposition of carbonate ions to form CO_2 (g).

The XRD spectra of the calcined powder P1 at 600, 800, 1000, and 1100 °C are reported in Fig. 8. In comparison with the as-dried powder, the crystallinity of the powders calcined at 800, 1000, and 1100 °C was raised and all the diffractions peaks correspond to the hydroxyapatite phase (ICDD N°96-900-2217). Low-intensity peaks at 37.44° and 54.55° in the DRX paterns of powder P1 calcined at 800, 1000 and 1100 °C are attibuted to CaO phase, which corresponds to the remained Ca $(OH)_2$ of the reaction or to the no total decomposition of carbonate ions present in the lattice of hydroxyapatite powder (Landi et al. 2004).

Fig. 7 FTIR spectra of the powder P2: as-dried and calcined at 600, 800, 1100 °C for 1 h at 5 °C/min

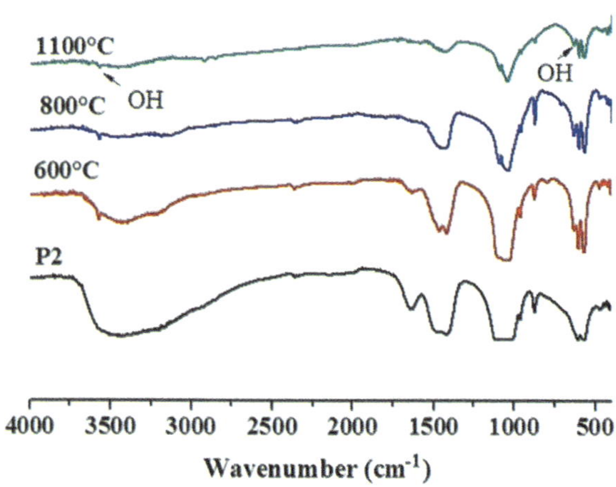

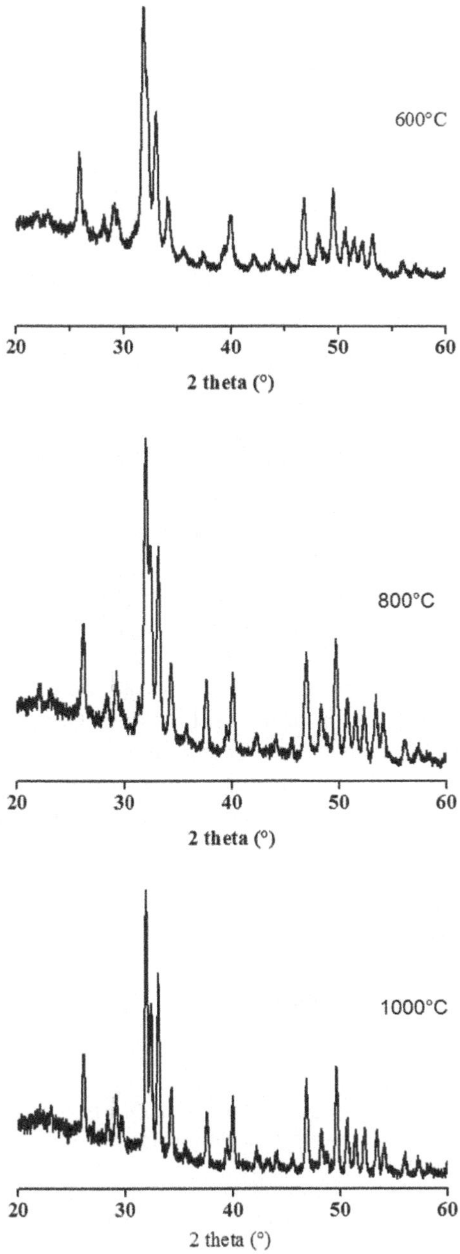

Fig. 8 DRX patterns of powder P1: as-dried and calcined at 600, 800, 1000, 1100 °C for 1 h at 5 °C/min

Fig. 8 (continued)

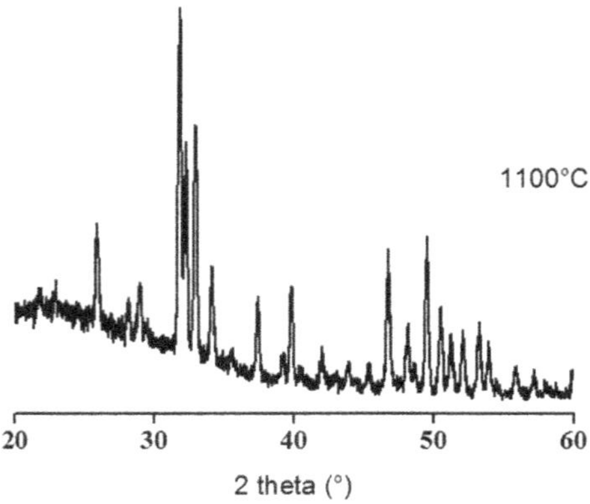

4 Conclusion and Future Perspective

Acid–base neutralization is an easy method to prepare hydroxyapatite powder by using $Ca(OH)_2$ and H_3PO_4 as precursors. The results of this work have shown that using rich calcium solution allows preparing hydroxyapatite at room temperature (23 °C) at reduced aging time (1 h) by the calcination of the obtained precipitate at 600 °C for 1 h. The particle size of the obtained powder depends on the concentration of the acidic solution. The evolution of pH of synthesis with time depends on the concentration of the acid solution as well. The obtained hydroxyapatite powder is stable even at high temperature of treatment. Depending on the application field of the synthesized powder, other properties are required to study in the future as the morphology of the particles, surface area, and porosity of the powder.

References

Azade Y, Suat Y (2018) Wet chemical precipitation synthesis of hydroxyapatite (HA) powders. Ceram Int 44:9703–9710

Blumenthal C, Posner AS, Holms JM (1972) Effect of preparation conditions on the propperties and transformation of amorphous calcium phosphate. Mat Res Bull 7:1181–1190

Doan PM, Haroun S, Ange N, Patrick S (2012a) Apatitic calcium phosphate: Synthesis, characterization and reactivity in the removal of lead (II) from aqueos solution. Chem Eng J 180–190

Doan PM, Nathalie L, Haroun S, Ange N, Sharrock P (2012b) Synthesis of calcium hydroxyapatite from calcium carbonate and different orthophosphate sources: a comparative study. Mater Sci Eng B 177:1080–1089

Doan PM, Sébastien R, Patrick S, Haroun S, Nathalie L, Ngoc DT, Mohamed R, Ange N (2014) Hydroxyapatite starting from calcium carbonate and orthophosphoric acid: synthesis, characterization, and applications. J Mater Sci 49(12):4261–4269

Elena L, Anna T, Giancarlo C, Lucia V, Monica S (2004) Influence of synthesis andsintering parameters on the characteristics of carbonate apatite. Biomaterials 25:1763–1770

Elhammari L, Merroun H, Coradin T, Cassaignon S, Laghzizil A, Saoiabi A (2007) Mesoporous hydroxyapatites prepared in ethanol–water media: structure and surface properties. Mater Chem Phys 104:448–453

Florin M, Aura-Catalina M, Catalina AD, Andreea M, Dan B, Andrei B, Stefan IV, Marian M, Vijay KT, Lucian TC (2017) Facile synthesis and characterization of hydroxyapatite particles for high value nanocomposites and biomaterials. Vacuum 146:614–622

Gema G, Cesar CV, Luis JB, Damarys S, Lisbeth L, Jose IC, Camilo JD, Luis L (2018) Effect of carbonates on hydroxyapatite self-activated photoluminescence response. J Lumin 195:385–395

Haifeng G, Changhua J, Zhenxuan X, Pingying L, Zhaofu F, Jia Z (2019) Synthesis of bitter gourd-shaped nanoscaled hydroxyapatite and its adsorption property for heavy metal ions. Mater Lett 241:176–179

Heughebaert J-C, Montel G (1977) Etude de l'évolution de l'orthophosphate tricalcique non cristalin en phosphate apatitique à la faveur d'une réaction chimique, à température ordinaire. Revue de physique appliquée. Tome 12

Kang S, Seo JT, Park S-H, Jung Y, Lee CY, Park J-W (2019) Qualitative analysis on crystal growth of synthetic hydroxyapatite influenced by fluoride concentration. Arch Oral Biol 104:52–59

Landi E, Tampieri A, Celotti G, Vichi L, Sandri M (2004) Influence of synthesis and sintering parameters on the characteristics of carbonate apatite. Biomaterials 25:1763–1770

Maya I, Madona L, Jean-M G, Jean-François L (2020) Hydroxyapatite, a multifunctional material for air, water and soil pollution control: a review materials. J Hazard 383:121–139

Mehdi SS, Mohammad-TK, Ehsan DK, Ahmad (2013) J Synthesis methods for nanosized hydroxyapatite with diverse structures. Acta Biomaterialia 9:7591–7621

Michele F, Sebastiano C, Marco S, Claudio E, Paolo C, Antonella G (2019) In-depth study of the mechanism of heavy metal trapping on the surface of hydroxyapatite. Appl Surf Sci 475:397–409

Moghimian P, Najafi A, Afshar S, Javadpour CJ (2012) Effect of low temperature on formation mechanism of calcium phosphate nano powder via precipitation method. Adv Powder Technol 23:744–751

Peipei W, Caihong L, Haiyan G, Xuerong J, Hongqiang W, Kaixing L (2010) Effects of synthesis conditions on the morphology of hydroxyapatite nanoparticles produced by wet chemical process. Powder Technol 203(2):315–321

Sangeetha K, Vidhya G, Vasugi G, Girija EK (2018) Lead and cademium removal from single and binary metal ion solution by novel hydroxyapatite/alginate/gelatin nanocomposites. J Environ Chem Eng 6(1):1118–1126

Souzaa FS, Matosb MJS, Galvãoa BRL, Arapiracaa AFC, Silvaa SN, Pinheiroa IP (2019) Adsorption of CO_2 on biphasic and amorphous calcium phosphates: an experimental and theoretical analysis. Chem Phys Lett714:143–148

Chapter 10
Bioactive Compounds and Antioxidant Activity of Ethanolic Extracts from the Leaves of *Tetraclinis Articulate* (Vahl) Masters

Bensaber Bensebia, Ouahida Bensebia, Ghania Belloumi, and Soumia Kouadri Mostefai

Abstract *Tetraclinis articulata* (Vahl) masters is an endemic plant from the south-western Mediterranean. This species was widely used in Algeria for its medicinal virtues. In this work, dried leaves of *T. articulata* have been analyzed for their polyphenol and flavonoid composition, and antioxidant activity. The influence of extraction time (30–240 min) and concentration of ethanol (40, 60 and 80%) on the extraction yield of phenolic and flavonoid compounds of extracts were investigated. The extraction yield of *T. articulata* ranged from 10.03 to 40.45% and the maximum extraction yield was obtained at 240 min with 60% of ethanol. The evaluation of total phenolic compounds was realized by the method of the Folin–Ciocalteu and the estimation of total flavonoid compounds was determined using a method of aluminum chloride. In addition, the antioxidant activity was evaluated by 2,2-diphenyl-1-picrylhydrazyl (DPPH) radical scavenging activity method. The highest extraction yield of total phenolic (95.28 mg GAE/g) and flavonoid (36.70 mg QE/g) compounds was obtained with 60% ethanol concentration for an extraction time of 210 min and 120 min, respectively. The best antioxidant activity was obtained from the *T. articulata* leaves extracted for an extraction time of 120 min with 60% ethanol concentration. The results demonstrated that an ethanol concentration of 60% was better to obtain extracts of *T. articulata* with greater amount of bioactive compounds and antioxidant activity.

B. Bensebia (✉)
Laboratoire "Chimie Verte-Eau", Faculté de Technologie,
Université Hassiba Benbouali, B.P. 151, 02000 Chlef, Algeria
e-mail: bensebiab@yahoo.fr

O. Bensebia
Laboratoire des Sciences du Génie des Procédés Industriels (LSGPI), Faculté de Génie Mécanique et de Génie des Procédés, Université des Sciences et de la Technologie-Houari Boumediene, BP 32, 16111 El Alia-Bab Ezzouar, Algeria
e-mail: bensebiab@yahoo.fr

G. Belloumi · S. K. Mostefai
Département Génie des Procédés, Faculté de Technologie,
Université Hassiba Benbouali, B.P. 151, 02000 Chlef, Algeria

D. Trache et al. (eds.), *Materials Research and Applications*,
Materials Horizons: From Nature to Nanomaterials,
https://doi.org/10.1007/978-981-15-9223-2_10

Keywords Polyphenol · Flavonoid · Antioxidant activity · *Tetraclinis articulate*

1 Introduction

Tetraclinisarticulate (Thuya) was a resinous species of the Cupressaceae family. This specie was widely distributed in North Africa region (Algeria, Morocco, and Tunisia) and a small stand in other countries such as Spain Morte and Honrubia (1996) and Letruch-Belarouci (1991) reported that thuya covered an area of 160,000 ha in Algeria in 1991, but this area has sharply declined due to anthropogenic action. According to the report established by Sahi et al. (2016), the surface area of thuya decreased to 124,000 ha in 2016. However, these numbers should be taken with precaution because the estimation of the areas of spontaneous plants is delicate and difficult to identify, especially if the sector is unstructured and poorly organized.In Algeria, *Tetraclinisarticulate* occupies the fourth place in terms of area, after Aleppo pine, green oak, and cork oak (Sahi et al. 2016). This species was used by the rural population in traditional medicine due to its pharmacological attributes. Actually, this endemic plant has been studied as a source among natural bioactive compounds. Plants synthesize a diverse array of secondary metabolites (phytochemicals); among these, the most important and abundant ones were phenolic, flavonoid, and tannin compounds (Manach et al. 2004). These compounds had several properties, such as antimicrobial and antioxidant activity, anti-inflammatory, antiallergenic, antiartherogenic, antithrombotic, cardioprotective, and vasodilatory effects (Middletonet al. 2000). The composition of the essential oil obtained from different organs of *Tetraclinis articulata* was investigated by several authors (Boussaïdet al. 2015; Chikhoune et al. 2013; Abi-Ayad et al. 2013; Djouahri et al. 2013). Many studies reported the antimicrobial (Ben-Ghnyaet al. 2016; Bahriet al. 2015; Bourkhiss et al. 2007), antioxidant (Djouahri et al. 2013; Bourkhiss et al. 2007), antitumor, anti-inflammatory (El Jemli et al. 2017; Djouahri et al. 2014), antibacterial (Bourkhiss et al. 2007; Djouahri et al. 2014; Sadiki et al. 2018; Abi-Ayada et al. 2011), and antifungal (Abi-Ayad et al. 2013; Tékaya-Karaoui et al. 2007) activities of the essential oil of *Tetraclinis articulata.* The few studies carried out on the extracts of *Tetraclinis articulate* have shown that this species is a rich source of bioactive phytochemicals, such as the polyphenol compounds. The phytochemical analysis of extracts, obtained from twigs of *T. articulata* using ethyl acetate and methanol as solvents, has revealed the presence of a number of polyphenolic compounds such as amentoflavone, myricitrin, and quercitrin (Zidane et al. 2014). The authors concluded that these compounds were responsible for the vasorelaxant activity of this species. In another work, Rached et al. (2018) evaluated several phytochemical properties (antioxidant, antibacterial, anti-inflammatory, and cytotoxic) of the aqueous extracts of the Algerian *tetraclinis articulate* leaves. Their results showed that the flavonoid compounds had a significant contribution in the bioactive properties of this species. The study realized by Herzi et al. showed that the best antioxidant activity of *T. articulate* leaves was obtained with a supercritical CO_2 extraction extract (Herziet al. 2013b). In addition, other biological activities have been identified in extracts of thuya. Dane et al. (2015) evaluated the insecticidal and antifungal activities of *T. articulate*. These authors concluded that these activities could be connected to flavonoids compounds such as quercetin-3-O-rhamnoside (31.70%). Ababsa et al. (2019) reported that butanol extracts of *T. articulate* leaves had an antidiarrhoeal and anti-ulcerative activities. The isolation of bioactive compounds was frequently carried out by solvent extraction technique. Nevertheless, the extract yields, the content of phenolic and flavonoid compounds, and the antioxidant activity of the plant were dependent on the operating conditions such as the method of extraction, nature of solvent, temperature, and extraction time. The choice of solvent extraction was very important for the recovery of bioactive products, with varied chemical characteristics and polarities. The polyphenol compounds were often extracted from a plant by the use of polar solvents. Peschel et al. (2006) reported that aqueous mixtures containing ethanol, methanol, acetone, and ethyl acetate were the most convenient solvents to the recovery of the polyphenols compounds. In this context, the present research aim was to determine the influence of the operating conditions of the extraction on the recovery of the polyphenol compounds and the antioxidant activity of Algerian *tetraclinis articulate* (Vahl) masters leaves.

2 Material and Methods

2.1 Plant Material

The leaves of *Tetraclinisarticulate* (Vahl) masters were gathered from Ain Merane situated in Chlef in the western center of Algeria. These leaves were dried for ten days and were grounded and sieved. The moisture content of *Tetraclinis articulate* leaves was 0.536 kg/kg (wet basis) and the mean average size was 573 μm.

2.2 Preparation of Extracts

Dried and ground leaves (10 g) were extracted using maceration at room temperature. Aqueous ethanol was chosen as solvent with a solid/liquid ratio 1/10 (w/v),for different times (30–240 min) and with different ethanol concentration (40, 60, 80%).The mixture was filtered and evaporated under reduced pressure using a rotary evaporator at 40 °C until a semi-solid sticky mass was obtained. Crude extracts were used for total phenolic, flavonoid, and antioxidant activity determinations.

2.3 Estimation of Total Phenol content (TPC)

The method of McDonald et al. (2001) was applied to estimate the total phenolic content (TPC) of the leaf extracts. 0.5 mL of each extract was mixed with 2.5 mL of Folin–Ciocalteu reagent (diluted 10 times in distilled water). After 4 min of incubation at room temperature, 2 mL of sodium bicarbonate solution (7.5%) was added to the mixture and then incubated in the dark for 30 min. After oxidation, the complex formed was measured at 765 nm using UV–visible spectrophotometer. The standard curve [$Y = 3.327 \times (R^2 = 0.999)$] was prepared using gallic acid, in the range of 0–200 mg/L. The concentrations of total phenol content (TPC) expressed as gallic acid equivalent (GAE) in % w/w of the extracts were calculated using the standard curve. All determinations were carried out in triplicate.

2.4 Estimation of Total Flavonoid Content (TFC)

The total flavonoids content (TFC) in extracts was estimated by the method described by Quettier-Deleuet al. (2000). 1.5 mL of extract sample was separately mixed with 1.5 mL of 2% aluminum chloride methanolic solution. After incubation at room temperature for 10 min, the absorbance of the extract was measured at 430 nm. Quercetin (0–40 mg/L) was used as standard to make the calibration curve. The calibration curve equation for quercetin was: $Y = 0.031 + 0.100 \times (R^2 = 0.995)$. The total flavonoid content was expressed in milligram as quercetin equivalent per gram of the extracts.

2.5 Estimation of Antioxidantactivity

The antioxidant activity was evaluated by the 1,1-diphenyl-2-picrylhydrazilradical (DPPH) free radical test according to Sánchez-Moreno et al. (1998). A volume of 50 μL of each extract dissolved in methanol at different concentrations (from 0.1 to 1 mg/mL) was added to 1.95 mL of the methanolic solution of DPPH (0.025 g/L). In parallel, a negative control is prepared by mixing 50 μL of methanol with 1.95 mL of the methanolic solution of DPPH. After incubation for 30 min in the dark at room temperature, the absorbance was measured at 517 nm, and the percentage of radical scavenging activity was calculated according to the equation $[1—(A_{sample}/A_{negatifcontrol})] \times 100$, where A_{sample} is the absorbance of the sample compounds and $A_{negatif\ control}$ is the absorbance of the control reaction. The antioxidant activity of extract expressed as the concentration for which 50% of DPPH disappears (efficient concentration $EC_{50}C$) was estimated from the graph of inhibition percentage plotted against extract concentration.

Table 1 Effect of extraction time and concentration of ethanol on the extraction yield of *Tetraclinis articulate* leaves

Time (min)	30	60	90	120	150	180	210	240
% Ethanol	Yield (%)							
80	20.90	26.16	30.33	30.98	34.06	37.42	39.28	40.18
60	29.13	31.47	32.97	33.76	35.97	39.90	40.25	40.45
40	10.03	18.70	19.45	23.20	23.24	24.38	31.51	37.55

3 Results and Discussion

3.1 *Effect of Extraction Time and Concentration of Ethanol on the Extraction Yield*

Extraction time is a critical parameter during extraction from plant matrix.Table1shows that the extraction yield of *T. articulata* ranged from 10.03 to 40.45%. The highest yield was obtained by the extraction method with 60% ethanol content and an extraction time of 240 min. It can be seen from Table 1 that the extraction yield increases as the time increases, and the highest extraction yield was obtained with 60% ethanol concentration. Djouahri et al. (2013) reported that the best extraction yield (35.9%) was obtained by cold maceration using methanol in Soxhlet extractor. However, a lower yield (17.64%) has been reported by Dane et al. (2015). This difference may be due to the type of plant used. In fact, these authors used in their study a cultivated plant and not a spontaneous one. In addition, the required time to obtain this yield was 48 h, which is quite high compared to the one optimized in our study. Vuong et al. (2011) reported that a long time may not be appropriate for all types of bioactive compounds, since it increased the oxidation and the degradation of these compounds.

3.2 *Effect of Extraction Time and Concentration on TPC*

The results displayed in Fig. 1 showed that the content of the phenolic compounds as extracts ranged from 37.09 to 95.29 mgGAE/g. These results exhibited that the content of phenolic compounds was dependent on the time extraction and the ethanol concentration.

The maximum extraction yield regarding TPC (95.29mgGAE/g) was achieved with 60% ethanol concentration for an extraction time of 210 min. Similar value (93.1 mg GAE/g) has been obtained using ethyl acetate fraction. However, butanol fraction and crude aqueous extract presented lower values (43.87 ± 0.02 and 21.2 ± 0.2 mg/g, respectively) compared with the ethyl acetate fraction (2018). Our results were in accordance with a previous work carried out by Zhang et al. (2007). These

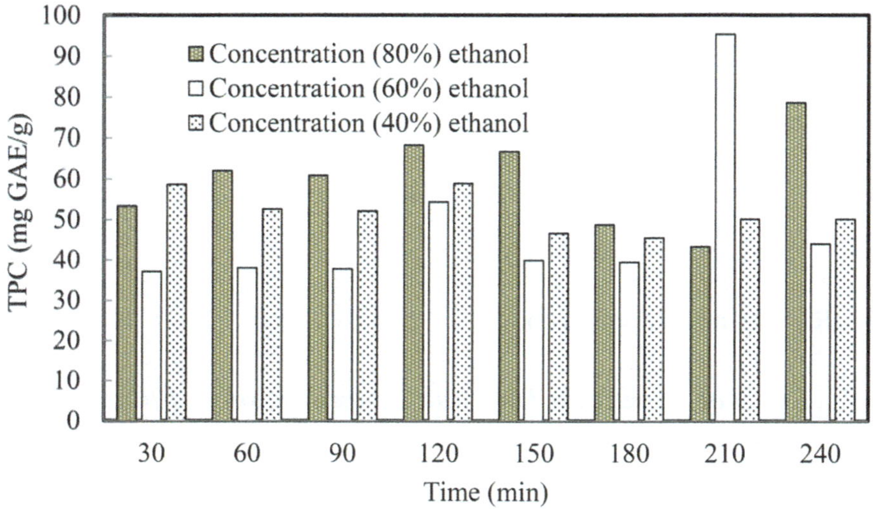

Fig. 1 Total phenol content (TPC) of *Tetraclinis articulate* extracted with different ethanol concentration (40, 60 and 80%) and at different extraction time (30–240 min)

authors suggested that anaqueous solution containing ethanol gave better results compared to water or pure ethanol in the extraction of phenolic compounds (Zhang et al. 2007).

The study carried out by Wang et al. (2004), dealing with the influence of aqueous solutions of ethanol on the quantity of extracted phenolic acids, showed that an optimal extraction yield was obtained with 30–60% solutions of ethanol. Gong et al. (2012) reported that the denaturation of protein in addition to the deceleration of the dissolution of polyphenol could be caused by a high concentration of ethanol. From the results depicted in Fig. 1, the extraction time did not have a very important impact on the levels of TPC with 40% ethanol concentration, where the values ranged from 45.43 to 58.89 mg GAE/g. This was not the case for extraction with 60 and 80% ethanol concentration (33.96–95.28 and 43.22–78.56 mg GAE/g, respectively). The highest yields of TPC were obtained with extraction time of 210 min, and thus, an extended contact time of the plant matrix with the solvent was required to the suitable recovery of the polyphenol compounds.

The results obtained by Djouahri et al. (2013) showed that the methanolic extracts of Algerian *T. articulate* were rich in phenolic compounds, and the extracts obtained by heat maceration contained a significant concentration of these compounds (156.63 mg GAE/g) compared to extracts obtained by cold maceration (133.5 mg GAE/g). Therefore, the quantity of total phenolic compounds was better when the extraction was done under reflux and with hot solvents. Rached et al. (2010) have studied the effect of different times for the harvesting of *T. articulate*. They reported that the highest (206.187 ± 16.612 mg GAE/g) amount of total phenolics (TP) was obtained when leaves were harvested in the spring (April). The results obtained by EL Jemli et al. (2016) using species from Morocco showed that the TPC in the extracts of *T. articulata* had a value of 175.67 ± 10.22 mg GAE/g. These results showed that the quantity of phenolic compounds present in extracts of Algerian and Moroccan T. *articulata* was similar. However, a significant difference (17.7 ± 0.7 to 30.1 ± 1.0 mg of GAE/g) was observed with the results reported by Sliti et al. (2016) and by Ben Jemia et al. (2013) (32.47 ± 1.89 mg of GAE/g) for various extracts of Tunisian *T. articulate.*

3.3 Effect of Extraction Time and Concentration on TFC

Figure 2 shows the effect of the extraction time and the ethanol concentration on TFC. The amount of the flavonoid compounds in *T. articulata* extracts ranged from 19.03 to 36.70 mgQE/g. The maximum extraction yield regarding TFC (36.70mgQE/g) was achieved with 60% ethanol concentration for an extraction time of 120 min. It was observed that the values of TFC increase as the extraction time increases from 30 to 120 min at fixed ethanol concentration. These values reached a maximum and subsequently decreased when the extraction time increased. This could be due to the increase in extraction time, which generated the degradation of flavonoid compounds. Thus, extraction time of 120 min was revealed enough to obtain the maximum TFC from *T. articulate* leaves at different ethanol concentrations. The amount of total flavonoid (TFC) investigated in the present study was in agreement with the previous report by Rached et al. (2010),where the flavonoid content ranged

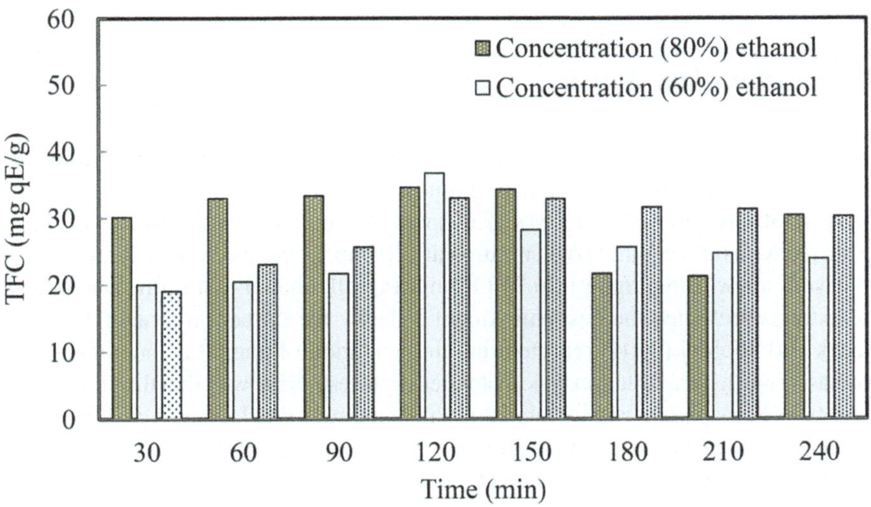

Fig. 2 Total flavonoid content (TFC) of *Tetraclinis articulate* extract with different ethanol concentration (40, 60 and 80%) and at different extraction time (30–240 min)

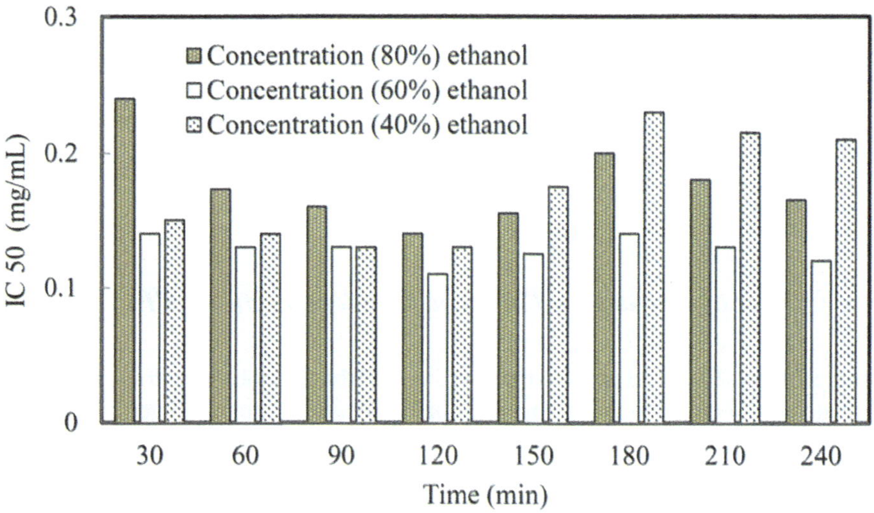

Fig. 3 EC_{50} values of *Tetraclinis articulate* extract with different ethanol concentration (40, 60 and 80%) and at different extraction time (30–240 min)

from 27.392 to 72.425 mg CE/g, but was higher than that reported by Ben Jemia et al. (2013) (17.97 ± 0.86 mg CE/g) and El Jemli et al. (2016) (11.78 ± 0.30 mg QE/g). This difference was because of the type of material, harvest period, extraction method, solvent used, and others operating conditions.

3.4 Antioxydant Activity

Figure 3 presents the values of EC_{50} of different extracts, which show that the antioxidant activity depends on the extraction time and the solvent concentration. The best antioxidant activity ($IC_{50} = 0.110$ mg/mL) was obtained from the *T. articulata* leaves extracted with 60% ethanol concentration at 120 min. In these conditions, the content of flavonoids was the highest(36.70 QE/mg).For all ethanol concentration studied, the extracts presented the best antioxidant capacity for extraction time of 120 min. Naczk and Shahidi (2004) reported that the oxidation of phenolic compounds was increased when the extraction was prolonged. Our results showed that the antioxidant activity of *T. articulate* can be related to the amount of the flavonoid present in this species. These results were in agreement with the study carried out by Rached et al. (2018), who concluded that the extracts of *T. articulata* were rich in flavonoids such as catechin, rutin, and quercetin. According to Pittela et al. (2009), the flavonoids are highly polar compounds with low molecular weight. Hence, as DPPH radicals preferentially react with low molecular weight phenolic compounds, these compounds are prone to react with DPPH radicals (Paixãoet al. 2007). Our results were similar to those reported by Herzi et al. (2013a), (EC_{50} = 108.16 μg/mL), and were higher than the results available in the literature (Rached et al. 2018; El Jemli et al. 2016; Sliti et al. 2016; Ben Jemia et al. 2013). Rached et al. (2018) reported that the lower EC_{50} value, namely the high antioxidant activity was observed with the ethyl acetate fraction (4.51 ± 0.15 μg/mL), compared to other studied samples (EC_{50} values: 8.65 ± 0.16 and 12.68 ± 0.67 μg/mL for butanol fraction and aqueous extract, respectively). The study conducted by Ben Jemia et al. (2013) showed that the antioxidant activity ($IC_{50} = 5.5$ μg/mL) of acetone extract was two times higher than the positive control (BHT).

4 Conclusion

The phenolic compounds concentration and the antioxidant activity of T. *articulate* were dependent on the extracting solvent and the extraction time. The highest yield was obtained by the extraction method with 60% ethanol concentration and an extraction time of 240 min. The optimal extraction conditions for optimized phenolic compounds recovery from T. *articulata* were 60% ethanol concentration at 210 min for total phenolic compounds and at 120 min for total flavonoids, respectively. The antioxidant capacity of the extracts was sensitive to the solvent concentration. For all ethanol concentrations studied, the extracts presented the best antioxidant capacity at the extraction time of 120 min. The best anti-radical activity, with values of IC_{50} of 0.110 mg/mL, was obtained from the *T. articulate* leaves extracted with 60% ethanol concentration at 120 min. With these conditions, the content of flavonoids was the highest (36.70 QE/mg). Finally, his work showed that the leaves of Algerian *tetraclinis articulate* (Vahl) mastersleaves, endemic plant from south-western Mediterranean, could be exploited in food and pharmaceutical industries as source of natural products.

References

Ababsa ZEA, Derouiche MT, Medjroubi K, Akka S (2019) In-vivo Antidiarrhoeal and anti-ulcerative activities of the Tetraclinisarticulata species of the Cupressaceae family. Acta Sci Nat 6:50–53

Abi-Ayad FZ, Abi-Ayad M, Lazouni AH, Rebiahi SA (2013) Evaluation of Tetraclinisarticulata essential oil from Algeria flora as a potential source of antifungal activity and study of its chemical composition. J Indian Acad Wood Sci 10:9–15

Abi-Ayada FZ, Abi-Ayada M, Lazzounia HA, Rebiahib S, Bessierec C (2011) Antibacterial activity of essential oil extracted from leaves of Tetraclinisarticulata (Vahl) Masters from Algeria flora. J Microbiol Biotechnol Res 1:1–6

Bahri F, Romane A, Höferl M, Wanner J, Schmidt E Jirovetz L (2015) Chemical composition and antimicrobial activity of essential oil of Algerian Tetraclinisarticulata (Vahl) Masters. J Essent Oil Res 28:42–48

Ben Jemia M, Chaabane S, Senatore F, Bruno M, Kchouk M E (2013) Studies on the antioxidant activity of the essential oil and extract of Tunisian Tetraclinisarticulata (Vahl) Mast. (Cupressaceae). Nat Prod Res: Formerly Nat Prod Letters 27(16):1419–1430

Ben-Ghnya A, Amri I, Hanana M, Gargouri S, Jamoussi B, Romana A, Hamrouni L (2016) Tetraclinisarticulata (Vahl.) masters essential oil from Tunisia: chemical characterization and herbicidal and antifugal activities assessment. Ind Crops Prod 83:113–117

Bourkhiss B, Ouhssine M, Hnach M, Amechrouq A, Chaouch A, Satrani B (2007) Chemical composition of the essential oil of Tetraclinisarticulata (Vahl) from Maroc. Phys Chem News 35:128–132

Boussaïd M, Bekhechi C, Beddou F, Sari DC, Bighelli A, Casanova J, Tomi F (2015) Chemical variability of the essential oil isolated from aerial parts of Tetraclinisarticulata from North-Western Algeria. Nat Prod Commun 10:1447–1452

Chikhoune A, Hazzit M, Kerbouche L, Baalioamer A, Aissat K (2013) Tetraclinis articulate (Vahl) Masters essential oils chemical composition and biological activities. J Essent Oil Res 25:300–307

Dane Y, Mouhouche F, Canela-Garayoa R, Delpino-Rius A (2015) Phytochemical analysis of Methanolic extracts of Artemisia absinthium L. 1753 (Asteraceae), Juniperusphoenicea L., and Tetraclinisarticulata (Vahl) Mast, 1892 (Cupressaceae) and evaluation of their biological activity for stored grain protection. Arab J Sci Eng 41:2147–2158

Djouahri A, Boudarene L, Sabaou N (2013) Phytochemical evaluation of antimicrobial activity of crudes extracts of leaves Tetraclinisarticulata (Vahl) Masters Algerian. Int J Pharmacogn Phytochem Res 5:45–50

Djouahri A, Sakaa B, Boudarene L (2014) In vitro synergistic/antagonistic antibacterial and anti-inflammatory effect of various extracts/essential oil from cones of Tetraclinisarticulata (Vahl) Masters with antibiotic and anti-inflammatory agents. Ind Crops Prod 56:60–66

El Jemli M, Kamal R, Marmouzi I, Zerrouki A, Cherrah Y, Alaoui K (2016) Radical-scavenging activity and ferric reducing ability of Juniperusthurifera (L.), J. oxycedrus (L.), J. phoenicea (L.) and Tetraclinisarticulata (L.). Adv Pharmacol Sci 2016:1–6

El Jemli H, Kamal R, Marmouzi I, Doukkali Z, Bouidida EH, Touati D, Nejjari R, El Guessabi L, Cherrah Y, Alaoui K (2017) Chemical composition, acute toxicity, antioxidant and anti-inflammatory activities of Moroccan Tetraclinisarticulata L. J Tradit Complement Med 7:281–287

Gong Y, Hou Z, Gao Y, Xue Y, Liu X, Liu G (2012) Optimization of extraction parameters of bioactive components from defatted marigold (Tageteserecta L.) residue using response surface methodology. Food Bioprod Process 90:9–16

Herzi N, Bouajila J, Camy S, Romdhane M, Condoret JS (2013) Comparison of different methods for extraction from Tetraclinisarticulata: yield, chemical composition and antioxidant activity. Food Chem 141:3537–4354

Herzi N, Camy S, Bouajila J, Destrac P, Romdhane M, Condoret JS (2013) Supercritical CO^2 extraction of Tetraclinisarticulata: Chemical composition, antioxidant activity and mathematical modeling. J Supercrit Fluids 82:72–82

Letreuch-Belarouci N (1991) Les reboisements en Algérie et leur perspective d'avenir. Vol. I. Alger, Office des Publications Universitaires (O.P.U.)

Manach C, Scalbert A, Morand C, Rémés C, Jiménez L (2004) Polyphenols: food sources and bioavailability. Am J Clin Nutr 79:727–747

McDonald S, Prenzler PD, Antolovich M, Robards K (2001) Phenolic content and antioxidant activity of olive extracts. Food Chem 73:73–84

Middleton E, Kandaswami C, Theoharides TC (2000) The effects of plant flavonoids on mammalian cells: implications for inflammation, heart disease and cancer. Pharmacol Rev 52:673–751

Morte MA, Honrubia M (1996) Biotechnology in agriculture and forestry. In: Bajaj (ed), vol. 35 Trees IV. Springer, Heidelberg

Naczk M, Shahidi F (2004) Extraction and analysis of phenolics in food. J Chormatogr A 1054:95–111

Paixão N, Perestrelo R, Marques JC, Câmara JS (2007) Relationship between antioxidant capacity and total phenolic content of red, rosé and white wines. Food Chem 105:204–214

Peschel W, Sanchez-Rabaneda F, Plescher WA, Gartzia I, Jimenez D, Lamuela-Raventos R, Buxaderas S, Condina C (2006) An industrial approach in the search of natural antioxidants from vegetable and fruit wastes. Food Chem 97:137–150

Pittela F, Dutra RC, Junior DD, Lopes MTB, Barbosa NR (2009) Antioxidant and Cytotoxic Activities of Centellaasiatica (L) Urb. Int J Mol Sci 10:3713–3721

Quettier-Deleu C, Gressier B, Vasseur J, Dine T, Brunet C, Luyckx M et al (2000) Phenolic compounds and antioxidant activities of buckwheat (FagopyrumesculentumMoench) hulls and flour. J Ethnopharmacol 72:35–42

Rached W, Benamer H, Bennaceur M, Marouf A (2010) Screening of the antioxidant potential of some Algerian indigenous plants. J Biol Sci 10:316–324

Rached W, Zeghada FZ, Bennaceur M (2018) Phytochemical analysis and assessment of antioxidant, antimicrobial, anti-inflammatory and cytotoxic properties of Tetraclinisarticulata (Vahl) masters leaves. Ind Crops Prod 112:460–466

Sadiki FZ, El Idrissi M, Sbiti M, Lemrhari A, Trifan A, Cioanca O, PostuPA HL (2018) Chemical composition and antibacterial activity of essential oil of Tetraclinis articulate (Vahl) masters branches of eastern Morocco. Chem Biol Technol Agric 24:2–6

Sahi L, Courivaud A, Chailan C (2016) Le marché des plantes aromatiques et médicinales: analyse des tendances du marché mondial et des stratégies économiques en Albanie et en Algérie. CIHEAM/FranceAgriMer, Montpellier

Sanchez-Moreno C, Larrauri JA, Saura-Calixto F (1998) A procedure to measure the antiradical efficiency of polyphenols. J Sci Food Agric 76:270–276

Sliti S, Ayadi S, Dumarçay S (2016) Evaluation of essential oil composition and antioxidant capacity of hydromethanolic extracts of Tetraclinisarticulata, depending on location and seasonal variations. J Mater Enviro Sci 7:968–980

Tékaya-Karaoui A, Ben Jannet H, Mighri Z (2007) Essential oil composition of terminal branches, cones and roots of Tetraclinisarticulata from Tunisia. Pak J Bio Sci 15:2495–2499

Vuong QV, Golding JB, Stathopoulos CE, Nguyen MH, Roach PD (2011) Optimizing conditions for the extraction of Catechins from green tea using hot water. J Sep Sci 34:3099–3106

Wang H, Provan GJ, Helliwell K (2004) Determination of rosmarinic acid and caffeic acid in aromatic herbs by HPLC. Food Chem 87:307–311

Zhang ZS, Li D, Wang LJ, Ozkan N, Chen XD, Mao ZH et al (2007) Optimization of ethanol–water extraction of lignans from flax seed. Sep Purif Technol 57:17–24

Zidane A, Tits M, Angenot L, Wauters JN, Frederich M, Dib I, MekhfiH AM, Bnouham M, Legssyer A, Ziyyat A (2014) Phytochemical analysis of Tetraclinisarticula in relation to its vasorelaxant property. J Mater Environ Sci 5:1368–1375

Chapter 11
Methylene Blue Dye Removal Through Adsorption Onto Amorphous BaO Nanoparticles Decorated MWCNTs

Arvind K. Bhakta, Sunita Kumari, Sahid Hussain, Preema C. Pais, Praveen Martis, Ronald J. Mascarenhas, Samir Belkhiri, Joseph Delhalle, and Zineb Mekhalif

Abstract This work presents an extensive study on the effect of various parameters (concentration, solvents, infrared irradiation, salts, and temperature variation) on the barium oxide nanoparticles decorated multi-wall carbon nanotubes. A simple and effective technique involving infrared irradiation and diazonium chemistry was utilized. The materials were characterized using X-ray photoelectron spectroscopy, Raman spectroscopy, energy dispersive X-ray spectroscopy, transmission electron microscopy, powder X-ray diffraction, thermogravimetric analysis and ultraviolet–visible spectrophotometry. A homogeneous distribution of amorphous barium oxide nanoparticles onto carbon nanotubes was observed. Particles size varies from 3–18 nm (Gaussian mean diameter ~11.7 nm). Thermogram showed that nanoparticles act as a catalyst to decompose nanotubes even at lower temperature. A common dye pollutant that is methylene blue was considered to evaluate the ability of the present composite for water purification for the first time. As our technique is also applicable to bulk synthesis, it can be worthful for wide range of nanotechnological applications.

Keywords Carbon nanotubes · Methylene blue dye · Barium oxide nanoparticles

A. K. Bhakta (✉) · J. Delhalle · Z. Mekhalif (✉)
Laboratory of Chemistry and Electrochemistry of Surfaces, NISM, University of Namur, B-5000 Namur, Belgium
e-mail: arvind.bhakta@unamur.be

Z. Mekhalif
e-mail: zineb.mekhalif@unamur.be

S. Kumari · S. Hussain
Department of Chemistry, Indian Institute of Technology Patna, Bihta, Bihar 801106, India

P. C. Pais · P. Martis
Department of Chemistry, St. Aloysius College, Mangalore 575003, India

R. J. Mascarenhas
Electrochemistry Research Group, St. Joseph's College, Bangalore 560027, India

S. Belkhiri
Ecole Militaire Polytechnique, BP17 Bordj El-Bahri, 16046 Algiers, Algeria

D. Trache et al. (eds.), *Materials Research and Applications*,
Materials Horizons: From Nature to Nanomaterials,
https://doi.org/10.1007/978-981-15-9223-2_11

1 Introduction

Carbon nanotubes (CNTs) have gained an utmost importance in the field of nanoscience and nanotechnology (Bianco et al. 2018) over the past two decades due to their uniqueness in properties such as structural flexibility, remarkable electrical conductivities, large surface area, chemical stability and exceptional tensile strength and thermal conductivities (Ajayan 1999). They have been employed as potential candidates over wide range of useful applications such as in tissue engineering (Kouser et al. 2018), pollutants removal (Tang et al. 2017), and electrochemical sensors (Erady et al. 2019), etc. Chemical inertness and insolubility of CNTs in most of the solvents limit their application in potentially diverse fields. These problems could be resolved by functionalization of CNTs. The functionalized CNTs are used as suitable templates for assembly of metal or metal oxide nanoparticles on their surface as it contributes to innumerable applications (Yadav et al. 2020). Barium oxide (BaO) exhibits superior catalytic performance among the alkaline oxides as it possesses stronger basic sites favoring transesterification reactions (Martinez-Guerra and Gude 2014). It also catalyzes isomerization and co isomerization reactions and has other applications (Song et al. 2017). Only two techniques have been described so far (Hasnahena et al. 2016; Bhakta et al. 2018) for the synthesis of barium oxide nanoparticles (BaO NPs) decorated multi-wall carbon nanotubes (MWCNTs). They have certain limitations such as low amount, non-homogeneous distribution, and large size of particles on the surface of CNTs. Thus, in the current work, efforts have been under taken to optimize the various parameters (concentration, solvents, infrared (IR) irradiation, salts, and temperature variation), to overcome the limitations reported by the previous methods. A common dye pollutant that is methylene blue (MB) was considered to evaluate the ability of the present composite for water purification. To the best of our knowledge, there is no work reported so far on any applications of this material except for sensing of NH_3 and CO gases (Hasnahena et al. 2016).

2 Materials and Methods

2.1 Chemicals

Barium acetate (99%), $NaNO_2$ (99%), 5-Amino-1,2,3-benzenetricarboxylic acid (97%), $HClO_4$ (70%), methylene blue, $(CH_3)_2CO$ (>99%), NaOH ($\geq$98%) and C_5H_{12} (99%) were purchased from MERCK, Fisher-Scientific (UK), ABCR, Merck, Sigma-Aldrich, ChemLab, ACS Reagent and Lab-Scan Analytical Sciences, respectively.

2.2 Apparatus

TEM (Tecnai 10 Philips microscope) study was carried out by maintaining it at an accelerating voltage of 80 kV. PXRD (PAN analytical diffractometer) analysis was performed with Cu Kα radiation (wavelength $\lambda = 1.5418$ Å). Other conditions like current and voltage were 30 mA and 45 kV, respectively. XPS studies were performed via Thermo Scientific K-Alpha spectrometer with a monochromatized Al Kα radiation (1486.6 eV) and hemispherical analyzer, operated at a pass energy of 50 and 200 eV for high resolution and survey spectra, respectively.

2.3 Synthesis of Material

Crude MWCNTs (4 g) were mixed with 400 ml of 12 M NaOH solution, refluxed at 170 °C for 12 h. Further, diluted with water, filtered, washed with water, then acetone and ultimately dried. This purified MWCNTs are referred to as p-MWCNTs. Afterwards, the combination of p-MWCNTs (1 g), 5- amino-1,2,3-benzenetricarboxylic acid (1.87 g), water (125 ml), sodium nitrite (0.57 g), and perchloric acid (862 μl,) was IR irradiated (1 h). The mixture was filtered; the residue was washed with pentane followed by acetone and dried in air (Bhakta et al. 2017, 2019a). The tricarboxylic aryl azo functionalized MWCNTs are referred to as p-MWCNTs-D3. 100 ml aqueous (optimum solvent) solution of 2.550 g (optimum amount) of barium acetate (BA) (optimum salt) and p-MWCNTs-D3 (0.0960 g) are dispersed by sonication (5 min) and then submitted to IR irradiation (5 h-optimum duration). The obtained residue is filtered, washed, and dried. They are denoted as p-MWCNTs-D3/BA. Afterwards, they are calcined at 400 °C (optimized temperature) for 2 h in inert atmosphere (argon gas) and labeled as p-MWCNTs-D3/BaO. This synthesis procedure is adopted from our previous work (Bhakta et al. 2018) after modification of some parameters.

2.4 Adsorption Tests

40 ppm MB stock solution is prepared by dissolving 10 mg MB in 250 ml water. Afterwards, 5 ml from stock solution is mixed with 5 ml of water. Finally, 3 mg of p-MWCNTs-D3/BaO were added and magnetically stirred in dark condition (to avoid the possibility of photodegradation). The adsorption process is controlled by measuring the UV-V is absorbance. For each interval of time, fresh solution is taken and then adsorbent is added and stirred. A prominent MB peak was observed at 664 nm. This experiment was also performed using p-MWCNTs. Adsorption isotherm was also obtained at different concentrations (10, 20, 30, 40, 50, and 60 ppm) for p-MWCNTs-D3/BaO.

3 Results and Discussion

3.1 Percentage Composition (By XPS and EDX) and Crystallinity Study (By PXRD)

Atomic % of different samples is shown in Table 1. Non-appearance of aluminium element in purified nanotubes (Fig. 1a) confirms the purification (complete removal of alumina impurity, which was used as catalyst support during MWCNTs synthesis) of MWCNTs. Further, occurrence of N1s and high intensity of O1s (Bhakta et al. 2017, 2018, 2019a) (Fig. 1b) indicate that MWCNTs are modified with carboxylic containing diazonium groups. Presence of BA in the impregnated sample was established by the occurrence of the typical Ba3d peak (Fig. 1c). This peak also remains

Table 1 Percentage composition of various samples obtained by (a) XPS and (b) EDX analysis

Material	C %	O %	N %	Ba %	Al %
(a) XPS					
Crude MWCNTs	97.19	1.96	–	–	0.85
p-MWCNTs	98.30	1.70	–	–	–
p-MWCNTs-D3	72.23	24.61	3.16	–	–
p-MWCNTs-D3/Ba	77.93	16.74	1.53	3.80	–
p-MWCNTs-D3/BaO	82.79	12.57	1.05	3.59	–
(b) EDX					
Crude MWCNTs	91.39	6.91	–	–	1.70
p-MWCNTs	98.34	1.62	–	–	0.04
p-MWCNTs-D3/BaO	82.65	13.79	–	3.55	–

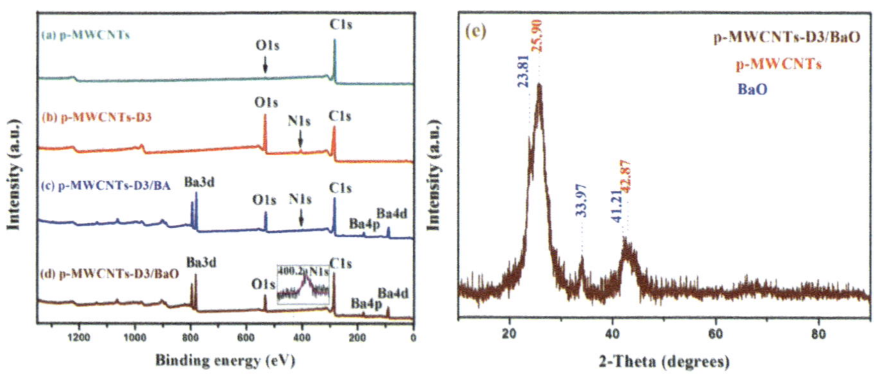

Fig. 1 a–d Left side-XPS survey spectra (Inset is the N1s high resolution XPS spectrum) and **e** right side-PXRD pattern of p-MWCNTs-D3/BaO

in the calcinated sample (Fig. 1d). The PXRD peak patterns obtained from the p-MWCNTs-D3/BaO sample (Fig. 1e) show similarities with those reported in the literature for BaO NPs decorated CNTs (Hasnahena et al. 2016). BaO NPs are amorphous in nature as the diffraction peak is not sharp. It is found that the BaO nanoparticles in its amorphous form are very interesting for nanoapplications (Song et al. 2017).

3.2 *Reaction Pathway and Materials Morphology by TEM*

The most common pathway for diazonium modification of MWCNTs surface is well reported in the literature (Bhakta et al. 2019a). The overall scheme of decoration is shown in Fig. 2f. TEM picture of impurified MWCNTs (Fig. 2a) has large species, i.e., alumina (contamination), completely removed in purified MWCNTs (Fig. 2b) indicating the efficiency of the applied basic treatment (purification step). Intact nanotubes after the purification process (Fig. 2b) and functionalization steps (Fig. 2c) point to effectiveness of the present method. This has a definite advantage over chemical oxidation method where CNTs shortenings, incorporation of unwanted groups, and severe damages to surface structure were observed (Kilinç 2016). Figure 2d shows the presence of homogeneously distributed BaO NPs (Fig. 2d) in the range of 3–18 nm (Fig. 2e). The particle size was measured using the ImageJ software. This

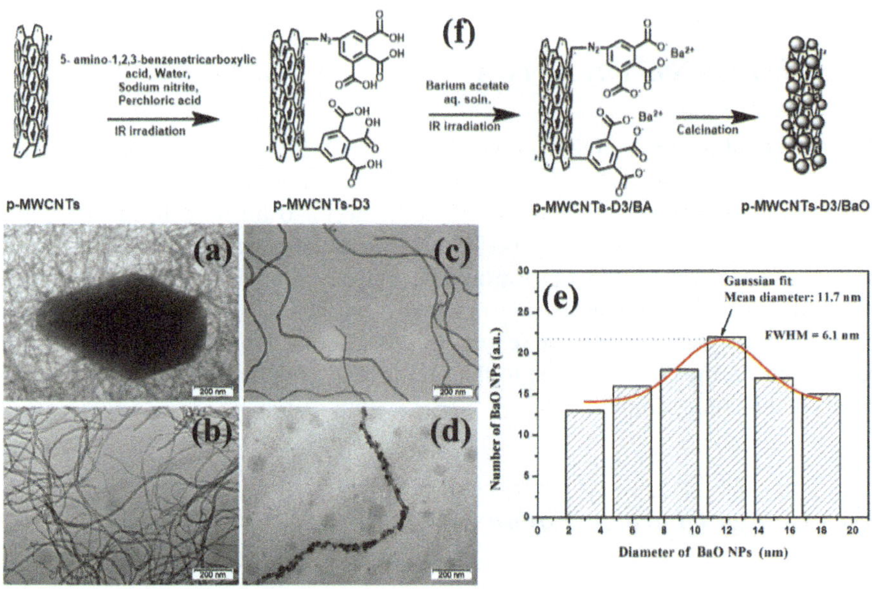

Fig. 2 **a–d** TEM images, **e** diameter distributions of the BaO-NPs, and **f** steps in the decoration of MWCNTs

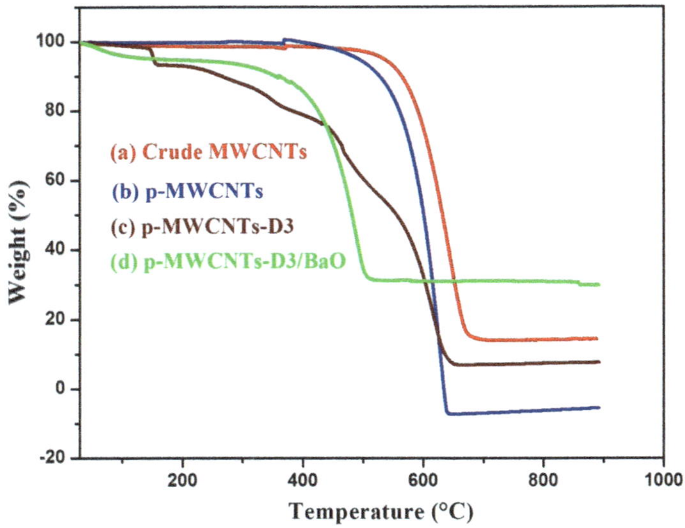

Fig. 3 TGA thermograms of **a** crude MWCNTs, **b** p-MWCNTs, **c** p-MWCNTs-D3 and p-MWCNTs-D3/BaO

proves (Fig. 2d) that not only the particles are homogeneously present but also the diazonium functions.

3.3 Thermal Stability by TGA

Thermal stability of prepared materials is depicted in Fig. 3. The weight loss above 500 °C in all the samples is due to MWCNTs decomposition (Silva et al. 2012). The thermogram (Fig. 3c) of the tricarboxylic aryl azo functionalized MWCNTs (p-MWCNTs-D3) between 150–500 °C is due to the moisture (Chakoli et al. 2014) as well as the decomposition of tricarboxylic aryl azo group. In the decorated sample (Fig. 3d), decomposition at lower temperature can be due to the presence of BaO NPs and its catalytic effect. One reason for the difference in the mass loss of crude and purified samples can be due to the presence of alumina (having high melting point) in the crude samples.

3.4 Raman Characterization of Materials

The important characteristics of Raman spectra (Fig. 4) are the *D* (~1342 cm^{-1}) and *G* (~1572 cm^{-1}) bands. Furthermore, a less prominent shoulder D' peak, related to symmetry loss of the carbon sp^2 network, is present at 1609 cm^{-1} (Dettlaff et al.

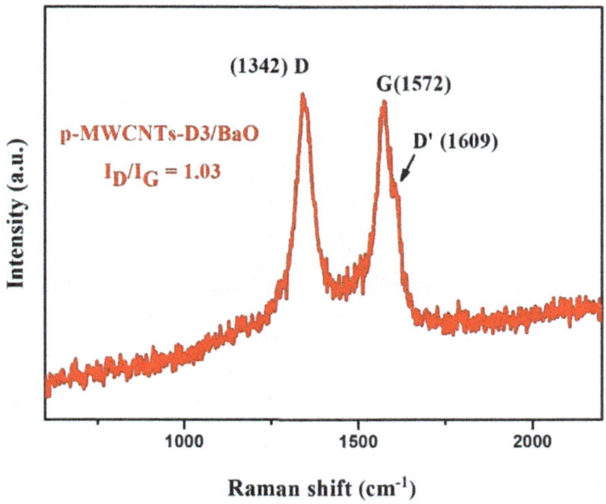

Fig. 4 Raman spectrum of p-MWCNTs-D3/BaO

2017). The I_D/I_G ratio is used as a parameter to assess the CNTs quality. The nanoparticles modified MWCNTs have a I_D/I_G equal to 1.03, which is less than the starting crude materials (Bhakta et al. 2019b), validating that the used method maintains the integrity of MWCNTs. Indeed, inherent defects which were produced during the synthesis procedure still exist. The present method, thus, proves the superiority over the other techniques (Sen et al. 2017) which suffer from defects problem after alteration with nanoparticles.

3.5 *Application*

It is experimentally observed that the MB peak intensity decreases as with a function of time indicating the MB dye adsorption onto the p-MWCNTs-D3/BaO surface (Fig. 5b). This has a faster rate compared to the adsorption at unmodified (Bhakta et al. 2019a) nanotubes (Fig. 5a). The quantification of adsorbed dye at time "*t*" was obtained by applying the following formula (Eq. 1):

$$q_t = \frac{V(C_0 - C_t)}{W} \tag{1}$$

$W(g)$ = mass of the adsorbent, $V(L)$ = volume of MB aqueous solution, C_0 and C_t (mg/L) are the initial and liquid-phase concentrations of MB at a given time t, respectively, and q_t (mg/g) represent the amount of dye adsorbed on the CNTs at time t.

Plot t/q_t versus t have a linear relationship with a R^2 of 0.9971 for p-MWCNTs-D3/BaO, pointing to a pseudo-second-order kinetics. Pseudo-second-order kinetics equation is given in (Eq. 2) (Kumari et al. 2019):

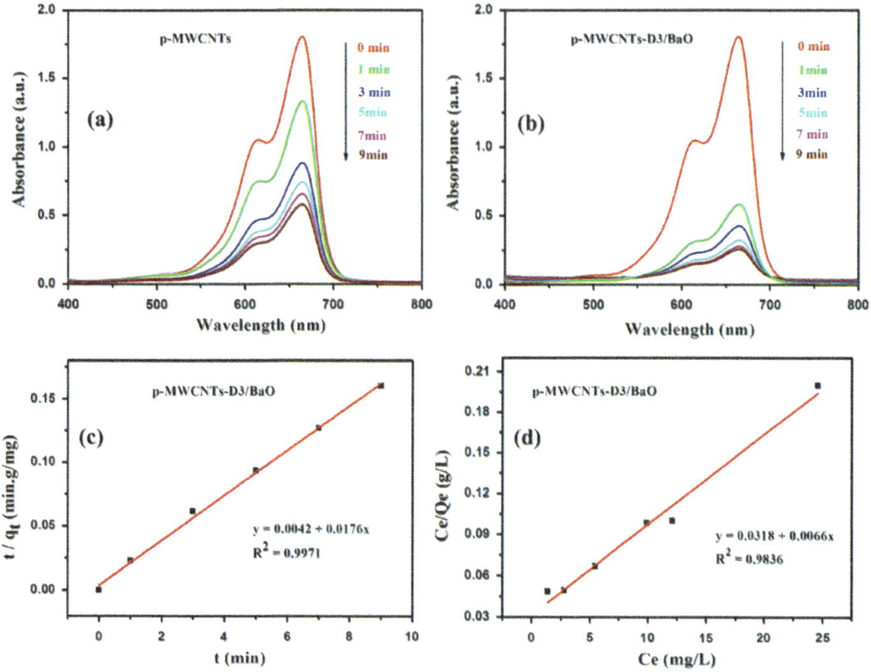

Fig. 5 UV-V is spectra of MB aqueous solution (20 mg L^{-1}) taken at different interval of time (0–9 min) of **a** p-MWCNTs, **b** p-MWCNTs-D3/BaO, **c** t/q_t versus t plot for MB adsorption at p-MWCNTs-D3/BaO, and **d** Langmuir adsorption isotherm

$$\frac{t}{q_t} = \frac{1}{K_2 q_e^2} + \frac{1}{q_e} t \quad (2)$$

where K_2 (g/mg/min) = pseudo-second-order rate constant and q_e (mg/g) = equilibrium adsorption capacity. K_2 is found to be 0.0737 g/mg/min. Initial fast adsorption process in both cases (Fig. 5a–b) is a result of monolayer adsorption of MB dye molecule. Then, a slow adsorption occurs which is a signature of the classical Langmuir adsorption (Mohammed et al. 2014). Thus, the adsorption isotherm obtained at different concentrations gives information about the adsorption mechanism. The resulting equilibrium adsorption data were obtained according to the Langmuir model (Langmuir 1918) (Eq. 3):

$$\frac{C_e}{Q_e} = \frac{1}{K_L Q_m} + \frac{C_e}{Q_m} \quad (3)$$

where Q_e (mg g^{-1}) and Q_m (mg g^{-1}) are the equilibrium and maximum adsorption capacities, respectively, and C_e (mg L^{-1}) is the equilibrium dye concentration. Binding energy associated with the adsorption process is denoted by K_L (Lm g^{-1}).

The value of Q_m is obtained from the slope of the linear plot $C_e/Q_e vs C_e$. The maximum adsorption capacity is found to be 151.52 mg g^{-1}. Comparison of this value with other similar adsorbents used for the MB dye removal in the literature (Ai et al. 2011; Wang et al. 2014) reveals that the present hybrid material exhibits high adsorption ability.

4 Conclusions and Future Perspectives

A simple, reproducible, and effective method involving diazonium chemistry and IR irradiation is applied to synthesize BaO nanoparticles decorated MWCNTs. Previously, only two works have been reported about this hybrid preparation but they have certain limitations (less particles loading, lower homogeneity, …) which are overcome by the present method. In addition, this material is utilized for the first time for water purification applications, the maximum adsorption capacity being 151.52 mg g^{-1}. A homogeneous distribution of barium oxide nanoparticles on the MWCNTs surfaces is observed, the nanoparticles size varying from 3 to18 nm (Gaussian mean diameter ~11.7 nm). p-MWCNTs-D3/BaO is an effective adsorbent for MB dye. It opens possible environmental as well as nanotechnological applications.

Conflict of Interest

All the authors confirm that there is no conflict of interest exist.

Acknowledgements A. K. Bhakta convey his sincere thanks to the University of Namur for awarding CERUNA doctoral fellowship.

References

Ai L, Zhang C, Liao F et al (2011) Removal of methylene blue from aqueous solution with magnetite loaded multi-wall carbon nanotube: kinetic, isotherm and mechanism analysis. J Hazard Mater 198:282–290. https://doi.org/10.1016/j.jhazmat.2011.10.041

Ajayan PM (1999) Nanotubes from carbon. Chem Rev 99:1787–1799. https://doi.org/10.1021/cr970102g

Bhakta AK, Detriche S, Martis P et al (2017) Decoration of tricarboxylic and monocarboxylic aryl diazonium functionalized multi-wall carbon nanotubes with iron nanoparticles. J Mater Sci 52:9648–9660. https://doi.org/10.1007/s10853-017-1100-z

Bhakta AK, Mascarenhas RJ, Martis P et al (2018) Multi-wall carbon nanotubes decorated with barium oxide nanoparticles. Synth Catal Open Access 3:1–4. https://doi.org/10.4172/2574-0431.100019

Bhakta AK, Kumari S, Hussain S et al (2019a) Synthesis and characterization of maghemite nanocrystals decorated multi-wall carbon nanotubes for methylene blue dye removal. J Mater Sci 54:200–216. https://doi.org/10.1007/s10853-018-2818-y

Bhakta AK, Kumari S, Hussain S et al (2019b) Differently substituted aniline functionalized MWCNTs to anchor oxides of Bi and Ni nanoparticles. J Nanostruct Chem 9:299–314. https://doi.org/10.1007/s40097-019-00319-8

Bianco A, Chen Y, Chen Y et al (2018) A carbon science perspective in 2018: current achievements and future challenges. Carbon N Y 132:785–801. https://doi.org/10.1016/j.carbon.2018.02.058

Chakoli AN, He J, Cheng W, Huang Y (2014) Enhanced oxidized regenerated cellulose with functionalized multiwalled carbon nanotubes for hemostasis applications. RSC Adv 4:52372–52378. https://doi.org/10.1039/C4RA07704K

Dettlaff A, Sawczak M, Klugmann-Radziemska E et al (2017) High-performance method of carbon nanotubes modification by microwave plasma for thin composite films preparation. RSC Adv 7:31940–31949. https://doi.org/10.1039/c7ra04707j

Erady V, Mascarenhas RJ, Satpati AK et al (2019) Carbon paste modified with Bi decorated multiwalled carbon nanotubes and CTAB as a sensitive voltammetric sensor for the detection of Caffeic acid. Microchem J 146:73–82. https://doi.org/10.1016/j.microc.2018.12.023

Hasnahena ST, Satpati B, Roy M (2016) Enhanced sensing of NH_3 gas by decorated multiwalled carbon nanotube. AIP Conf Proc 1731:0500951–0500953. https://doi.org/10.1063/1.4947749

Kilinç E (2016) γ-Fe_2O_3 magnetic nanoparticle functionalized with carboxylated multi walled carbon nanotube: synthesis, characterization, analytical and biomedical application. J Magn Magn Mater 401:949–955. https://doi.org/10.1016/j.jmmm.2015.11.003

Kouser R, Vashist A, Zafaryab M et al (2018) Biocompatible and mechanically robust nanocomposite hydrogels for potential applications in tissue engineering. Mater Sci Eng C 84:168–179. https://doi.org/10.1016/j.msec.2017.11.018

Kumari S, Khan AA, Chowdhury A et al (2019) Efficient and highly selective adsorption of cationic dyes and removal of ciprofloxacin antibiotic by surface modified nickel sulfide nanomaterials: Kinetics, isotherm and adsorption mechanism. Colloids Surfaces A Physicochem Eng Asp 586:124264. https://doi.org/10.1016/j.colsurfa.2019.124264

Langmuir I (1918) The adsorption of gases on plane surfaces of glass, mica and platinum. J Am Chem Soc 40:1361–1403. https://doi.org/10.1021/ja02242a004

Martinez-Guerra E, Gude VG (2014) Transesterification of used vegetable oil catalyzed by barium oxide under simultaneous microwave and ultrasound irradiations. Energy Convers Manag 88:633–640. https://doi.org/10.1016/j.enconman.2014.08.060

Mohammed MI, Razak AAA, Al-Timimi DAH (2014) Modified multiwalled carbon nanotubes for treatment of some organic dyes in wastewater. Adv Mater Sci Eng 2014

Sen B, Kuzu S, Demir E et al (2017) Highly monodisperse RuCo nanoparticles decorated on functionalized multiwalled carbon nanotube with the highest observed catalytic activity in the dehydrogenation of dimethylamine-borane. Int J Hydrogen Energy 42:23292–23298. https://doi.org/10.1016/j.ijhydene.2017.06.032

Silva WM, Ribeiro H, Seara LM et al (2012) Surface properties of oxidized and aminated multiwalled carbon nanotubes. J Braz Chem Soc 23:1078–1086. https://doi.org/10.1590/S0103-505 32012000600012

Song Y, Wang W, Ge L et al (2017) Rational design of a water-storable hierarchical architecture decorated with amorphous barium oxide and nickel nanoparticles as a solid oxide fuel cell anode with excellent sulfur tolerance. Adv Sci 4:17003371–17003378. https://doi.org/10.1002/advs.201700337

Tang Y, Tian J, Malkoske T et al (2017) Facile ultrasonic synthesis of novel zinc sulfide/carbon nanotube coaxial nanocables for enhanced photodegradation of methyl orange. J Mater Sci 52:1581–1589. https://doi.org/10.1007/s10853-016-0452-0

Wang P, Cao M, Wang C et al (2014) Kinetics and thermodynamics of adsorption of methylene blue by a magnetic graphene-carbon nanotube composite. Appl Surf Sci 290:116–124. https://doi.org/10.1016/j.apsusc.2013.11.010

Yadav A, Upadhyaya A, Gope J et al (2020) Silver-decorated multiwall carbon nanotubes: synthesis characterization and application in polymer composite-based devices. J Mater Sci Mater Electron 31:1451–1460. https://doi.org/10.1007/s10854-019-02659-1

Chapter 12
Bioremediation of Oily Sludge Contaminated Site—A Pilot-Scale Study

Kahina Ighilahriz, Mohamed Khodja, Samia Haddadi, Amina Benchouk, and Hichem Boutamine

Abstract The present paper treats the bioremediation feasibility of hydrocarbon-contaminated soil from oily quagmire generated during drilling operation. The experience took place in the region of *Hassi Messaoud*, located in the southeast of Algeria. Through which, methods of natural attenuation, remediation-using poultry manure and bentonite were subject of evaluation and comparison. A homogenized oily sludge sample was inoculated with adequate amount of organic and mineral stimulant, in microcosm having dimensions of $1\ m \times 2\ m \times 1\ m$ for length, breadth and depth respectively. During the treatment, variation of microbial growth was followed, where the highest microbial growth was obtained in the case of poultry manure (2.46×10^6 CFU/mL). The undertaken bioremediation process resulted in 85% removal of total petroleum hydrocarbon (TPH) after 5 months of treatment compared to 14% removal of TPH in control plots. Evaluation of operating costs indicated that the treatment cost per m^3 of pollutant would be approximately \$20/m^3.

Keywords Bioremediation · Oily sludge · Total petroleum hydrocarbons · Arid conditions

1 Introduction

One operation of drilling oil, in the *Hassi Messaoud* oil field, generates on average 1500 m^3 of solid wastes, composed mainly of residual fluids and cuttings stored in quagmire before treatment. It is to note that the diesel oil-based mud constitutes 90% of used drilling fluids adjusted with several chemical and mineral additives. Therefore, drilling wastes are composed of hydrocarbons and heavy metals posing environmental issues for industry and human health (Leonard and Stegemann 2010). Over

K. Ighilahriz (✉) · M. Khodja · A. Benchouk · H. Boutamine
Central Directorate Research and Development, SONATRACH, Boumerdes, Algeria
e-mail: kahina.ighilahriz@sonatrach.dz; k_ighilahriz@hotmail.com

S. Haddadi
Department of Bacterial Corrosion, Laboratory Division, SONATRACH, Boumerdes, Algeria

D. Trache et al. (eds.), *Materials Research and Applications*,
Materials Horizons: From Nature to Nanomaterials,
https://doi.org/10.1007/978-981-15-9223-2_12

the years, several remediation processes have been applied such as solidification and stabilization (Tuncan et al. 2000), incineration (Zhou et al. 2009), microwave heating technique (Shang et al. 2006), and surfactant-bio-surfactant washing (Hu et al. 2013; Lai et al. 2009). However, these techniques are expensive, less effective and may form a secondary pollution that requires subsequent treatment (Haque et al. 2008). The biological curing process of drilling waste has been attracting attention since the 1980s, and it continues to raise interest. Later studies classified the bioremediation as one of the most effective methods of soil remediation, ecofriendly, and economic (Wolińska et al. 2016). Different strategies were developed, natural attenuation, bio-stimulation, and bio-augmentation. Natural attenuation is frequently considered as the initial remediation mechanism, based on the use of natural processes to reduce the concentration of contaminants (Declercq et al. 2012; Mulligan et al. 2004). Bio-stimulation consists of adding nutrients such as nitrogen, phosphorus, urea, sawdust, compost, bio-solid, manure, and wastewater sludge in order to stimulate the activity and growth of native soil microbial populations (Cho et al. 1997; Namkoong et al. 2002; Rosenberg et al. 1992; Walworth et al. 1995; Williams et al. 1999; Agarry et al. 2010). Bio-augmentation includes the addition of strains or consortia of microorganisms having a metabolic ability to degrade hydrocarbons. Application of poultry manure can contribute to the degradation of hydrocarbons; poultry manure could be used for a double purpose for bioremediation and soil amendment. It is a rich source of nutrient and has a large number of microorganisms with biodegradable capacity (Semple et al. 2001; Wyss et al. 2006). Furthermore, biodegradation of organic pollutants can be enhanced by clay minerals (Knaebel et al. 1994), and clay minerals have a role in the biosorption of organic and inorganic compounds as well as microorganisms (Lin and Puls 2000) which provide an effective habitat and substrate for microorganisms (Southam et al. 2012). Consequently, better degrading microbe's activity and growth will be achieved by using this interaction, which in turn leads to better bioremediation of contaminants (Naidu et al. 2013).

The present study aims to investigate hydrocarbon degradation influenced by adding poultry manure and bentonite and to validate the performance of the method through a pilot-scale experiment under desert climate.

2 Materials and Methods

2.1 Site Description

The pilot site is situated in Hassi Messaoud (31° 40′ 45″ N, 6° 01′ 56″ E), which is located in the southeast of Algeria. Hassi Messaoud is considered to have a desert climate; the average annual temperature and rainfall are 22.4 °C and 40 mm, respectively. The experience was conducted from April to October, the meteorological conditions were unstable characterized by strong sand winds during April to June, extremes temperatures which reached 49 °C during July and August and severe changes of temperature between the day and the night occurred during September to October. We constructed four piles covered with a polyethylene waterproof membrane to avoid infiltration. The pollutant was transported from another site, and we conducted an ex-situ treatment.

2.2 *Experimental Procedure*

A volume of 7 m^3 oily sludge from the quagmire was dispersed through different parcels dimensioned as follows: 2 m × 1 m × 1 m in width, length, and depth, respectively, and each parcel received approximately 1000 tons of waste. A summary of treatment and control trays tested is presented in Table 1. Poultry manure was used as an amendment and a source of microbes, obtained from a farm located in Ouargla, Algeria. Bentonite was employed as a mineral stimulant; it was provided from a commercial supplier. Poultry manure was added at the beginning, and watering of the parcels to retain humidity and tilling was performed simultaneously to each parcel every 20 days expect *S*4, where no operation was applied on it. The pollutant was obtained from 6-month-old quagmire in Hassi Messaoud, and it was then characterized for initial concentrations of TPH, physical and chemical properties (Table 2). The sampling dates were 0 and 22 weeks after the beginning of treatment and kept at 4 °C before analysis (Figs. 1 and 2).

Microbial count (CFU/mL) was determined using a standard serial dilution and Tryptic Soy Agar (TSA) medium; the samples were incubated for 24 h at 37° C, and after incubation, the counting was carried out according to the following formula:

$$\text{Colony Forming Units/mL} = \frac{\text{colony number}}{\text{volume} \times \text{dilution}} \tag{1}$$

Toxicity tests were conducted on sludge at the beginning and after the treatment for the scenario who gives the best rate of TPH degradation. Soil toxicity tests were conducted using *Triticum durum Desf* and *Hordeum vulgare*(ISO 11269-2, OECD). The seeds were stored at 4 °C and in the dark before germination tests. Before use, the seeds were examined; damaged or small seeds are put aside. They were then disinfected in a bath of 5% sodium hypochlorite solution during 5 min and then rinsed with distilled water several times to remove all traces of chlorine. Petri dish

Table 1 Experimental setup for the treatment

Scenario	Treatment units
*S*1	Oily sludge + bentonite (1.5% (w/w) + water (10% v/v)
*S*2	Oily sludge + poultry manure (5% w/w) + bentonite (2% w/w) + water (30% v/v)
*S*3	Oily sludge + poultry manure (8% w/w) + water (40% v/v)
*S*4	Oily sludge (control)

Table 2 Initial composition of oily sludge

Parameters	Value	Analysis standard
Physical and chemical analysis		
pH	8.7	ISO 10390
Chemical oxygen demand (mg/L)	277	ISO 6060
Biological oxygen demand$_5$(mg/L)	8.5	ISO 5815
Total organic carbon (mg/L)	69.85	ISO 10694
Nitrate (mg/L)	5.5	ISO 14238
Nitrite (mg/L	0.066	
Phosphate (mg/L)	5.8	
TPH (g/kg)	87.1	API 13B-2
Saturated hydrocarbure (%)	68.71	NF T 60-115
Aromatic hydrocarbure (%)	8.48	
Resins (%)	10.52	
Asphalten (%)	12.29	
Total bacteria		
Microbial count (CFU/mL)	1.93×10^6	ISO 8199

The term 5 indicates the reaction time

Fig. 1 Biological treatment operation on site. (1): Mixture deposit, (2): parcels, (3): control

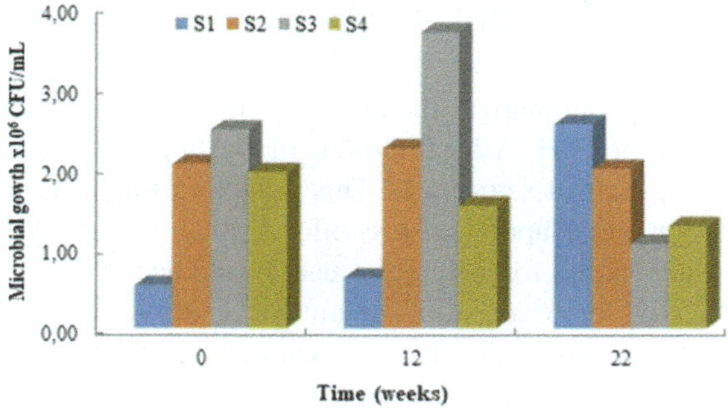

Fig. 2 Microbial count as function of time. *S*1: oily sludge + bentonite; *S*2: oily sludge + poultry manure + bentonite; *S*3: oily sludge + poultry manure; *S*4: control (Results represent the means of three separate experiments)

containing 20 g of contaminated, treated sludge, and uncontaminated soil was seeded in three replicates, containing 10 seeds in each essay. After 5 days of incubation in the dark, at room temperature, the seeds were examined to determine whether they had germinated or not (Plaza et al. 2005). The percentage of germination is expressed by the following formula (Saterbak et al, 1999; Rahman et al. 2002):

$$\text{Germination rate}(\%) = \frac{\text{number of germinated seeds}}{\text{number of seeds sown}} \times 100. \quad (2)$$

3 Results and Discussion

3.1 *Microbial Count*

An initial hydrocarbon-degrading microbial population of about 1.93×10^6, 5.22×10^5, 2.03×10^6, and 2.46×10^6 CFU/mL was observed at the beginning of the treatment for *S*4, *S*1, *S*2, and *S*3, respectively. Maximum microbial population, 2.23×10^6, 3.68×10^6 CFU/mL of sludge was observed under *S*2 and *S*3, respectively, after 2 months of experiment. Higher microbial population in *S*3 may be due to the nutrient induced by poultry manure addition. After 5 months, decline in microbial population occurred in *S*3 and *S*4, probably due to the lack of nutrient and degradable organic carbon (Martínez et al. 2017). On the other hand, adding bentonite (*S*1) resulted in the reduction of microbial population and then slowly increased after 2 months of reaction to attain 2.54×10^6 CFU/mL at the end of treatment. Probably, the microbial community took time to adapt to the new environmental conditions caused by the addition of bentonite. Bacteria that are attached faster to clay in the clay-microbial system can degrade hydrocarbons more effectively (Crocker et al. 1995; Wick et al. 2013).

3.2 TPH Degradation

Figure 3 summarizes the effect of different bioremediation strategies on the degradation of TPH. After 5 months of treatment, the lowest yield (14%) was obtained in the control setup (*S*4). This reduction might be a result of both abiotic degradation and evaporation loss effects. Shabir et al. (2008) found similar results. The results varied for different scenarios studied; *S*2 and *S*3 were found to be the most effective bioremediation treatment, resulting in 87 and 77% of TPH degradation. The previous studies have shown that poultry manure has microbial consortia that are proficient to degrade pollutants, as well as many nutrients, which makes it well suited for supporting comfortable microbial growth (Ezenne et al. 2014). In fact, the plate counts of bacteria revealed that poultry manure contained 2.32×10^6 CFU/mL. Similarly, Ezenne et al. (2014) determined the role of poultry droppings in the process of reducing the concentrations of TPH and poly-aromatic hydrocarbon (PAH) in crude-oil-contaminated soils. The application rate of animal manure had significant reduction effects on the concentrations of TPH and PAH in the contaminated soils, and 94.0% of TPH degradation occurred in contaminated soil samples. Otherwise, Hesnawi et al.(2013) reported that the addition of poultry manure to diesel fuel contaminated soil led to increase the degradation of TPH with a rate of 72.2% compared to a degradation in the control soil (52.4%), after 5 months of treatment. Besides, *S*2 gives better degradation yield compared to *S*3, and this is due to the presence of bentonite. Later, studies classified mineral clay as an effective support material; it offers a protective habitat for microorganisms by forming a biofilm (Bhabananda et al. 2015). In the other part, clay mineral possesses a high specific surface area and cationic exchange capacity (Churchman et al. 2006), and this encourages the adhesion of pollutant on clay surface and clumping of microorganisms (Vieira and Melo 1995). When using clay mineral alone in the case of *S*1, the degradation rate was not significant, and it achieved only 50% after 5 months of treatment. This can be explained by the fact that clay enhances the degradation activity of microorganisms contained in the contaminated sludge or in the case of microorganisms lack, the clay would serve as a vector for delivering the selective microorganisms to the sludge (Sowers et al. 2012).

3.3 Toxicity Test

The results of seed germination before and after bioremediation (*S*2) are exposed in Fig. 4. In the uncontaminated soils, seed germination of wheat and barley species achieved 100% germination rate. In contrast, seed germination in the contaminated sludge for both wheat and barley was negative. The oil may seep into the seed, affect the metabolic reactions, and can disrupt or block embryogenesis or even kill the embryo. Besides, the oil film formed around the seeds can act as a physical barrier, reducing both the transfer of water and gases assimilated by the seeds; thus,

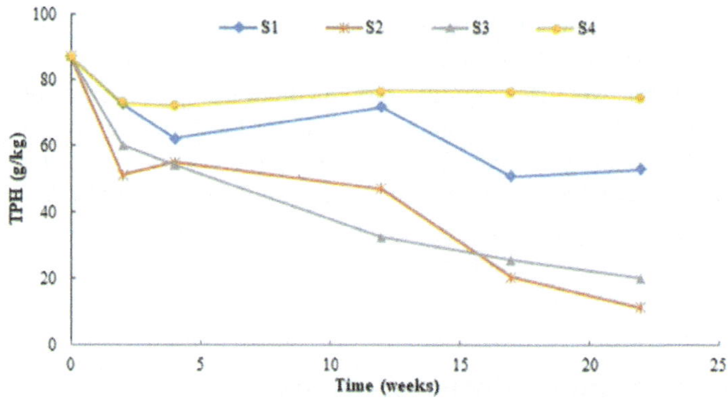

Fig. 3 TPH as function of time for different strategies. *S*1: oily sludge + bentonite; *S*2: oily sludge + poultry manure + bentonite; *S*3: oily sludge + poultry manure; *S*4: control ((Results represent the means of three separate experiments)

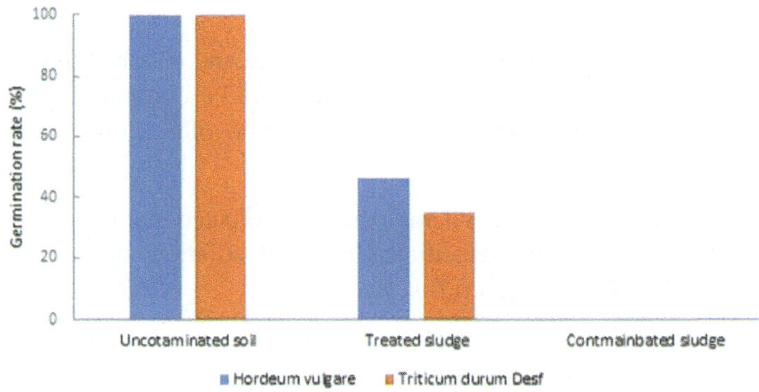

Fig. 4 Germination rate for *Hordeum vulgare* and *Triticum durum Desf* (results represent the means of three separate experiments)

the germination can be inhibited (Chaineau et al. 1997). In the treated sludge, we obtained more than 40% of the germination rate with a better trend for barley.

3.4 Cost Analysis

The decision to industrialize a process is based primarily on the assessment of its economic interest and therefore the calculation of its profitability. The cost of a bioremediation process depends on the methodologies used, the location of site treatment, the quantity, and characteristics of the pollutant treated. A cost analysis of the present test comprises the cost of amendments \$12/m^3, excavation and a liner cost about \$5/m^3, and the transportation of amendments cost \$3/m^3, and thus, the treatment cost per m^3 of sludge would be approximately \$20/m^3. Later studies showed that a bioremediation test can cost \$22.5 to \$230.5 per m^3 of contaminant (Line et al. 1996), which classes the bioremediation as a cost-effective process. In general, the cost of bioremediation depends on the time required for the test, the amount of pollutant treated which will affect the cost of excavation and liner (Troquet and Troquet 2002) and the methodology (in situ or ex situ). In case of in situ bioremediation, the treatment of pollutant where it was found can be considered cheaper compared to ex situ bioremediation.

4 Conclusions and Future Perspectives

From the present study, it can be concluded that:

- Bioremediation is a suitable process for oily sludge from oil drilling quagmire.
- The oily sludge contained a variety of indigenous microorganisms (1.93×10^6 CFU/mL).
- Using poultry manure and bentonite can stimulate and enhance the degradation process, and the degradation rate was improved by73% compared to control test.
- The cost analysis showed that the bioremediation of the oily sludge can be remediated for \$20/m^3.

The results from this pilot-scale demonstration indicated that the oily sludge can be effectively and economically remediated by bioremediation under arid conditions and has a potential to be used in a full-scale.

References

Agarry SE, Owabor CN, Yusuf RO (2010) Studies on biodegradation of kerosene in soil under different bioremediation strategies. Bioremediat J 14:135–141

Bhabananda B, Binoy S, Ruhaida R, Ravi N (2015) Bioremediation of PAHs and VOCs: advances in clay mineral-microbe interaction. Environ Int 85:168–181

Chaineau CH, Morel JL, Oudot J (1997) Phytotoxicity and plant uptake of fuel oil hydrocarbons. Journal OfEnvironmental Quality 26:1478–1483

Cho BH, Chino H, Tsuji H, Kunito T, Nagaoka K, Otsuka S, Yamashita K, Matsumoto S, Oyaiz H (1997) Laboratory-scale bioremediation of oil-contaminated soil of Kuwait with soil amendment materials. Chemosphere 35:1599–1611

Churchman GJ, Gates WP, Theng BKG, Yuan G (2006) Clays and clay minerals for pollution. Develop Clay Sci 1:625–675

Crocker FH, Guerin WF, Boyd SA (1995) Bioavailability of naphthalene sorbed to cationic surfactant-modified smectite clay. Environ Sci Technol 29:2953–2958

Declercq I, Cappuyns V, Duclos Y (2012) Monitored natural attenuation (MNA) of contaminated soils: state of the art in Europe-a critical evaluation. Sci Total Environ 426:393–405

Ezenne GI, Nwoke OA, Ezikpe DE, Obalum SE, Ugwuishiwu BO (2014) Use of poultry droppings for remediation of crude-oil-polluted soils: effects of application rate on total and poly-aromatic hydrocarbon concentrations. Int Biodeteriorat Biodegradat 92:57–65

Haque N, Peralta-Videa JR, Jones GL, Gill TE, Gardea-Torresdey JL (2008) Screening the phytoremediation potential of desert broom (Braccharissarathroides Gray) growing on mine tailings in Arizona, USA. Environ Pollut 153:362–368
Hesnawi RM, Adbeib MM (2013) Effect of nutrient source on indigenous biodegradation of diesel fuel contaminated soil. APCBEE Procedia 5:557–561
Hu G, Li J, Zeng G (2013) Recent development in the treatment of oily sludge from petroleum industry: a review. J Hazard Mater 261:470–490
Knaebel DB, Federle TW, McAvoy DC, Vestal JR (1994) Effect of mineral and organic soil constituents on microbial mineralization of organic compounds in a natural soil. Appl Environ Microbiol 60:4500–4508
Lai CC, Huang YC, Wei YH, Chang JS (2009) Biosurfactant-enhanced removal of total petroleum hydrocarbons from contaminated soil. J Hazard Mater 167:609–614
Leonard SA, Stegemann JA (2010) Stabilization/solidification of petroleum drill cuttings: leaching studies. J Hazard Mater 174:484–491
Lin Z, Puls RW (2000) Adsorption, desorption and oxidation of arsenic affected by clay minerals and aging process. Environ Geol 39:753–759
Line MA, Garland CD, Crowley M (1996) Evaluation of landfarm remediation of hydrocarbon-contaminated soil at the Inveresk Railyard, Lauceston, Australia. Waste Manage 7:567–570
Martínez Álvarez LM, Ruberto L, Lo Balbo A, Mac Cormack WP (2017) Bioremediation of hydrocarbon-contaminated soils in cold regions: development of a pre-optimized biostimulation biopile-scale field assay in Antarctic. Sci Total Environ 15:194–203
Mulligan CN, Yong RN (2004) Natural attenuation of contaminated soils. Environ Int 30:587–601
Naidu R (2013) Recent advances in contaminated site remediation. Water Air Soil Pollut 224:1–11
Namkoong W, Hwang E, Park J, Choi J (2002) Bioremediation of diesel-contaminated soil with composting. Environ Pollut 119:23–31
Plaza G, Grzegorz NJ, Krzysztof U, Robin L, Brigmon, (2005) The application of bioassays as indicators of petroleum-contaminated soil remediation. Chemosphere 59:289–296
Rahman KSM, Banat IM, Thahira J, Thayumanavan T, Lakshmanaperumalsamy P (2002) Bioremediation of gasoline contaminated soil by a Microbial consortium amended with poultry litter, coir pith, and rhamnolipidbiosurfactant. Bioresources Technol 1:25–32
Rosenberg E, Legmann R, Kushmaro A, Taube R, Adler E, Ron EZ (1992) Petroleum bioremediation—a multiphase problem. Biodegradation 3:337–350
Saterbak A, Toy RJ, Wong Diana CL, Mcmain Bruce J, Williams MP, Dorn Philip B, Brzuzy Louis P, Chai Eric Y, Salanitro Joseph P (1999) Ecotoxicological and analytical assessment of hydrocarbon-contaminated soils and application to ecological risk assessment. Environ Toxicol Chem 18(7):1591–1607
Semple KT, Reid BJ, Fermor TR (2001) Impact of composting strategies on the treatment of soils contaminated with organic pollutants. Environ Pollut 112:269–283
Shabir G, Muhammad A, Anwar F, Tahseen R, Khalid ZM (2008) Biodegradation of kerosene in soil by a mixed Microbial culture under different nutrient conditions. Int Biodeteriorat Biodegradat 61:161–166
Shang H, Snape CE, Kingman SW, Robinson JP (2006) Microwave treatment of oil contaminated North Sea drill cuttings in a high power multimode cavity. Sep Purif Technol 49:84–90
Southam G (2012) Mineral as substrates for life: the prokaryotic view. Elements 8:101–106
Sowers KR, Kjellerup B, Ghosh U (2015) Organic biofilm substrata as a microbial inoculum delivery vehicle for bioaugmentation of persistent organic pollutants in contaminated sediments and soils. US Patent, No. US8945906B2
Troquet J, Troquet M (2002) Economic aspects of polluted soil bioremediation. WIT Press, Ashurst Lodge, Southampton, SO40 7AA, UK
Tuncan A, Tuncan M, Koyuncu H (2000) Use of petroleum contaminated drilling wastes as subbase material for road construction. Waste Manage Res 18:489–505
Vieira MJ, Melo LF (1995) Effect of clay particles on the behavior of biofilms formed by Pseudomonas fluorescens. Water Sci Technol 32:45–52

Walworth JL, Reynolds CM (1995) Bioremediation of a petroleum- contaminated cryic soil: effects of phosphorus, nitrogen, and temperature. J Soil Contaminat 4:299–310
Wick LY, Springael Harms H(2013) Microbial strategies to improve the bioavailability of hydrophobic organic. In: Treatment of Contaminated Soil: Fundamentals, Analysis, Applications, Stegmann et al. (Eds.), Springer, 203–217
Williams CM, Grimes JL, Mikkelsen RL (1999) The use of poultry litter as co-substrate and source of inorganic nutrients and microorganisms for the ex situ biodegradation of petroleum compounds. Poult Sci 78:956–964
Wolińska A, Kuźniar A, Szafranek-Nakonieczna A, Jastrzębska N, Roguska E, Stępniewska Z (2016) Biological activity of autochthonic microbial community in oil-contaminated soil. Water Air Soil Pollut 227:130
Wyss A, Boucher J, Montero A, Marison I (2006) Micro-encapsulated organic phase for enhanced bioremediation of hydrophobic organic pollutants. Enzyme Microbial Technol 40:25–31
Zhou L, Jiang X, Liu J (2009) Characteristics of oily sludge combustion in circulating fluidized beds. J Hazard Mater 170:175–179

Chapter 13
Pretreatment of Seawater Using Precipitation Agents to Retard Scale Formation in Multi-Stage Flash Evaporators

Khedidja Dahmani, Djamal Eddine Kherroub, and Mohamed Belloul

Abstract The objective of this study is to investigate the formation of scale at the level of multi-stage flash evaporators (MSF). Scale (tartar) is the result of the precipitation of insoluble species such as calcium sulfate, calcium carbonate, and magnesium hydroxide. As a solution to this problem, we propose to pre-treat seawater before introducing it into evaporators. Pre-processing by the precipitation process is a method that protects evaporators (MSF). Our strategy consists of pre-treating seawater by three different ways using potassium hydroxide at a temperature of 25 °C, sodium carbonate at different temperatures 25, 35, 50, and 100 °C, and the combination of potassium hydroxide and sodium carbonate at 25 and 100 °C. As a result, we noticed that the treatment of 1 L of seawater heated at 100 °C using 10 g of sodium carbonate completely eliminated the calcium ions and reduced the concentration of magnesium ions from 747.22 to 142.8 mg/L. The treatment of the filtered solution of 10 g/L of sodium carbonate heated to 100 °C with 3 g of potassium hydroxide, however, allowed the removing of the remaining magnesium ions. On the other hand, the pre-treatment of seawater carried out with a mixture of sodium carbonate and potassium hydroxide and heated at 100 °C gave the best result through the total elimination of both ions of sodium carbonate and magnesium.

Keywords Seawater · Temperature · Evaporators · Alkaline tartar · Sodium carbonate · Potassium hydroxide

---

K. Dahmani (✉)
Institute of Sciences and Technology, University Center Ahmed Zabana of Relizane, BP 48000 Relizane, Algeria
e-mail: khedidja.dahmani@hotmail.fr; khadidjadahmani76@yahoo.fr

D. E. Kherroub
Polymer Chemistry Laboratory, Department of Chemistry, Faculty of Exact and Applied Sciences, University of Oran, 1 Ahmed Ben Bella, BP 1524 El'Menouer, 31000 Oran, Algeria

K. Dahmani · M. Belloul
University Abdelhamid Ibn Badis of Mostaganem, 27000 Mostaganem, Algeria

D. Trache et al. (eds.), *Materials Research and Applications*,
Materials Horizons: From Nature to Nanomaterials,
https://doi.org/10.1007/978-981-15-9223-2_13

1 Introduction

The multi-stage flash (MSF) is a thermal distillation process, which represents one of the most efficient technologies for the desalination of seawater, due to the better quality of production of freshwater obtained (Nigim and Eaton 2017; Roy et al. 2017). Distilled water produced from this technique generally has a total dissolved solid (STD) content of less than 50 ppm (Bandi et al. 2016). Seawater expands in the evaporators due to the increase in temperature and depressurization inside the evaporators. Then, the rising vapors condense to produce distilled water (El-Ghonemy 2017). The material structure of this process mainly consists of heat exchangers and brine heater. The MSF process is applied in many countries of the world, where it represents around 64% of the total world capacity for the production of distilled water (Shatat and Riffat 2014). Algeria uses this process to produce distilled water from seawater. The seawater desalination plant called Kahrama consists of three units for which each unit produces distilled water with a rate of 29.629 m^3/day.

However, the high temperature in the MSF process causes the formation of scale on the surface of evaporators, which leads to a reduction in the production of distilled water, and sometimes even the shutdown of an entire unit at the factory (Al-Hamzah and Fellows 2015; Al-Rawajfeh 2016). Scale formation is caused by the crystallization of calcium carbonate, magnesium hydroxide salts as well as calcium sulfate (Ghani and Al-Deffeeri 2010; Al-Jaroudi et al. 2010). The formation of alkaline tartar of calcium carbonate and magnesium hydroxide depends on the temperature, concentration of calcium, magnesium, bicarbonate and carbonate ions, and pH (Alsadaie and Mujtaba 2017). On the other hand, at the MSF evaporators, the non-alkaline tartar of sulfates results from the direct crystallization of anhydrite ($CaSO_4$), hemihydrate ($CaSO_4 . ½H_2O$), or gypsum ($CaSO_4 . H_2O$) (Bystrianský et al. 2016). Several studies have been devoted to combating the precipitation of mineral salts on the hot surfaces of the MSF system. Among the prevention methods used, one can cite the addition of acid to prevent decomposition of bicarbonate, incorporation of anti-tartar agents such as tartar inhibitors (Ghani and Al-Deffeeri 2010), and periodic cleaning to limit the formation of scale at the evaporators. Nanofiltration is a pre-treatment process used to remove mineral ions from seawater, during the supply of thermal power plants. Nanofiltration integrated into thermal desalination allows this process to operate at high temperatures. Other methods use reverse osmosis as a pre-treatment step for the feed water of MSF desalination plants to reduce the concentration of mineral ions (Thabit et al. 2019). Al Khadra et al. (2020) used shock electrodialysis to desalt seawater. They chose sodium citrate as a reagent to avoid scaling and showed that 99.99% of the magnesium has been eliminated.

The brine temperature depends on the concentration of calcium and magnesium ions. The upper brine temperature reaches 160 °C after using a graphene oxide membrane as a pre-treatment step to remove these substances (Abdelkader et al. 2019). Pre-treatment by precipitation is an important strategy, which is based on the injection of carbon dioxide with the addition of sodium hydroxide to remove calcium from seawater (Zhao et al. 2013). In 2016, the same authors (Zhao et al. 2016) studied the mechanism of the pre-treatment by precipitation. They found that calcium and magnesium precipitate quickly with the addition of sufficient soda (NaOH). In a previous work, Ayoub et al. (2014) have shown that pre-treatment of seawater using precipitation agents at a pH of 12 could completely remove calcium and magnesium ions.

The present work, however, aims to eliminate calcium and magnesium ions in seawater by precipitation before introducing it into desalination evaporators, using the following precipitation agents: sodium carbonate, potassium hydroxide, and coupling sodium carbonate–potassium hydroxide at the same time. The evolution of the elimination rate as a function of operating conditions, such as temperature and concentrations, has also been followed.

2 Methods and Materials

2.1 Properties of Seawater

In order to give more socioeconomic impact to our study, the seawater used in our study is the same one used by Kahrama. It is a factory located at the seaside in Oran, intended for the desalination of seawater. The water used is characterized by a very high hardness reaching 4600 mg/L, which means a strong presence of calcium and magnesium ions. The strong presence of these substances is directly responsible for the formation of tartar. So we propose a technique for pre-treating seawater before feeding it to multi-stage flash evaporators. Precipitants such as potassium hydroxide, sodium carbonate, and their combination are used. The composition of the seawater used by Kahrama is shown in Table 1.

Table 1 Seawater composition of Kahrama desalination plant

Parameters	Value
Total hardness, mg/L	4600.00
Ca^{2+}, mg/L	610.00
Mg^{2+}, mg/L	747.22
Cl^{-}, mg/L	20,412.50
TA, mg/L	25.00
TAC, mg/L	85.00
TDS, at 180 °C, mg/L	38,942.00
Na^{+}, mg/L	13,013.00
pH	7.89
Temperature, °C	25.00

2.2 *Pre-Treatment of Seawater: Elimination of Calcium and Magnesium*

Before starting the study of the precipitation pre-treatment process, we measured the following characteristic parameters of seawater such as pH, the concentration of calcium (Ca^{+2}), magnesium (Mg^{+2}), bicarbonates (HCO^{3-}), carbonates ($CO_3{}^{2-}$), and hydroxide ions, as well as the total hardness (TH).

To measure the concentration of calcium and magnesium ions, the complexometric titration method (EDTA) was used. The pH value of the seawater and the filtered solutions was determined by using an electrode of the Starter 210 pH meter. The measurement of these parameters was carried out at 25 °C.

Experiment 1: In order to determine the optimal conditions for precipitation with potassium hydroxide (KOH), an amount of KOH varied from 1 to 9 g was introduced into a series of nine vials containing 1 L of seawater. Each mixture (seawater and KOH) was stirred for 5 min at a temperature of 25 °C and then was filtered and analyzed. The pH, the total hardness (TH), and the concentration of calcium and magnesium were then determined.

Experiment 2: In order to determine the optimal quantity of sodium carbonate to use, we introduced different quantities of sodium carbonate ranging from 1 to 13 g into 13 beakers; each beaker contains 1 L of seawater. Each mixture (seawater and sodium carbonate) was well stirred for 5 min at a speed of 250 rpm. It was then heated to different temperatures: 25, 35, 50, and 100 °C.

Experiment 3: In the third process, to study the pre-treatment by precipitation of seawater, different quantities of potassium hydroxide ranging from 1 to 6 g were introduced into a series of six bottles for which each containing 1 L of a filtered solution previously treated with 10 g/L of sodium carbonate. The mixture was then stirred for 5 min at a speed of 250 rpm at a temperature of 25 and 100 °C, separately. All filtered solutions were analyzed at 25 °C.

3 Results and Discussion

3.1 *Elimination of Calcium and Magnesium Ions by the Addition of KOH*

The variation in the concentrations of calcium and magnesium ions in seawater versus the amount of KOH used is presented in Fig. 1. When the amount of KOH added increases in the seawater solution, the concentrations of calcium and magnesium ions are decreased from 610 mg/L and 747.22 mg/L to 140 mg/L and 36.46 mg/L, respectively. The addition of 9 g of KOH to 1 L of the solution seawater has led to the almost total removal of magnesium ions and to a significant reduction in the concentration of calcium ions.

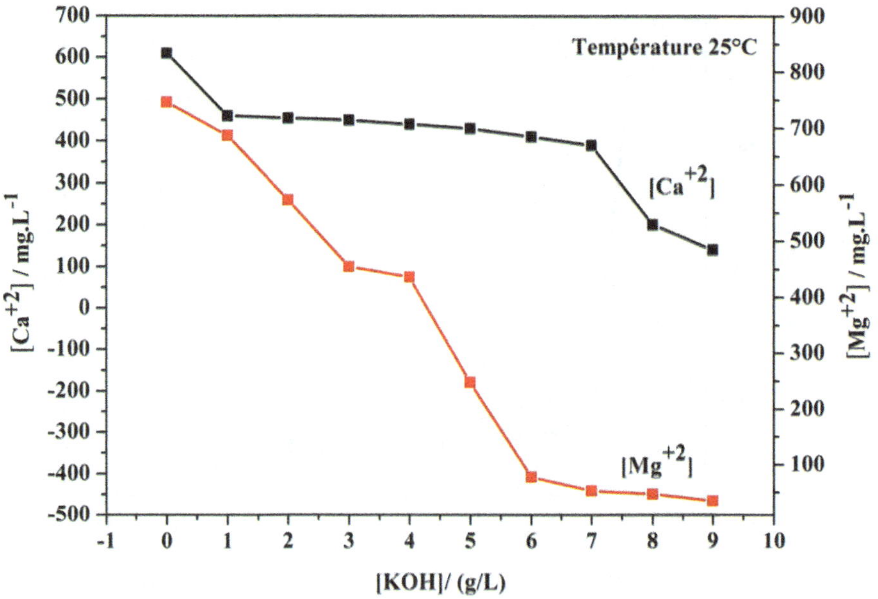

Fig. 1 Calcium and magnesium content as a function potassium hydroxide concentration at 25 °C

3.2 *Evaluation of the Influence of the Added Amount of Sodium Carbonate (Na_2CO_3)*

Figures 2 and 3, respectively, show the evolution of the concentration of calcium and magnesium as a function of the amount of sodium carbonate added to seawater at different temperatures of 25, 35, 50 and 100 °C. It is clearly remarkable the decrease in the concentrations of calcium and magnesium ions with the increase in the amount of sodium carbonate used. For example, the concentrations of calcium and magnesium ions have been reduced to 5 mg/L and 379.91 mg/L, respectively, after the addition of 13 g of sodium carbonate. This result means that sodium carbonate promotes the precipitation of calcium. Similar result was mentioned by the Jeldres and Arancibia-Bravo (2017), where the calcium concentration decreased to 0 mg/L after the addition of 0.05 M sodium carbonate and stirring for 5 min at room temperature. The results also exhibit the temperature effect since the precipitation of calcium and magnesium ions is more favored at high temperatures. The addition of 10 g of sodium carbonate to 1 L of seawater at 25 °C led to a decrease of the concentration of calcium and magnesium ions from 610 mg/L and 747.22 mg/L to 40 mg/L and 413.10 mg/L, respectively. A similar amount of sodium carbonate added to 1 L of seawater heated to 100 °C can lead to a decrease of the concentration of calcium and magnesium from 610 mg/L and 747.22 mg/L to 2.5 mg/L and 142.7 mg/L, respectively. On the other hand, one can observe that beyond this quantity of 10 g of sodium carbonate, the concentrations of calcium and magnesium have remained constant. Hence, 10 g

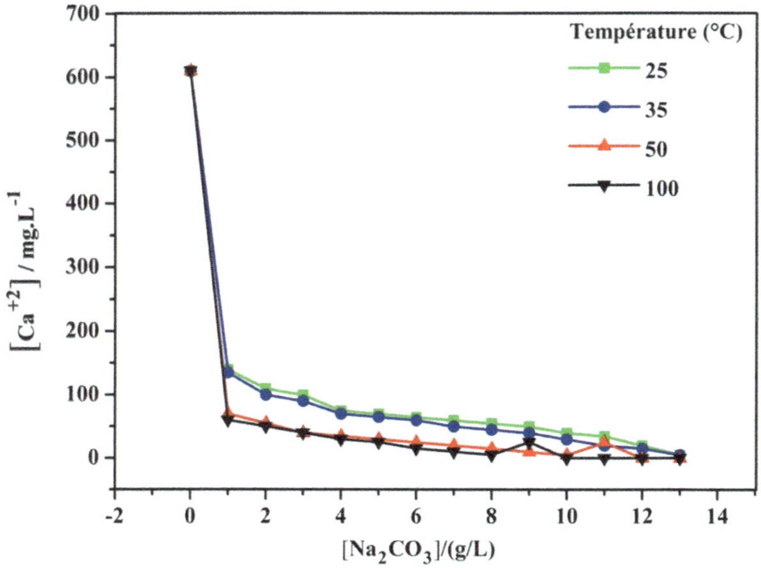

Fig. 2 Ca^{+2} concentrations in filtered seawater solution after adding various amounts of soda ash (Na_2CO_3) at 25, 35, 50, and 100 °C

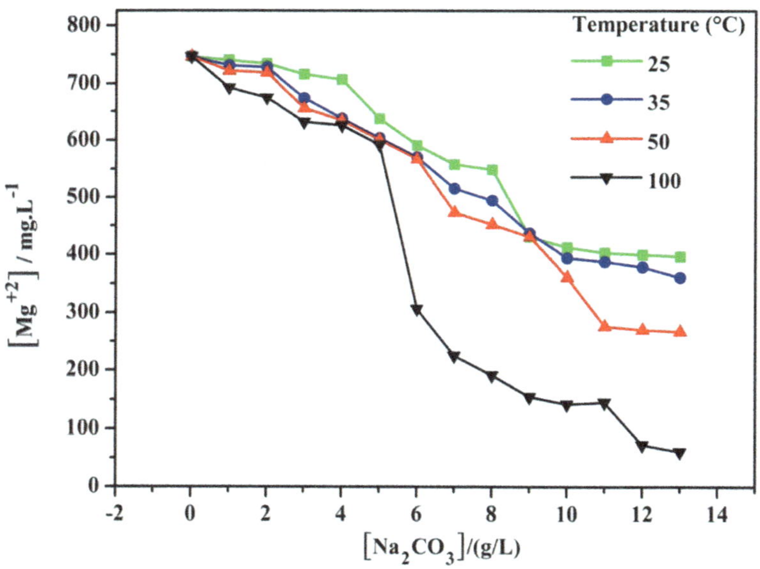

Fig. 3 Mg^{+2} concentrations in filtered seawater solution after adding various amounts of soda ash (Na_2CO_3) at 25, 35, 50, and 100 °C

of sodium carbonate added to 1 L of seawater heated to 100 °C can be assumed as the optimal amount to use.

When sodium carbonate is added to seawater, it dissolves, producing sodium ions [Na^+] and carbonate ions [CO_3^{-2}] (reaction 1). The carbonate ions react afterward with the calcium and magnesium ions present in seawater to give insoluble precipitates of calcium carbonate [$CaCO_3$] and magnesium carbonate [$MgCO_3$], respectively (reactions 2 and 3).

$$\mathbf{Na_2CO_3 \rightarrow 2Na^+ + CO_3^{-2}} \quad (1)$$

$$\mathbf{Ca^{+2} + CO_3^{-2} \rightarrow CaCO_3} \quad (2)$$

$$\mathbf{Mg^{+2} + CO_3^{-2} \rightarrow MgCO_3} \quad (3)$$

After that, at a moderate temperature, the bicarbonate ions in seawater decompose into carbonates by the following reaction (reaction 4).

$$\mathbf{HCO_3^- + OH^- \rightarrow CO_3^{-2} + H_2O} \quad (4)$$

The carbonate ions produced by reaction (4) allows the precipitation of calcium carbonate. As the temperature increases, the carbonate ions hydrolyze even more by the following reaction (reaction 5).

$$\mathbf{CO_3^{-2} + H_2O \rightarrow OH^- + CO_2} \quad (5)$$

Consequently, the production of hydroxyl groups causes the precipitation of magnesium hydroxide ($Mg(OH)_2$).

3.3 Evaluation of the Influence of Potassium Hydroxide Added to the Filtered Solution of Sodium Carbonate at 25 and 100 °C

Sodium carbonate provides carbonate ions to the seawater solution, which then reacts with calcium and magnesium to precipitate as calcium carbonate and magnesium carbonate, respectively. Adding KOH to the filtered solution previously treated with sodium carbonate at 25 °C (Fig. 4) leads to an increase in the pH of the solution. This consequently increases the concentration of hydroxide ions, which allows the decomposition of bicarbonate ions into carbonates. Therefore, the concentrations of calcium and magnesium ions were reduced to zero after the addition of 6 g of KOH to a liter of filtered seawater solution previously treated with 10 g of sodium carbonate at 25 °C. Sodium carbonate ions that exist at relatively high pH produce carbonate and

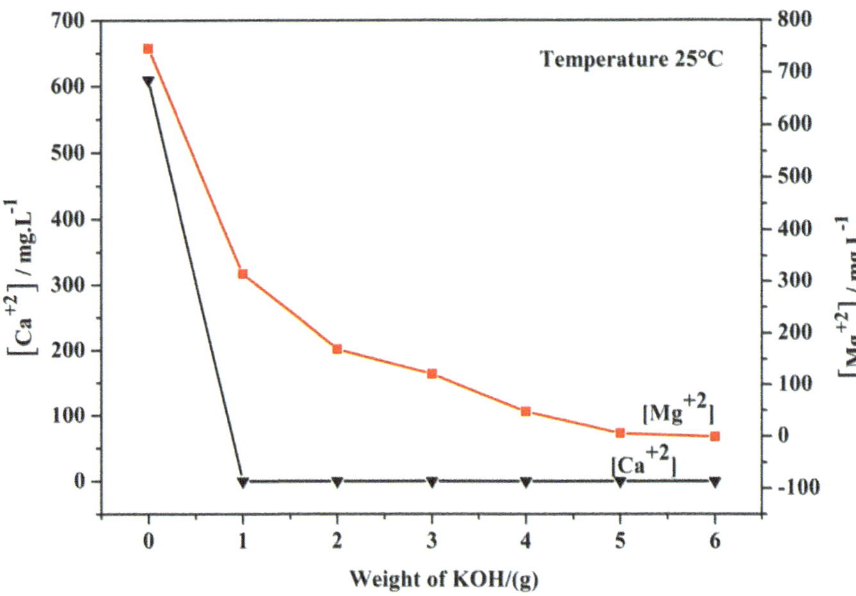

Fig. 4 Ca^{+2} and Mg^{+2} concentrations in filtered solution at 25 °C after adding various amounts of potassium hydroxide at a concentration of soda ash of 10 g/L

hydroxide ions. The two ions CO_3^{-2} and OH^- react with calcium and magnesium to form calcium carbonate and magnesium hydroxide, respectively. Adding KOH to a filtered solution previously treated with 10 g of sodium carbonate at 100 °C (Fig. 5) ensures the continued precipitation of the remaining magnesium ions. The calcium and magnesium ions were completely eliminated after the incorporation of 3 g of KOH to a filtered solution previously treated with 10 g of sodium carbonate at 100 °C. The high temperature hydrolyzes carbonate ions to generate hydroxide ions. Consequently, the hydroxide ions react with the remaining magnesium ions from the filtered solution and precipitate as magnesium hydroxide.

3.4 Alkalinity and PH behavior

The concentrations of HCO^{3-}, CO_3^{2-}, OH^- ions, and the pH value after adding different amounts of KOH to a filtered solution of seawater previously treated with 10 g of sodium carbonate at 100° C are shown in Fig. 6. At a temperature of 100 °C, the addition of 1 g of KOH to the filtered solution of seawater previously treated with 10 g of sodium carbonate is completely sufficient for the total disappearance of the bicarbonate ions for which the concentration has become 0 mg/L. On the other hand, the concentrations of carbonate and hydroxide ions increased with the increase in the added amount of KOH. The treatment at 100 °C, for example, by adding 3 g of

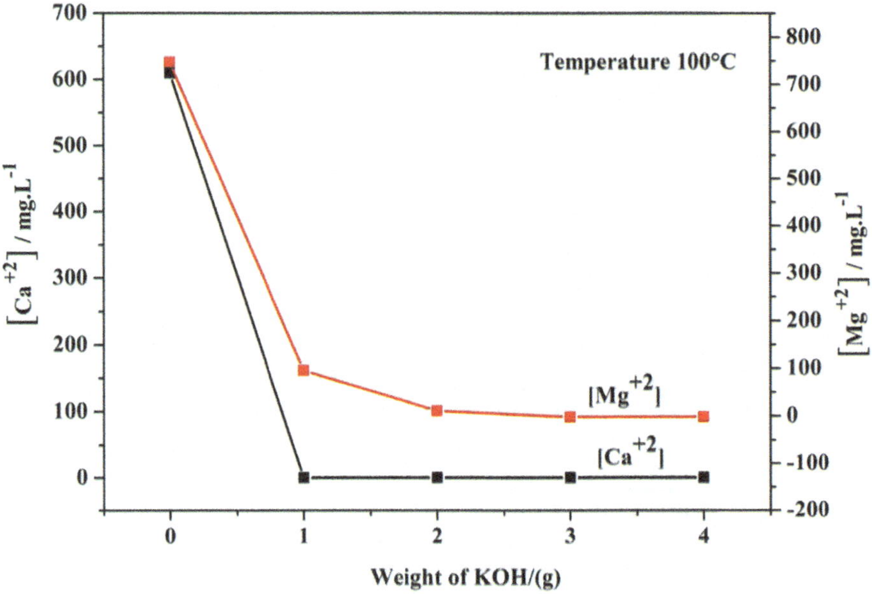

Fig. 5 Ca^{+2} and Mg^{+2} concentrations in filtered solution at 100 °C after adding various amounts of potassium hydroxide at a concentration of soda ash of 10 g/L

KOH changed the concentrations of bicarbonate, carbonate, and hydroxide ions from 40, 50, and 0 mg/L to 0, 4670, and 4680 mg/L, respectively. The pH of seawater at 25 °C is 7.89. The pH curve (Fig. 6) shows the change in pH of the filtered solution of seawater, preliminarily treated with 10 g of sodium carbonate at 100 °C. The pH value increased with the addition of KOH for which the value reached 11.8 after adding 3 g of KOH. This result shows the total elimination of calcium and magnesium ions by precipitation with potassium hydroxide and sodium carbonate can be carried out at pH 11.8.

4 Conclusion

In order to prevent scale formation at the multi-stage flash evaporators, the pretreatment of seawater seems a very effective approach before feeding. Precipitation is a technique used in the present work to remove calcium and magnesium ions suing different precipitation agents such as sodium carbonate, potassium hydroxide, and their combination. At a temperature of 25 °C, KOH promotes the precipitation of magnesium ions and reduces its concentration from 747.225 mg/L to 36.45 mg/L.

The precipitation with sodium carbonate is useful for reducing calcium ions more than magnesium ions. Thus, the addition of sodium carbonate to seawater heated at 100 °C decreased the concentration of calcium and magnesium ions with 95.90% and

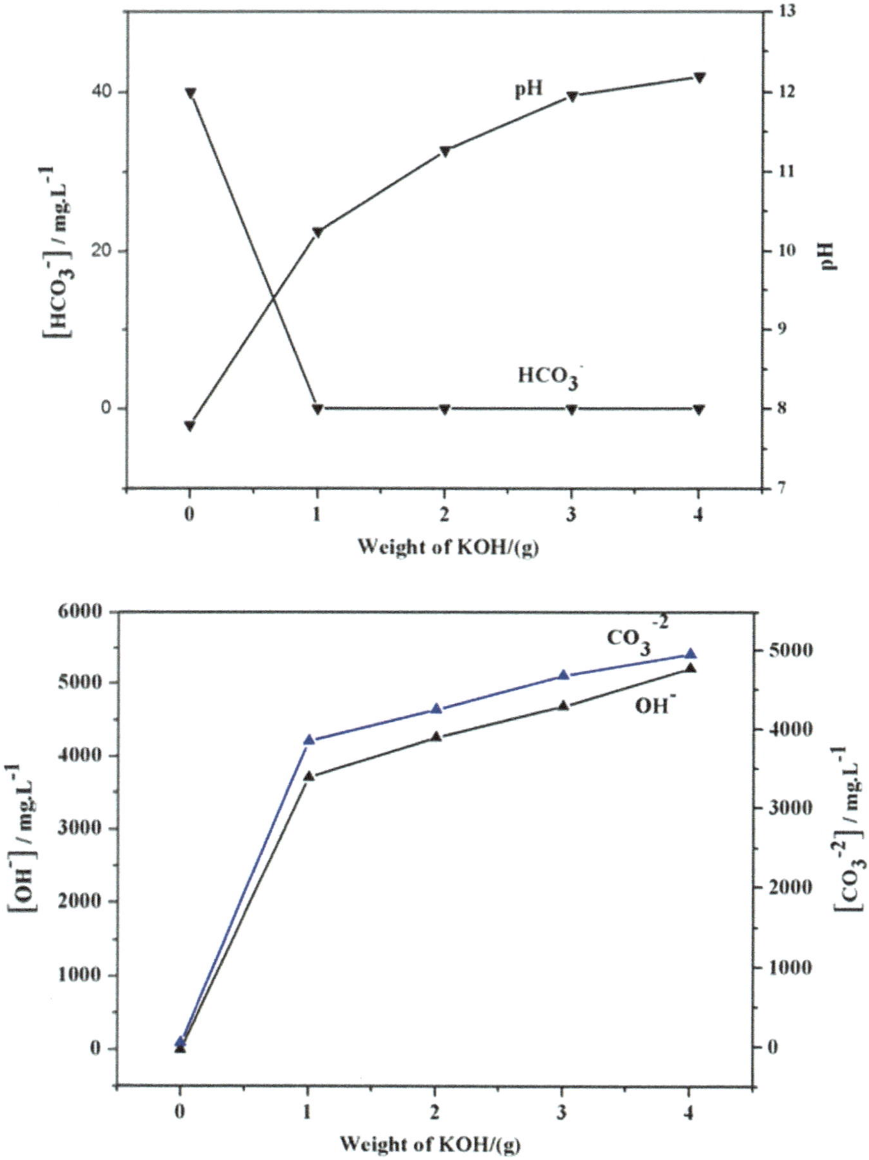

Fig. 6 pH and concentration of HCO_3^-, CO_3^{-2}, and OH^- in filtered solution at 100 °C after adding various amounts of KOH at a concentration of Na_2CO_3 of 10 g/L

80.89%, respectively. At a temperature of 100 °C, the incorporation of KOH to the filtered solution, previously treated with 10 g of sodium carbonate, completely eliminated the calcium and magnesium ions. Raising the temperature and combining KOH and Na_2CO_3 ensured the best result for the removal of Ca^{+2} and Mg^{+2} ions. Therefore, we recommend this precipitation process as a pretreatment step of seawater before introducing it into multi-stage flash evaporators to delay the formation of tartar.

References

Abdelkader BA, Antar MA, Laoui T, Khan Z (2019) Development of graphene oxide-based membrane as a pretreatment for thermal seawater desalination. Desalination 465:13–24

Al-Hamzah AA, Fellows CM (2015) A comparative study of novel scale inhibitors with commercial scale inhibitors used in seawater desalination. Desalination 359:22–25

Al-Jaroudi SS, Ul-Hamid A, Al-Matar JA (2010) Prevention of failure in a distillation unit exhibiting extensive scale formation. Desalination 260:119–128

Alkhadra MA, Gao T, Conforti KM, Tian H (2020) Small-scale desalination of seawater by shock electrodialysis. Desalination 476:114–219

Al-Rawajfeh AE (2016) Nanofiltration pretreatment as CO_2 deaerator of desalination feed: CO_2 release reduction in MSF distillers. Desalination 380:12–17

Alsadaie SM, Mujtaba IM (2017) Crystallization of calcium carbonate and magnesium hydroxide in the heat exchangers of once-through multistage flash process desalination. Comput Aided Chem Eng 40:349–354

Ayoub GM, Zayyat RM, Al-Hindi M (2014) Precipitation softening: a pretreatment process for seawater desalination. Environ Sci Pollut Res 21:2876–2887

Bandi CS, Uppaluri R, Kumar A (2016) Global optimization of MSF seawater desalination processes. Desalination 394:30–43

Bystrianský M, Nir O, Šír M, Honzajková Z, Vurm R, Hrychová P, Bervic A, Bvd B (2016) The presence of ferric iron promotes calcium sulphate scaling in reverse osmosis processes. Desalination 393:115–119

El-Ghonemy AMK (2017) Performance test of a sea water multi-stage flash distillation plant: case study. Alexandria Eng J 57:2401–2413

Ghani S, Al-Deffeeri NS (2010) Impacts of different antiscalant dosing rates and their thermal performance in multi stage flash (MSF) distiller in Kuwait. Desalination 250:463–472

Jeldres RI, Arancibia-Bravo MP (2017) The impact of seawater with calcium and magnesium removal for the flotation of copper-molybdenum sulphide ores. Miner Eng 109:10–13

Nigim TH, Eaton JA (2017) CFD prediction of the flashing processes in a MSF desalination chamber. Desalination 420:258–272

Roy Y, Thiel GP, Antar MA, Lienhard VJH (2017) The effect of increased top brine temperature on the performance and design of OT-MSF using a case study. Desalination 412:32–38

Shatat M, Riffat SB (2014) Water desalination technologies utilizing conventional and renewable energy sources. Int J Low-Carbon Technol 9:1–19

Thabit MS, Hawari AH, Ammar MH, Zaidi S, Zaragoza G, Altaee A (2019) Evaluation of forward osmosis as a pretreatment process for multi stage flash seawater desalination. Desalination 461:22–29

Zhao Y, Yuan J, Zhang J, Xie L, Ji Z, Su M, Chen J (2013) A different approach for seawater decalcification pretreatment using carbon dioxide as precipitator. Desalination 322:15–158

Zhao Y, Cao H, Xie Y, Yuan J, Ji Z, Yan Z (2016) Mechanism studies of a CO_2 participant softening pretreatment process for seawater desalination. Desalination 393:166–173

Chapter 14
Conversion of Methane Over $LaNiZnO_3$ Perovskite Catalysts

Djamila Sellam, Kahina Ikkour, Sadia Dekkar, and Mourad Halouane

Abstract Carbon dioxide reforming of methane reaction (DRM) is an important process to produce synthesis gas at low H_2/CO ratio and to consume the two main greenhouse gases CH_4 and CO_2. Nickel-based catalysts are the most frequently used in dry reforming of methane. However, deactivation of these catalysts is a major problem due to carbon deposition and metal sintering. In order to minimize the deactivation phenomena of the nickel-based catalysts, a great attention is given to the use of the perovskite oxide (ABO_3) catalysts. These oxides have a defined structure that yields small metallic particles highly active and quite stable during DRM reaction. This work deals dry reforming of methane to produce synthesis gas (H_2/CO), using the mixed oxides $LaNi_{1-x}Zn_xO_3$ ($x = 0, 0.2, 0.8$ and 1) perovskite-type catalysts. The auto-combustion using glycine as ignition promoter was the method for the synthesis of these catalysts, and then, they were characterized by using XRD, FTIR, and N_2 physisorption. XRD analysis of the solids $LaNi_{1-x}Zn_xO_3$ ($x = 0, 0.2, 0.8$) shows the formation of perovskite structure at 2θ between 32° and 33°. At high Zn content ($x = 1$), the catalyst showed the presence of individual metal oxides ZnO and La_2O_3. These oxides are confirmed by FTIR characterization with observation of bands in the range 450–700 cm^{-1} which are attributed to M–O-M of the perovskite structure and M–O of the simple oxide. The effect of partial substitution of Ni by Zn on the catalytic performance has been studied. $LaNi_{0.8}Zn_{0.2}O_3$ was the most active toward the CH_4 and CO_2 conversions and was selective for synthesis gas products at 750 °C. This activity was attributed to high dispersion of the small metallic nickel particles (19 nm) and the high surface area (27 m^2/g).

Keywords Methane · Catalyst · Perovskite · $LaNiZnO_3$

D. Sellam (✉) · S. Dekkar · M. Halouane
Laboratoire de Chimie Appliquée et de Génie Chimique, Université Mouloud Mammeri (UMMTO), 15000 Tizi-Ouzou, Algeria
e-mail: sellam_dj@yahoo.fr

K. Ikkour
Laboratoire de Physico-Chimie Des Matériaux et Catalyse (LPCMC), Faculté de Technologie, Université de Bejaia, 06000 Bejaia, Algeria

D. Trache et al. (eds.), *Materials Research and Applications*,
Materials Horizons: From Nature to Nanomaterials,
https://doi.org/10.1007/978-981-15-9223-2_14

1 Introduction

Dry reforming of methane reaction ($CH_4 + CO_2 \rightarrow 2H_2 + 2CO$ $\Delta H^{o}_{r\,298}$ = 247 kJ/mol) has received attention from an environmental protection point of view, because the conversion harmful gas to useful synthesis gas is possible, which is an important intermediary to produce chemicals and fuels and at the same time, to avoid the emission of CH_4 and CO_2 in atmosphere that causes the greenhouse effect (Subramani and Gangwal 2008; Liu et al. 2010; Raudaskoski et al. 2009). The CO_2 reforming of methane reaction gives H_2/CO molar ratio around unity, which is more suitable for Fischer–Tropsch synthesis to produce long-chain hydrocarbons (Zhang and Verykios 1996; Kapokova et al. 2011). DRM is a highly endothermic reaction requiring high reaction temperatures at which coke formation and sintering of catalysts become serious problems (Valderrama et al. 2008). Catalysts based on group VIIIB metals (Rh, Pt, Ni, Co, Fe, Ir …) are known to be more or less active and selective. The catalysts based on noble metals are very active for this reaction (Solymosi et al. 1991). Despite the high activity and stability of the noble metals, the use of these metals remains limited because of their high cost. From an industrial point of view, nickel is the most suitable catalyst for the reaction due to its availability and low cost. In comparison with noble metals, nickel-based catalysts showed similar activities and selectivities to those reported for these metals (Goula et al. 2006; Enger et al. 2008). However, their major problem is deposition of carbon on active species and metal sintering which can cause catalyst deactivation. Several works have been done and reported on understanding of deactivation phenomena of Ni-based catalysts, and several efforts have been made in order to improve their stability in CO_2 reforming of methane (Zhang et al. 2007). Numerous investigations attempted to reduce these problems by insertion the metal into well-defined structures as perovskite-type oxides. These oxides with general formula ABO_3 (where A^{x+} and B^{y+} are two cations having sum of charges compensating the charge of the three oxygen) have attracted significant interest as precursors in the catalytic DRM. These oxides present high thermal stability and produce very small metallic particles, highly dispersed at the surface of a basic support under reduction process of the B site elements, which can improve the catalytic activity and stability and inhibit carbon formation (Goldwasser et al. 2005). Another advantage of perovskite-type structures is the possibility of total or partial substitution of its cations (A and B) with other metals could modify their oxidation state, mobility of oxygen lattice, and redox properties. These characteristics can modify the catalytic activity and stability of the catalyst (Valderrama et al. 2005, 2008, 2010; Gallego et al. 2008). Partial substitution at the B site creates isostructural crystals $AB_{1-x}B_xO_3$, which can lead to structural distortions and electronic defects, by stabilization of unusual oxidation states of the B components. It can also produce highly dispersed B metallic particles upon reduction treatments and reduce metal sintering during reaction conditions, which strongly affects the catalytic activity (Provendier et al. 1999). Moradi et al. (2014) studied the effect of partial substitution of Ni by Zn in $LaNi_{1-x}Zn_xO_3$ perovskites (x = 0 to 1) on the CO_2 reforming of methane (CH_4/CO_2 /He=1/1/1, 750 °C). The results showed that the conversion rates of CO_2 and CH_4 increased as Ni was partially substituted by Zn. For low x value ($x \leq 0.4$), the mixed oxides are efficient for the production of synthesis gas. Among all the catalysts, $LaNi_{0.8}Zn_{0.2}O_3$ showed superior activity in terms of high CH_4 and CO_2 conversions and resistance to carbon deposition. TPR results reveal that the catalyst reduced at high temperature indicates a more stable of the perovskite structure, and this can reduce the Ni sintering during the reaction. It indicates that Zn stabilizes the perovskite structure. For this reason, $LaNi_{0.8}Zn_{0.2}O_3$ was more resistant to coke formation in comparison with $LaNiO_3$ during 75 h time on stream stability test. Sutthiumporn et al. (2012) studied the effect of substitution of (M=Bi, Co, Cr, Cu, Fe) in the $La_{0.8}Sr_{0.2}Ni_{0.8}M_{0.2}O_3$ perovskite system. They found enhanced mobility of lattice oxygen. This oxygen species was found to be the reason for the removal of coke formed during the DRM reaction. Song et al. (2016) synthesized Fe substituted La_2NiO_4 and $LaNiO_3$ perovskites and used them as catalysts for CO_2 reforming of methane. Fe partial substitution enhanced significantly the stability of the perovskite structure and increased carbon resistance in these catalysts, which is attributed to the small Ni metallic particles highly dispersed, resulted from the strong metal–support interaction. A stable Ni-based catalyst with high carbon resistance for dry reforming of methane is obtained through the $LaNi_{0.5}Fe_{0.5}O_3$ precursor. Gallego et al. (2009) studied the effect of substitution of the A site in $LaNiO_3$ by Pr and Ce. After reduction treatment, the average diameters of nickel metallic obtained from $LaNiO_3$, $La_{0.9}Ce_{0.1}NiO_3$ and $La_{0.9}Pr_{0.1}NiO_3$ were 15 nm, 9 nm, and 6 nm, respectively. $La_{0.9}Pr_{0.1}NiO_3$ catalyst showed high catalytic activity without carbon deposition after 100 h of reaction. This stability was attributed to the small nickel metallic particles as well as to the redox property of Pr_2O_3. It is reported that during the dry reforming of methane on the $LaNiO_3$-based catalysts, the CO_2 adsorbs on La_2O_3 and generates the formation of the oxycarbonate $La_2O_2CO_3$ species, and these latter stabilize the nickel particles inhibiting sintering (Su et al. 2014). In the present work, Zn was chosen for partial substitution of Ni in $LaNiO_3$ perovskite. Therefore, a series of $LaNi_{1-x}Zn_xO_3$ ($x = 0$, 0.2, 0.8 and 1) perovskites catalysts were prepared by auto-combustion method. The aim of the study was to investigate the catalytic performances of these materials with respect to the dry reforming of methane into synthesis gas.

2 Experimental

2.1 *Preparation*

The oxides perovskites $LaNi_{1-x}Zn_xO_3$ ($x = 0$, 0.2, 0.8 and 1) were prepared by the auto-combustion method using stoichiometric amounts of $La(NO_3)0.6H_2O$, $Ni(NO_3)_2,6H_2O$, and $Zn(NO_3)_2,6\ H_2O$ nitrates as starting material (Batiot-Dupeyrat et al. 2005). In a first step, nitrate precursors with a mass ratio of lanthanum nitrate to nickel nitrate and zinc nitrate of 3:2 were dissolved in a limited volume of distilled water (20 mL). After complete dissolution at room temperature, glycine was then added to the nitrate solution with NO_3^-/glycine molar ratio of 1 under constant agitation for about 2 h. The resulting solution was slowly evaporated at 100 °C until a vitreous gel was obtained. Then, glycine auto-ignition occurred on increasing the temperature to 280 °C. The powder obtained was calcined by increasing the temperature from ambient to steady temperature of 750 °C at a rate of 1 °C/min. It is maintained at that temperature for 4 h in order to obtain the perovskite structure and to eliminate the residual carbonaceous.

2.2 *Characterization*

The catalysts were characterized by X-ray powder diffraction (XRD) on D8-DVANCE diffractometer using filtered Cu-Kα radiation ($\lambda = 1.506$ Å) at 2θ between 10° and 80° with a 0.020° step and an acquisition time of 0.80 s at each step for crystalline phase detection. We used database to identify the crystalline phases. The crystallite size was calculated from the full width at half maximum of the most intense diffraction peaks using Scherer's equation. Infrared spectroscopy (FTIR) was performed by diluting with KBr in a pellet on spectrometer with transform of Fourier FTIR 8400 marks SHIMADZU in order to determine various bands, which characterize the structure. Textural properties were obtained by N_2-sorption experiment at −196 °C on a Micromeritics ASAP 2420 apparatus. Prior to analysis, the samples were degassed at a temperature of 250 °C. The BET method was employed to measure the specific surface areas.

2.3 *Catalytic Activity*

CH_4/CO_2 reforming reaction was performed at atmospheric pressure in a U quartz reactor (internal diameter of = 6.0 mm) containing 25 mg of catalyst. Before reaction, the catalysts were reduced with pure hydrogen (20 mL.min^{-1}) at 600 °C for 1 h. Then, the reactant mixture (CH_4/CO_2=1/1, at a GHSV of 600 L.h^{-1}.g^{-1}) was introduced at 750 °C. After stabilization of the reacting flow, the activity was measured each 20 min for 4 h. The reactants (CH_4 and CO_2) and the products formed (CO and H_2) were analyzed by an online gas chromatograph (SHIMADSU, GC–14B) equipped with a TCD using a Carbosieve column.

3 Results and Discussions

3.1 Physicochemical Characterization

Figure 1 summarizes the results of the XRD analysis of the calcined solids. The patterns of $LaNi_{1-x}Zn_xO_3$ (x = 0, 0.2 and 0.8) catalysts present a crystalline phase which corresponds to a structure perovskite similar to $LaNiO_3$. The position of the characteristic peak of $LaNiO_3$ was located at $2\theta = 32.9°$ (Batiot-Dupeyrat et al.2005). According to DRX patterns, the obtained perovskite seems pure because of formation of some crystalline such as La_2O_3 and NiO. However, In the case of $LaZnO_3$, the structure perovskite was not formed, and two phases were observed, corresponding to the simple oxides of ZnO and La_2O_3. The infrared spectrum of calcined catalysts is shown in Fig. 2. It reveals the existence of the following bands: A broad absorption band at 3480 cm^{-1} corresponds to the vibration of the O–H bond of the hydroxyl groups and to the water molecules at the catalyst's surface. Relatively more intense bands detected around 1384 and 1650 cm^{-1} are attributed to free CO_2 vibration modes. Bands in the range 450–700 cm^{-1} were also observed which are attributed to M–O–M of the perovskite structure (Firdous et al. 2009) and M–O of the simple oxide (Nakamato 1986; Fernandes et al. 2002). These results confirmed those obtained with XRD analysis.

Table 1 shows the specific surface areas and particle sizes of our solids $LaNi_{1-x}Zn_xO_3$ (x = 0, 0.2, 0.8, and 1). From these results, we recorded specific surface areas between 20 and 31 m^2/g. This result is consistent with several studies on perovskites prepared by the auto-combustion method (Batiot-Dupeyrat et al. 2005). The specific surface area values of our solids are much higher than the other prepared

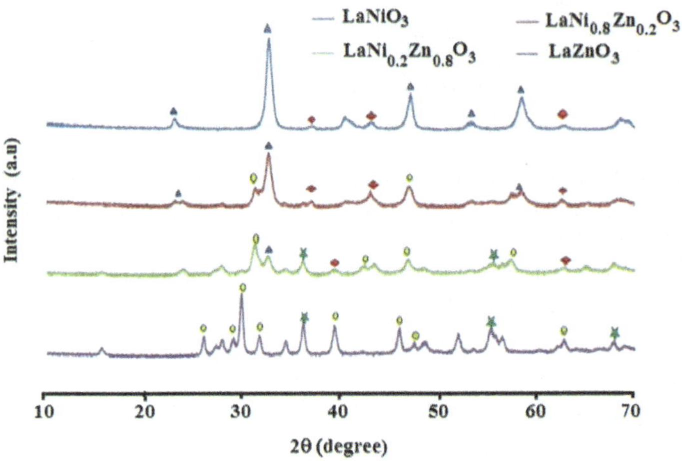

Fig. 1 XRD patterns of the prepared catalysts $LaNi_{1-x}Zn_xO_3$ (x = 0, 0.2, 0.8 and 1) ▲$LaNiO_3$ ◆NiO ○La_2O_3 ☆ZnO

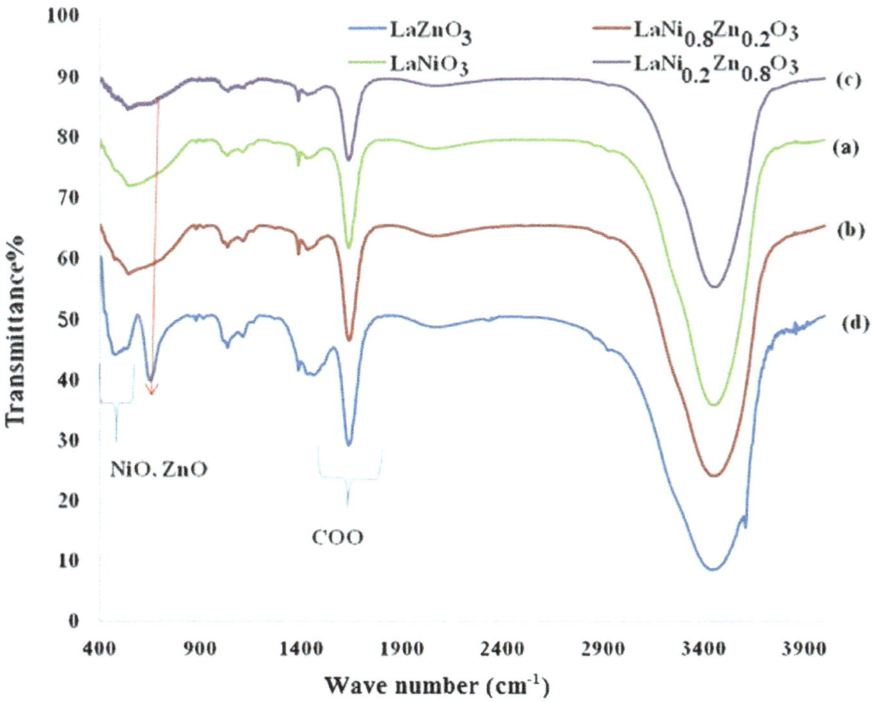

Fig. 2 FTIR spectrum of the prepared catalysts: **a** $LaNiO_3$, **b** $LaNi_{0.8}Zn_{0.2}O_3$, **c** $LaNi_{0.2}Zn_{0.8}O_3$, **d** $LaZnO_3$

Table 1 Specific surface areas and particle sizes of prepared $LaNi_{1-x}Zn_xO_3$ (x = 0, 0.2, 0.8, and 1) catalysts

$LaNiO_3$	$LaNi_{0.8}Zn_{0.2}O_3$	$LaNi_{0.2}Zn_{0.8}O_3$	$LaZnO_3$
Surface area (m^2/g)20	27	25	31
Crystal size (nm)		231,920	25

by sol–gel and hydrothermal method (Sagar et al. 2014). With this method of preparation, we also recorded a decrease in particle sizes from 23 nm of $LaNiO_3$ to 19 nm of $LaNi_{0.8}Zn_{0.2}O_3$.The effect of substitution of Ni by Zn on the specific surface areas and the particle sizes of the catalysts is outstanding; the increase in specific surface areas and the decrease in particle sizes are probably due to structural defects.

3.2 *Catalytic Activity*

Figures 3 and 4 summarize the results of the activity of $LaNi_{1-x}Zn_xO_3$ catalysts under the following test conditions: (CH_4/CO_2=1/1, GHSV of 600 $L.h^{-1}.g^{-1}$, Tr =

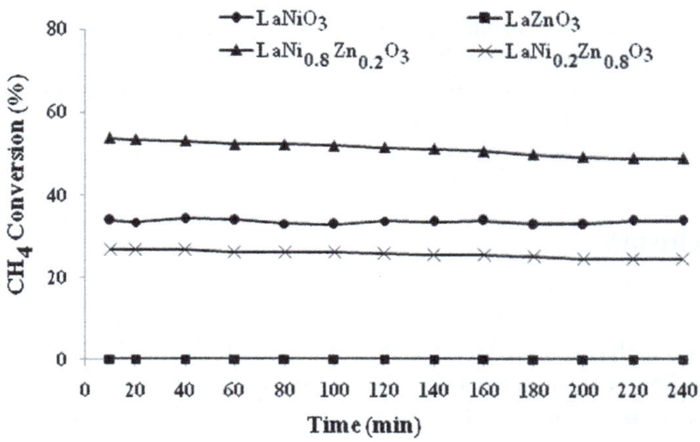

Fig. 3 Conversion of CH_4 over reduced $LaNi_{1\text{-}x}Zn_xO_3$ (x = 0, 0.2, 0.8 and 1) catalysts. Feed: CH_4/CO_2 = 50/50 mL.min^{-1}, GHSV = 600 L.h^{-1}.g^{-1}, T = 750 °C

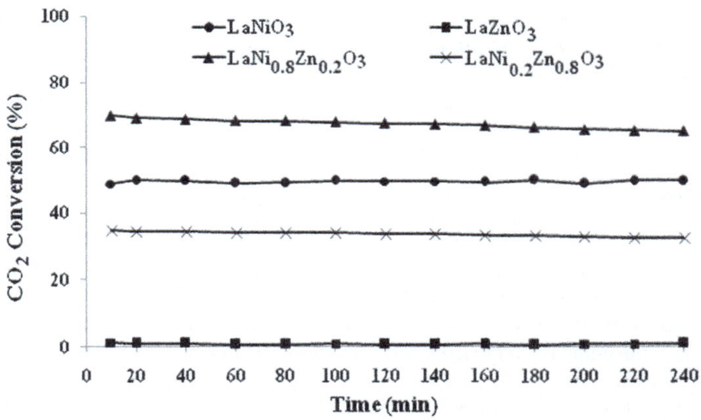

Fig. 4 Conversion of CO_2 over reduced $LaNi_{1\text{-}x}Zn_xO_3$ (x = 0, 0.2, 0.8 and 1) catalysts. Feed: CH_4/CO_2 = 50/50 mL.min^{-1}, GHSV = 600 L.h^{-1}.g^{-1}, T = 750 °C

750 °C). According to the obtained results, we note that the $LaZnO_3$ is inactive. Numerous authors reported that the active species in dry reforming of methane is the metallic Ni° particles present at the catalyst's surface (Liu and Au 2003). However, the $LaNi_{0.8}Zn_{0.2}O_3$ catalyst showed higher catalytic activity than the $LaNiO_3$ one. After 3 h of time on stream, methane and carbon dioxide conversions remained constant, respectively, at around 50% and 65%. During the reaction, the conversion of CO_2 is higher than the conversion of methane, even the values of H_2/CO ratio are near to 0.8 less than 1 and the formation of water, suggesting that besides methane dry reforming reaction, parallel reactions may be present, and among these reactions, reverse water–gas shift reaction ($CO_2 + H_2 \rightarrow CO + H_2O$) is occurring. In general, CO_2 reforming

of methane reaction is typically accompanied by the simultaneous occurrence of reverse water–gas shift reaction (Rabelo-Neto et al. 2018). We attributed the better activity of this catalyst to the small metallic nickel particles (19 nm) high dispersed at the surface and the high surface area (27 m^2/g).

4 Conclusion

The perovskite oxides and $LaNi_{1-x}Zn_x$ O_3 (x = 0, 0.2, 0.8, and 1) were prepared by the auto-combustion method. Their activity was tested in the dry reforming of methane at 750 °C. The formation of the solid solution La-Ni-Zn–O (x = 0, 0.2, and 0.8) with perovskite structure is confirmed by XRD and FTIR analysis. The results showed that the partial substitution of nickel by zinc in $LaNiO_3$ perovskite improved the physicochemical and catalytic properties of the mixed oxide catalysts. The $LaNi_{0.8}Zn_{0.2}O_3$ catalyst presented a relatively high surface area (27m^2/g) and the small particle sizes (19 nm). Among the synthesized catalysts, $LaNi_{0.8}Zn_{0.2}O_3$ exhibited the highest catalytic activity, which is probably attributed to the small Ni^0 particle size (19 nm) highly dispersed at the surface of the catalyst and the high surface area (27 m^2/g).

References

Batiot-Dupeyrat C, Sierra GA, Mondragon F, Barrault J, Tatibouet JM (2005) CO_2 reforming of methane over $LaNiO_3$ as precursor material. Catal Today 107:474–480

Enger BC, Lødeng R, Holmen A (2008) A review of catalytic partial oxidation of methane to synthesis gas with emphasis on reaction mechanisms over transition metal catalysts. Appl Catal A: Gen 346:1–27

Fernandes JDG, Melo DMA, Zinner LB, Salustiano CM, Silva ZR, Martinelli AE, Cerqueira M, AlvesJúnior C, Longo E, Bernardi MIB (2002) Low-temperature synthesis of single-phase crystalline $LaNiO_3$ perovskite via Pechini method. Mater Lett 53:122–125

Firdous A, Quasim I, Ahmad MM, Kotru PN (2009) Studies on gel-grown pure and strontium-modified lanthanum tartrate crystals. J Cryst Growth 311:3855–3862

Gallego GS, Batiot-Dupeyrat C, Barrault J, Florez E, Mondragon F (2008) Dry reforming of methane over LaNi1−yByO3±δ (B=Mg, Co) perovskites used as catalyst precursor. Appl Catal A: Gen 334:251–258

Gallego J, Batiot-Dupeyrat C, Barrault J, Mondragón F (2009) severe deactivation of a $LaNiO_3$ perovskite-type catalyst precursor with H_2S during methane dry reforming. Energy Fuels 23:4883–4886

Goldwasser MR, Rivas ME, Pietri E, Pérez-Zurita MJ, Cubeiro ML, Grivobal-Constant A, Leclercq G (2005) Perovskites as catalysts precursors: synthesis and characterization. J Mol Catal A: Chem 228:325–331

Goula G, Kiousis V, Nalbandian L, Yentekakis IV (2006) Catalytic and electrocatalytic behavior of Ni-based cermet anodes under internal dry reforming of CH_4+ CO_2 mixtures in SOFCs. Solid State Ionics 177:2119–2123

Kapokova L, Pavlova S, Bunina R, Alikina G, Krieger T, Ishchenko A, Rogov V, Sadykov V (2011) Dry reforming of methane over $LnFe_{0.7}Ni_{0.3}O_{3-\delta}$perovskites: Influence of Ln nature. Catal Today 164:227–233

Liu BS, Au CT (2003) Carbon deposition and catalyst stability over $La_2NiO_4/\gamma\text{-}Al_2O_3$ during CO_2 reforming of methane to syngas. Appl Catal A Gen 244:181–195

Liu D, Cheo WNE, Lim YWY, Borgna A, Lau R, Yang Y (2010) A comparative study on catalyst deactivation of nickel and cobalt incorporated MCM-41 catalysts modified by platinum in methane reforming with carbon dioxide. Catal Today 154:229–236

Moradi GR, Rahmanzadeh M, Khosravian F (2014) The effects of partial substitution of Ni by Zn in $LaNiO_3$ perovskite catalyst for methane dry reforming. J CO_2 Utilization 6: 7–11

Nakamato K (1986) Infrared and raman spectra of inorganic and coordination compound, 4th edn. Wiley, New York, USA

Provendier H, Petit C, Estournes C, Libs S, Kiennemann A (1999) Stabilisation of active nickel catalysts in partial oxidation of methane to synthesis gas by iron addition. Appl Catal A: Gen 180:163–173

Rabelo-Neto RC, Sales HBE, Inocêncio CVM, Varga E, Oszko A, Erdohelyi A, Noronha FB, Mattos LV (2018) CO_2 reforming of methane over supported $LaNiO_3$ perovskite-type oxides. Appl Catal B: Environ 221:349–361

Raudaskoski R, Turpeinen E, Lenkkeri R, Pongrácz E, Keiski RL (2009) Catalytic activation of CO_2: Use of secondary CO_2 for the production of synthesis gas and for methanol synthesis over copper-based zirconia-containing catalysts. Catal Today 144:318–323

Sagar TV, Sreelatha N, Hanmant G, Surendar M, Lingaiah N, Rama Rao KR, Satyanarayana CVV, Reddy IAK, Sai Prasad PS (2014) Influence of method of preparation on the activity of La–Ni–Ce mixed oxide catalysts for dry reforming of methane. RSC Adv 4:50226–50232

Solymosi F, Kutsán G, Erdöhelyi A (1991) Catalytic reaction of CH_4 with CO_2 over alumina-supported Pt metals. Catal Lett 11:149–156

Song X, Dong X, Yin S, Wang M, Li M, Wang H (2016) Effects of Fe partial substitution of $La_2NiO_4/LaNiO_3$ catalyst precursors prepared by wet impregnation method for the dry reforming of methane. Appl Catal A: Gen 526:132–138

Su YJ, Pan KL, Chang MB (2014) Modifying perovskite-type oxide catalyst $LaNiO_3$ with Ce for carbon dioxide reforming of methane. Int J Hydrogen Energy 39:4917–4925

Subramani V, Gangwal SK (2008) A review of recent literature to search for an efficient catalytic process for the conversion of syngas to ethanol. Energy Fuels 22:814–839

Sutthiumporn K, Maneerung T, Kathiraser Y, Kawi S (2012) CO_2 dry-reforming of methane over $La_{0.8}Sr_{0.2}Ni_{0.8}M_{0.2}O_3$perovskite (M=Bi Co, Cr, Cu, Fe): Roles of lattice oxygen on C–H activation and carbon suppression. Int J Hydrogen Energy 37:11195–11207

Valderrama G, Goldwasser MR, de Navarro CU, Tatibouët JM, Barrault J, Batiot-Dupeyrat C, Martínez F (2005) Dry reforming of methane over Ni perovskite type oxides. Catal Today 107:785–791

Valderrama G, Kiennemann A, Goldwasser MR (2008) Dry reforming of CH_4 over solid solutions of $LaNi_1{-}xCo_xO_3$. Catal Today 133:142–148

Valderrama G, Kiennemann A, Goldwasser MR (2010) La-Sr-Ni-Co-O based perovskite-type solid solutions as catalyst precursors in the CO_2 reforming of methane. J Power Sour 195:1765–1771

Zhang Z, Verykios XE (1996) Mechanistic aspects of carbon dioxide reforming of methane to synthesis gas over Ni catalysts. Catal Lett 38:175–179

Zhang J, Wang H, Dalai AK (2007) Development of stable bimetallic catalysts for carbon dioxide reforming of methane. J Catal 249:300–310

Chapter 15
Homogenous Photo-Fenton Catalytic Oxidation of Olive Mill Wastewater Using Plackett and Burman's Design

Hamida Iboukhoulef, Hocine Kadi, and Ramdane Moussaoui

Abstract The main purpose of this study was to investigate the removal of the phenolic compounds from olive mill wastewater (OMW) by combination of catalytic and UV light processes. It is assumed that several factors are involved; thus, an experimental design of considered experiments is required as an effective statistical technique for optimizing them. The effect of the independent parameters such as type of catalyst, the stirring rate, pH, dilution of OMW, and UV light on phenolic removal was assessed. It has been found that the homogenous photo-Fenton method could efficiently reduce phenolic compounds. The results showed that all the factors, namely the type of catalyst, the stirring speed, the pH and the dilution are influential on the degradation of the phenolic compounds of olive mill wastewater.

Keywords Olive mill wastewater (OMWs) · Homogenous photo-Fenton · Phenolic compounds · Experimental design

1 Introduction

Olive mill wastewater (OMW) and olive pomace constitute the main by-products of the olive oil mills. The estimated annual volume of OMWs generated by the Mediterranean region is approximately 30 million m^3, and this huge amount is creating a serious ecological and environmental issue resulting in severe consequences for the soil and water quality (Gudina et al. 2016). OMWs are a cloudy aqueous liquid, reddish-brown to black in color depending on the state of degradation of the phenolic components and the olive from which they come. It has a pleasant smell and a bitter taste. Their physico-chemical properties are relatively different, according to the weather, the type of olives, time of storage, and the extraction method. OMWs have a very polluting activity: a biological oxygen demand (BOD) of 12–63 g L^{-1} and a chemical oxygen demand (COD) of 80–200 g L^{-1}. These values are 200–400 times

H. Iboukhoulef (✉) · H. Kadi · R. Moussaoui
Laboratoire de Chimie Appliquée et Génie Chimique, Université Mouloud Mammeri, Tizi-Ouzou, Algérie
e-mail: bhamidam@yahoo.fr; hamida.iboukhoulef@ummto.dz

D. Trache et al. (eds.), *Materials Research and Applications*,
Materials Horizons: From Nature to Nanomaterials,
https://doi.org/10.1007/978-981-15-9223-2_15

higher than municipal waters. The acidity of the margins (pH<5), the high salinity and the high concentration of phenolic compounds has harmful effects on the environment in general (0.5–24 g L^{-1}) cause the destruction of the soil microFlore and lead to toxic effects (Iboukhoulef et al. 2018). Various methods of treatment have been developed, settling with clay and/or chemical oxidation (Hodaifa et al. 2019), electrocoagulation (Esteves et al. 2019), agronomic uses for fertirrigation (Malvis et al. 2019), and biological treatment coupled with electrochemical methods (Markou et al. 2017).

Advanced oxidation process (AOP) constitutes a suitable method for OMWs treatment because these processes rely on the effects of chemical initiators (radicals) and energy to destroy the pollutants. These methods are referred to oxidation phenomenal where hydroxyl radical (OH·) is formed in situ from oxidant compounds such as O_2, O_3, H_2O_2 with/without catalyst and/or in the presence of energetic resources (electrochemical, UV, US, etc.). Although these systems use various mechanisms, they all result around the generation of hydroxyl radicals. AOPs applied to OMWs include ozonation (Iboukhoulef et al. 2019), combination of UV and H_2O_2 (Hodaifa et al. 2019), conductive diamond electro-oxidation (Barbosa et al. 2016), UV/TiO_2 (Cordova et al. 2017), and catalytic wet air oxidation (Guneysu and Arayıcı 2019).

Photolysis is a major naturally occurring degradation processes. It is a chemical reaction by which some functional groups of molecules are activated by the absorption of light (irradiation) to reach an excited state resulting in breaking down (of molecular bonds) into smaller units. Organic compounds do not absorb all irradiation, which makes photolysis a slow reaction. To increase the global degradation yield, some combinations are applied, such as the addition of H_2O_2 and a catalyst (Domingues et al. 2018). The hydroxyl radicals have been selected among the strongest oxidants that can be applied to wastewater depollution. They are among the most powerful oxidants known, with a redox potential of 2.73 V/ENH in acid solution. In alkaline medium, these radicals exist in their basic conjugated form: the oxygen anion radical $O\cdot^-$ (Iboukhoulef et al. 2019). The efficiency of photolysis depends on several parameters such as pH, temperature, type of pollutant, oxidant concentration, UV light intensity, and parasite reactions consuming hydroxyl radicals.

In our previous work, the effect of H_2O_2, temperature, and time has been evaluated. They are known to be influential on the degradation of phenolic compounds in OMWs, and this is why we have chosen to work at $[H_2O_2] = 12$ M, temperature of 30 °C for 30 min (Iboukhoulef et al. 2018, 2019).

In this study, we applied homogenous photo-Fenton process to the treatment of OMWs. The experimental design methodology was undertaken to investigate the effect of catalyst, agitation speed, pH, dilution, and UV light. A basic aim of the evaluation of the results is to develop an empirical model that links the response to the more significant variables.

2 Experimental Part

2.1 *OMW Characteristics*

The OMWs were tracked from a continuous olive oil processing plant located in Tizi Ouzou (Algeria). The OMWs were stored in a covered plastic container and kept at 4 °C. Experiments were done on the diluted OMW with pH 4.7, COD of 74 g/L, and total polyphenols of 2.7 g/L. All chemicals used were of analytical quality.

2.2 *Photochemical Oxidation*

The tests were conducted in a 500 mL vessel surrounded by a cylindrical glass shell, allowing water cooling of the reactor. In all experiments, the volume of OMW treated was 100 mL. To the diluted OMW (60 and 80) with adjusted pH (2 and 5) by addition of hydrochloric acid (2 M), 10 mL of solution catalyst (iron and copper) at 0.5 g L^{-1} was introduced followed by 10 mL of H_2O_2 (12 M). The batch was stirred at a given agitation rate (300 and 700 rpm) under total reflux at 30 °C for 30 min. These tests are carried out (with and without) UV lamp. Phenolic compounds were measured by the colorimetric method (Folin Ciocalteu method): to 0.5 mL OMW diluted in 10 mL distilled water, 0.5 mL Folin–Ciocalteu reagent is added. After 3 min, 1 mL of saturated sodium carbonate solution was introduced. After 30 min in the dark, the absorption was taken at 751 nm with a UV/VIS spectrophotometer (Shimadzu). Gallic acid was used for calibration, and the results of the analyses are given in gallic acid equivalent with $R^2 = 0.997$.

2.3 *Plackett–Burman Experimental Design*

Several factors are involved in this study; thus, it is difficult to adjust them. Experimental design is an effective statistical technique for optimizing this process. This methodology reduces significantly the number of experiments for evaluating the effects of many independent variables on response and for process modeling.

A Plackett–Burman design was used to screen the significant independent variable for photocatalytic OMWs treatment. Five variables are selected: type of catalyst, agitation speed, pH, dilution, and UV irradiation. The selected levels are listed in Table 1.

Each variable was represented at two levels, high and low symbolized by (+1) and (−1), respectively. A design matrix in coded and real values is given in Table 2. Eight combinations were organized according to applied design; in order to obtain the experimental variance, the number of experiments N carried out was doubled ($N = 16$).

Table 1 Experimental domain for screening design

Factors	Coded values	Unit	Level	
			(−1)	(+1)
Catalyst	$X1$	$g\ L^{-1}$	Fe(II)	Cu(II)
Agitation speed	$X2$	rpm	300	700
pH	$X3$		2	5
Dilution	$X4$		60	80
UV	$X5$	nm	Without	With (365)

The coded values $X1$, $X2$, $X3$, $X4$, and $X5$ correspond to real variables: type of catalyst, agitation speed, pH, dilution, and the UV irradiation, respectively. The selected experiments allowed the estimation of the coefficients of the model given by Eq. (1), where Y was the response (residual phenolic content); b_0 was the medium response; b_1, b_2, b_3, b_4, and b_5 were the linear coefficients (principal effects) of variables $X1$, $X2$, $X3$, $X4$, and $X5$.

$$y = b_0 + b_1X1 + b_2X2 + b_3X3 + b_4X4 + b_5X5 \tag{1}$$

The effect (b_j) of each variable was calculated using Nemrodw software.

The significance of effect of each variable was determined according to Student's t-test. To estimate the main effects of the factors, we must calculate the standard deviation (Eq. 4) using Eqs. (2) and (3), t of Student (Eq. 5), and significance level (%) for each coefficient. It is accepted that if the significance level is less than 5% (confidence levels greater than 95%), the effects are significant (Berkane et al. 2019).

$$S^2 = \frac{1}{N-p}\sum e_i^2 \tag{2}$$

$$e_i = Y_{exp} - Y_{the} \tag{3}$$

S: Estimator, N: Number of trials, p: Number of model coefficient, e_i: residues.

$$\sigma_i^2 = \frac{S^2}{N} \tag{4}$$

$$t_{\exp} = \frac{|b_j|}{\sigma(b_j)} \tag{5}$$

σ: standard deviation, $t_{\exp}$: t value.

Table 2 Design matrix and results obtained for residual phenolic compounds (Y)

N° exp	Coded variables					Real variables					Y (g L^{-1})
	$X1$	$X2$	$X3$	$X4$	$X5$	Catalyst	Agitation speed	pH	Dilution	UV	
1	1	1	1	−1	1	Cu	700	5	60	With UV	1.70
2	1	1	1	−1	1	Cu	700	5	60	With UV	1.69
3	−1	1	1	1	−1	Fe	700	5	80	Without UV	4.64
4	−1	1	1	1	−1	Fe	700	5	80	Without UV	4.66
5	−1	−1	1	1	1	Fe	300	5	80	With UV	3.14
6	−1	−1	1	1	1	Fe	300	5	80	With UV	3.02
7	1	−1	−1	1	1	Cu	300	2	80	With UV	2.12
8	1	−1	−1	1	1	Cu	300	2	80	With UV	1.66
9	−1	1	−1	−1	1	Fe	700	2	60	With UV	2.17
10	−1	1	−1	−1	1	Fe	700	2	60	With UV	2.43
11	1	−1	1	−1	−1	Cu	300	5	60	Without UV	1.52
12	1	−1	1	−1	−1	Cu	300	5	60	Without UV	1.36
13	1	1	−1	1	−1	Cu	700	2	80	Without UV	2.00
14	1	1	−1	1	−1	Cu	700	2	80	Without UV	2.26
15	−1	−1	−1	−1	−1	Fe	300	2	60	Without UV	2.75
16	−1	−1	−1	−1	−1	Fe	300	2	60	Without UV	2.88

3 Results and Discussions

3.1 Statistical Analysis

The experimental results of the degradation of phenolic compounds obtained from an experimental Plackett–Burman design are reported in the last column of Table 2.

Nemrodw software enabled to calculate the model coefficients. Model coefficients and statistical analysis including *t* values and *p* values are given in Table 3. All selected variables were identified as influents at 5% significant level. The type of catalyst is considered to be the most significant factor because it has the highest coefficient in absolute value (b = 0.711). In addition, agitation speed, pH, and dilution had a positive effect on the treatment, while the type of catalyst and UV had a negative effect. These coefficients values are represented in Fig. 1. This graph shows at the same time the magnitude and the importance of the effects. Statistical tools (Table 3 and Fig. 1) show that all the main effects were significant. According to these results, the effect of the type of catalyst is approximately 3.6 times greater than that of the stirring speed, 3.2 times that of pH, 1.62 times that of dilution, and 2.75 times greater than that of UV effect.

Table 3 Estimates and statistics of coefficients for residual phenolic compounds (IF = 1, $\sigma b_j = 0.04$)

Name	Coefficient	t_{exp}	Significance (%)
b_0	2.500	62.92	<0.01***
b_1	−0.711	−17.90	<0.01***
b_2	0.194	4.88	0.123**
b_3	0.216	5.44	0.0614***
b_4	0.438	11.01	<0.01***
b_5	−0.259	−6.51	0.0186***

*Significant at the level of 95%, **Significant at the level of 99%, ***Significant at the level of 99.9%, IF: Inflation factor, σb_j: Standard deviation of b_j, $t_{exp} = (bj/\sigma b_j)$: *t* of Student

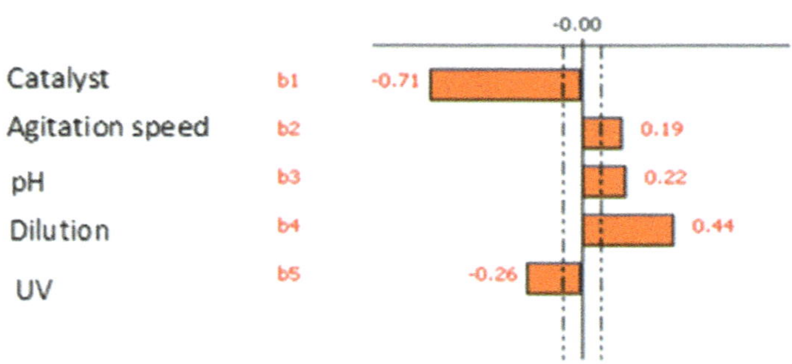

Fig. 1 Graphic study of the factor effects on the residual phenolic content

The data presented in this method are adapted to the real values of the variables, and the regression coefficient is calculated with a valid determination coefficient ($R^2 = 0.91$), which indicates that the model can explain more than 91% of the responses (Berkane et al. 2019). Figures 2 and 3 show that the residues are randomly distributed around zero and that the experimental values are in accordance with the values predicted by this model. This is further evidence that the model defined the region studied. This is proof still that the model describes well the region studied.

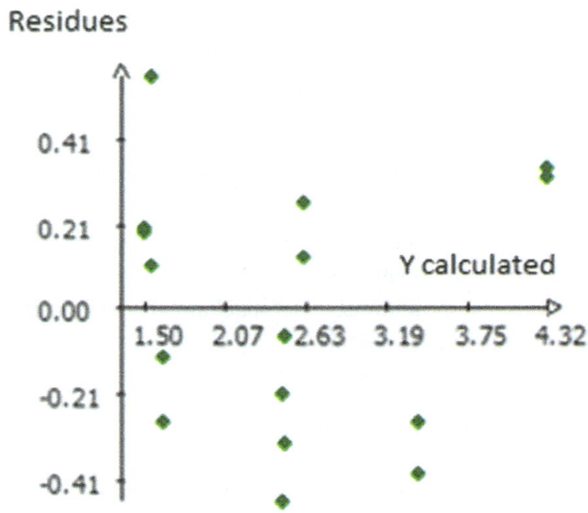

Fig. 2 Residual distribution analysis

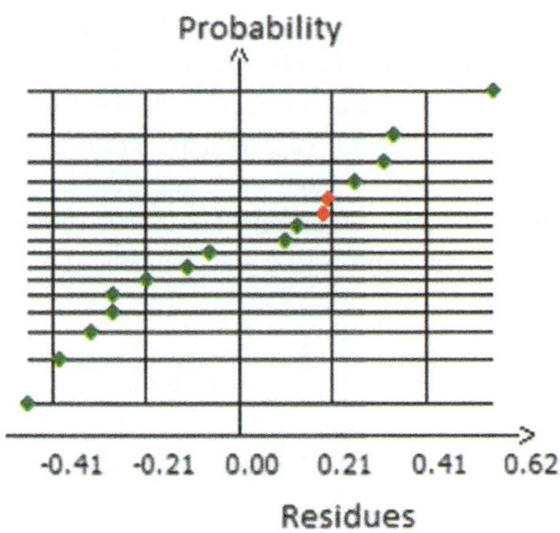

Fig. 3 Diagram of effect probability

3.2 Degradation Mechanism

The photolysis process increases the influence of hydrogen peroxide. UV light is utilized to break the O–O bond in H_2O_2 to produce the hydroxyl radical. Thus, more hydroxyl radicals can be produced with a small amount of hydrogen peroxide, which promotes the degradation of phenolic compounds in OMWs. The reactions are given below (Guneysu and Arayıcı 2019):

$$H_2O_2 \overset{h\vartheta}{\rightarrow} 2HO\cdot \tag{1}$$

$$H_2O_2 + HO\cdot \rightarrow HOO\cdot + H_2O \tag{2}$$

$$H_2O_2 + HOO\cdot \rightarrow HO\cdot + H_2O + O_2 \tag{3}$$

The Fenton and Fenton-like reactions are a combination of catalyst and hydroxyl radicals formed in reactions (1–3); they are known as a strong oxidizing agent for organic pollutants. The mechanism, with iron as catalyst, is related below (Domingues et al. 2018):

$$Fe^{2+} + H_2O_2 \rightarrow Fe^{3+} + OH^- + HO\cdot \tag{4}$$

$$Fe^{3+} + H_2O \rightarrow Fe^{2+} + H^+ + HO\cdot \tag{5}$$

The Fenton photo process (H_2O_2/Fe^{2+}/UV) leads the generation of HO· by photolysis of hydrogen peroxide. In the case of UV light, the ferric ions (Fe^{3+}) produced in reaction (4) are photocatalyzed into ferrous ions (Fe^{2+}), with an additional formation of hydroxyl radicals (reaction 5) (Hodaifa et al. 2019).

Among various catalysts available, copper has been successfully used in Fenton-type systems used to treat recalcitrant compounds, such as dyes (Abubakar and Gaya 2019), cyanide (Kamrani et al. 2019), or phenol derivatives (Kartal Temel and Gürkan 2016). The Fenton-like reactions with copper as catalyst are listed (Iboukhoulef et al. 2018):

$$H_2O_2 + Cu^{2+} \rightarrow Cu^+ + H^+ + HOO\cdot \tag{6}$$

$$H_2O_2 + Cu^+ \rightarrow Cu^{2+} + OH^- + HO\cdot \tag{7}$$

HOO· is a very unstable species, and perhydroxyl radicals can be removed by reaction with another copper (reaction 8) (Kartal Temel and Gürkan 2016):

$$HOO\cdot + Cu^{2+} \rightarrow Cu^+ + O_2 + H^+ \tag{8}$$

The hydroxyl radicals generated interact directly with the organic compounds, causing them to oxidize (reaction 9).

$$\text{OMWs} + \left(\text{HO}\cdot, \text{HOO}\cdot, \text{O}_2\cdot^-\right) \rightarrow \text{by products} \quad (9)$$

Since the photo-Fenton reaction is accelerated by light, it results in faster degradation rates than the conventional Fenton process, which is agreed with the literature (Domingues et al. 2018; Hodaifa et al. 2019).

4 Conclusion

This study demonstrated that the Plackett–Burman experimental designs can be used to determine significant independent variables for phenolic compounds degradation from OMWs by homogeneous photocatalytic process. This kind of treatment depends significantly on the type of catalyst, the stirring rate, pH, dilution of OMW, and UV light.

The amount of phenolic compounds in the OMWs decreases by combining copper as a catalyst with UV light ($b < 0$). This rate increases with stirring speed, pH, and dilution ($b > 0$).

The best degradation of phenolic compounds is obtained by using copper as a catalyst in the presence of UV radiation, stirring speed: 300 rpm, pH: 2, and dilution: 60 times. In perspective, in order to complete this study, we will have to calculate the value of the optimum using the response surfaces.

References

Abubakar A, Gaya U (2019) Box-Behnken-optimized Cu(I) photo-Fenton-like degradation of Basic Violet 3. J Mater Environ Sci 10(1):15–21

Barbosa J, Fernandes A, Ciríaco L, Lopes A, Pacheco MJ (2016) Electrochemical treatment of olive processing wastewater using a boron-doped diamond anode. CLEAN Soil Air Water44(9):1242–1249

Berkane N, Meziane S, Aziri S (2019) Optimization of Congo red removal from aqueous solution using Taguchi experimental design. Sep Sci Technol 55(2):278–288

Cordova R, Nagel-Hassemer N, Matias W, Muller J, Castilhos Junior A (2017) Removal of organic matter and ammoniacal nitrogen from landfill leachate using the UV/H_2O_2 photochemical process. Environ Technol 1–14

Domingues E, Gomes J, Quina M, Quinta-Ferreira R, Martins R (2018) Detoxification of olive mill wastewaters by Fenton's process. Catalysts 8(12):662

Esteves B, Rodrigues C, Maldonado-Hódar F, Madeira L (2019) Treatment of high-strength olive mill wastewater by combined Fenton-like oxidation and coagulation/flocculation. J Environ Chem Eng 7(4):103252

Gudina E, Rodrigues J, Freitas J, Azevedo Z, Teixeira J, Rodrigues L (2016) Valorization of agro-industrial wastes towards the production of rhamnolipids. Bioresour Technol 212:144–150

Guneysu S, Arayıcı S (2019) Wet peroxidation of olive oil mill wastewater for sludge minimization. In: Balkaya N, Guneysu S (eds) Recycling and reuse approaches for better sustainability. Environmental science and engineering. Springer, Cham

Hodaifa G, Gallardo P, García C, Kowalska M, Seyedsalehi M (2019) Chemical oxidation methods for treatment of real industrial olive oil mill wastewater. J Taiwan Inst Chem Eng 97:247–254

Iboukhoulef H, Amrane A, Kadi H (2018) Optimization of phenolic compounds abatement in olive mill wastewater by Fenton's like treatment with H_2O_2/Cu under microwave using experimental design. Environ Eng Manag J 17(5):1267–1274

Iboukhoulef H, Douani R, Amrane A, Chaouchi A, Elias A (2019) Heterogeneous Fenton like degradation of olive mill wastewater using ozone in the presence of $BiFeO_3$ photocatalyst. J Photochem Photobiol A Chem 383:112012

Kamrani M, Seifpanahi K, Seyed Hakimi A, Al G, Agarwa S, Gupta V (2019) Degradation of cyanide from gold processing effluent by H_2O_2, NaClO and $Ca(ClO)_2$ combined with sequential catalytic process. Bul Chem Commun 51(3):384–393

Kartal Temel N, Gürkan R (2016) Photocatalytic decolourization of bromophenol blue in aqueous solution with Cu(II)-peroxo complexes in presence of SDS. Anal Chem Lett 6(4):435–447

Malvis A, Hodaifa G, Halioui M, Seyedsalehi M, Sanchez S (2019) Integrated process for olive oil mill wastewater treatment and its revalorization through the generation of high added value algal biomass. Water Res 151:332–342

Markou V, Kontogianni M, Frontistis Z, Tekerlekopoulou A, Katsaounis A, Vayenas D (2017) Electrochemical treatment of biologically pre-treated dairy wastewater using dimensionally stable anodes. J Environ Manage 202(1):217–224

Chapter 16 Electrochemical Impedance Spectroscopy as a Powerful Assessment Tool for the Electropolishing Quality of AISI 304 Stainless Steel

Simon Detriche, Jean-François Vanhumbeeck, Joseph Delhalle, and Zineb Mekhalif

Abstract The electrochemical impedance spectroscopy behavior of AISI 304 stainless steel in a sulfuric/phosphoric acid electrolyte is described. Comparison with the literature is carried out, and common points are established. A new EIS behavior is reported for this stainless steel. It is related to the presence of nickel in this particular stainless steel. A parallel relationship has been established between the EIS characteristics, the surface state after polishing, and the applied potential.

Keywords Electrochemical impedance spectroscopy · Stainless steel · Electropolishing

1 Introduction

Electropolishing has been used since almost a century and proved to be a powerful and efficient method for surface treatments of several metals and alloys (Landolt 1987). This method allows the polishing of surfaces with complex morphologies and leads to a corrosion resistance and passivation enhancement. From a mechanistic point of view, it is widely known that electropolishing occurs via both anodic leveling and brightening. The leveling is a mass-transport dependent process resulting from a higher dissolution rate of the surface peaks caused by an accumulation of charges due to point effect. Brightening occurs when the dissolution process becomes independent of the crystallographic structure of the metal or the alloy (Datta and Landolt 1975; Clerc and Landolt 1987).

S. Detriche (✉) · J. Delhalle · Z. Mekhalif (✉)
Laboratory of Chemistry and Electrochemistry of Surfaces, University of Namur, Rue de Bruxelles, 61, 5000 Namur, Belgium
e-mail: simon.detriche@gmail.com

Z. Mekhalif
e-mail: zineb.mekhalif@unamur.be

J.-F. Vanhumbeeck
CRM Group, Avenue Bois Saint-Jean, 21, 4031 Liège, Belgium

D. Trache et al. (eds.), *Materials Research and Applications*, Materials Horizons: From Nature to Nanomaterials,
https://doi.org/10.1007/978-981-15-9223-2_16

Electrochemical impedance spectroscopy (EIS) is a powerful method of in situ investigation of electrochemical processes (Orazem and Tribollet 2008). It allows the understanding of complex interfacial mechanisms in particular the electropolishing processes for simple materials. However, only a few studies have been devoted to the EIS analyses of steels in electropolishing conditions. The group of Bojinov has reported on the EIS response in simple electrolytes like sulfuric or phosphoric acids of several alloys, mimicking simple or complex steels (Bojinov et al. 1999; Betova et al. 2002, 2003; Bojinov and Tzvetkoff 2003). Along with such mechanistic studies, the group of Landolt (1987) has highlighted the link between the EIS signal and the effectiveness of the electropolishing process (Magaino et al. 1993; Matlosz et al. 1994). More recently, Rotty (2018) reported a new mechanism using interpretation of EIS data on AISI 316 in sulfuric/phosphoric acid electrolyte.

In this work, the electrochemical impedance spectroscopy is used to assess, in commonly used electropolishing conditions, the behavior of AISI 304 stainless steel (containing nickel as additional alloying element) comparatively to the AISI 430 (nickel free) previously studied by Landolt et al. (Magaino et al. 1993; Matlosz et al. 1994). A new and typical EIS signature, only observed in bad AISI 304 surface electropolishing, and due to the presence of nickel, is reported.

2 Experimental

Electrochemical impedance spectra and current/tension curves have been recorded for the two stainless steels, AISI 430 (17% Cr) and AISI 304 (19% Cr and 8% Ni) used as working electrode in a spot cell (controlled sample surface area of 0.28 cm^2) with platinum and saturated calomel as counter and reference electrodes. The electrolyte is a mixture of 18 ml sulfuric acid (98 wt.%) and 102 ml phosphoric acid (70 wt.%).

EIS spectra have been recorded at various constant applied potentials ranging from 0.8 to 2.8 V versus SCE. Samples have been polarized during 20 min before each EIS measurement. A 10 mV sinusoidal variation has been applied, with a frequency ranging from 10,000 to 0.1 Hz. Electrochemical data have been recorded and analyzed using VersaStudio and EIS Analyzer softwares.

For some potential values with characteristic EIS responses, AISI 304 samples have been polarized 30 min, washed with ethanol, and their morphology characterized using scanning electron microscopy (SEM) (JEOL 7500).

3 Results and Discussion

Figure 1 shows the intensity/potential curves for AISI 430 (17% Cr) and AISI 304 (19% Cr and 8% Ni) stainless steels. The main features of these curves correspond to a classical behavior of all stainless steels in concentrated acids. A potential increase leads to an increase in current values during the so-called transpassive dissolution step. This step ranges from 1.2 to 1.55 V versus SCE for AISI 304 and 1.65 V for AISI 430. Beyond this value, the current decreases due to the secondary passivation step, followed by the oxygen evolution step starting from 1.90 V.

For each applied potential ranging from 0.8 to 2.8 V versus SCE, the EIS has been recorded. The Nyquist plots (Fig. 2) show five different typical behaviors denoted region I to IV and highlighted in Fig. 1. The transpassive dissolution step and the oxygen evolution are each characterized by a typical Nyquist plot (regions I and IV,

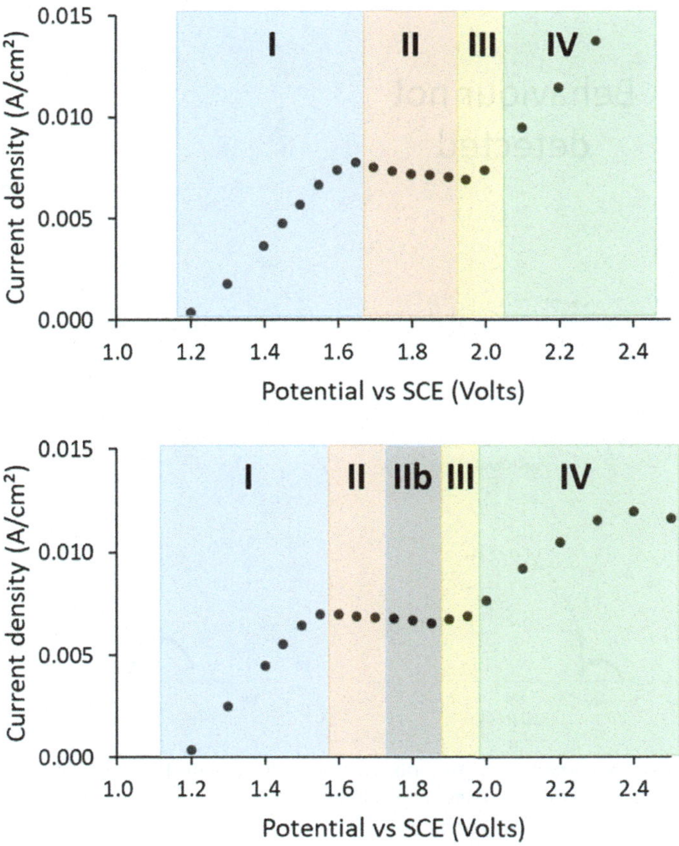

Fig. 1 Intensity/potential curves for AISI 430 (top) and AISI 304 (down)

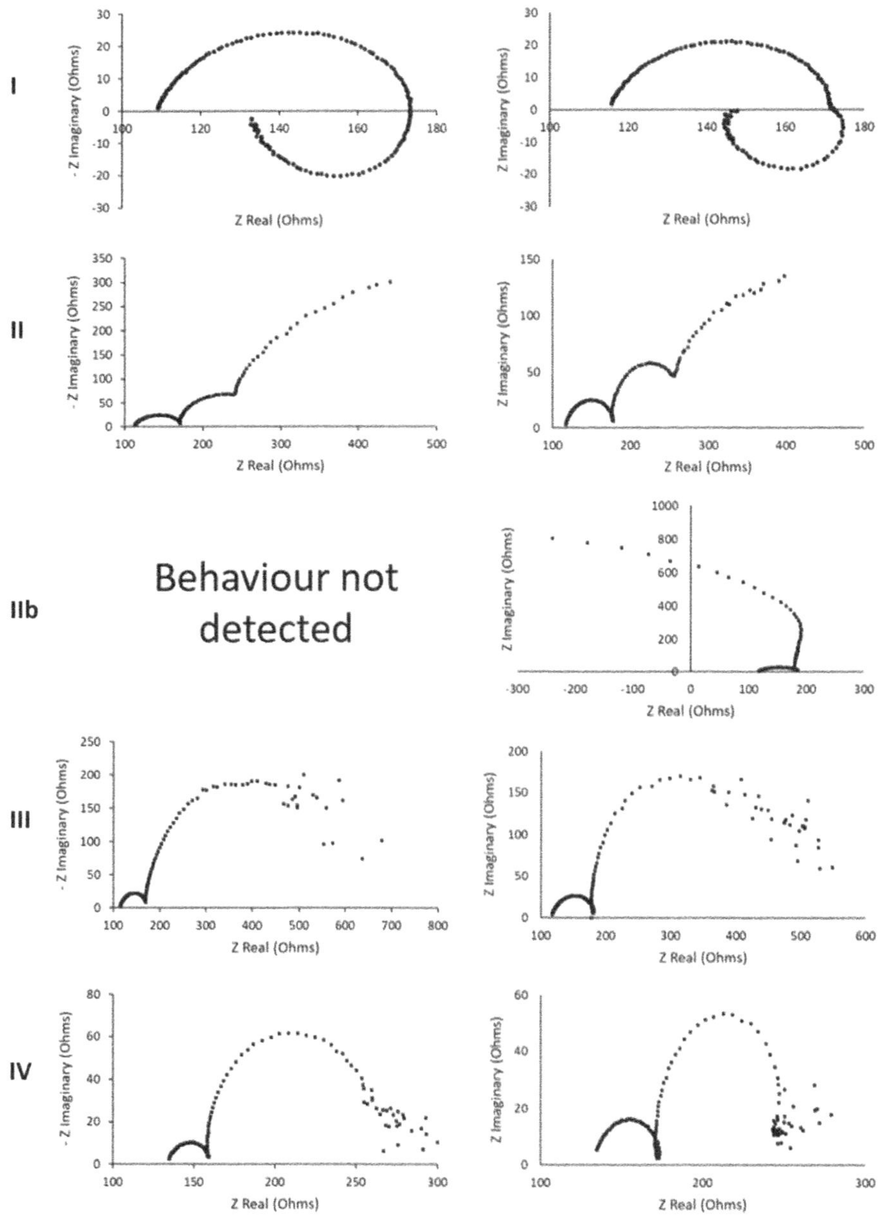

Fig. 2 EIS behavior corresponding to various regions defined in the I/V curves of Fig. 1. AISI 430 in the left column and AISI 304 in the right column

respectively) whatever the applied potential, while the secondary passivation step exhibits two or three different EIS behaviors, depending on the nature of stainless steel (AISI 304 and AISI 430).

In order to evidence the relationship between the EIS spectra and the surface state resulting from the application of different potentials for the steel of interest (AISI 304), SEM has been performed on representative samples in each region I–IV (Fig. 3).

The EIS spectra in region I display a capacitive loop at high frequency and an inductive one at low frequency identical to those obtained for various alloys (Fe/Cr, Cr/Ni and Fe/Cr/Mo) in both sulfuric and phosphoric acids (Bojinov et al. 1997, 1999; Bojinov and Tzvetkoff 2003). They are usually described as the fusion of two capacitive and two pseudo-inductive loops with close values of time constants. The high-frequency capacitive loop(s) is(are) ascribed to an impedance of the passive film and a charge-transfer reaction at the passive film/solution interface. The low-frequency loop(s) correspond(s) to the relaxation of negative charge carriers and of the surface coverage of an adsorbate (Bojinov et al. 1997, 1999; Bojinov and Tzvetkoff 2003). On the other hand, Landolt et al. have proven that the transpassive dissolution typically results in a bad polishing of the surface (Magaino et al. 1993). This is also observed for AISI 304 stainless steel in our conditions.

The EIS in the second region is characterized by an additional new capacitive loop at lower frequency (Betova et al. 2002, 2003; Bojinov and Tzvetkoff 2003), which has been assigned to the relaxation of the surface coverage by an intermediate deceleration of the dissolution rate leading to a current decrease (Fig. 1, region II). The existence and/or stability of these species seems to be related to the presence of phosphoric acid (Bojinov and Tzvetkoff 2003).

Similar EIS results (region II) have been reported by Landolt et al. (1987) and explained by a three-step process (Magaino et al. 1993; Matlosz et al. 1994). At middle frequency, the capacitive loop is related to a conversion reaction producing an insoluble intermediate product which covers the surface. The high-frequency capacitive loop is related to a charge-transfer reaction depending on the remaining free surface at the electrode. At low frequency, a capacitive loop is due to a complexation reaction between an acceptor molecule and the insoluble intermediate. The best electropolishing results are obtained for potentials exhibiting these EIS features. This is attributed to the equilibrium between the complexation reaction and the insoluble intermediate production which induces a slow but continuous etching of the surface. Moreover, the complexation reaction is independent of the crystallographic orientation of the alloy.

Our results on the AISI 304 stainless steel are in line with Landolt (1987) observations. A similar correlation exists between the electropolishing effectiveness (smooth and homogenous surface, low roughness) and the occurrence of three capacitive loops in the EIS spectra (Figs. 2 and 3, region II). This region width is narrower for AISI 304 than AISI 430 (100 mV width instead of 250 mV).

Region IIb is characterized by a capacitive loop at high frequency and another one with a negative resistance value at low frequency. This kind of behavior observed for AISI 304 has never been reported previously on stainless steels. In the case of other

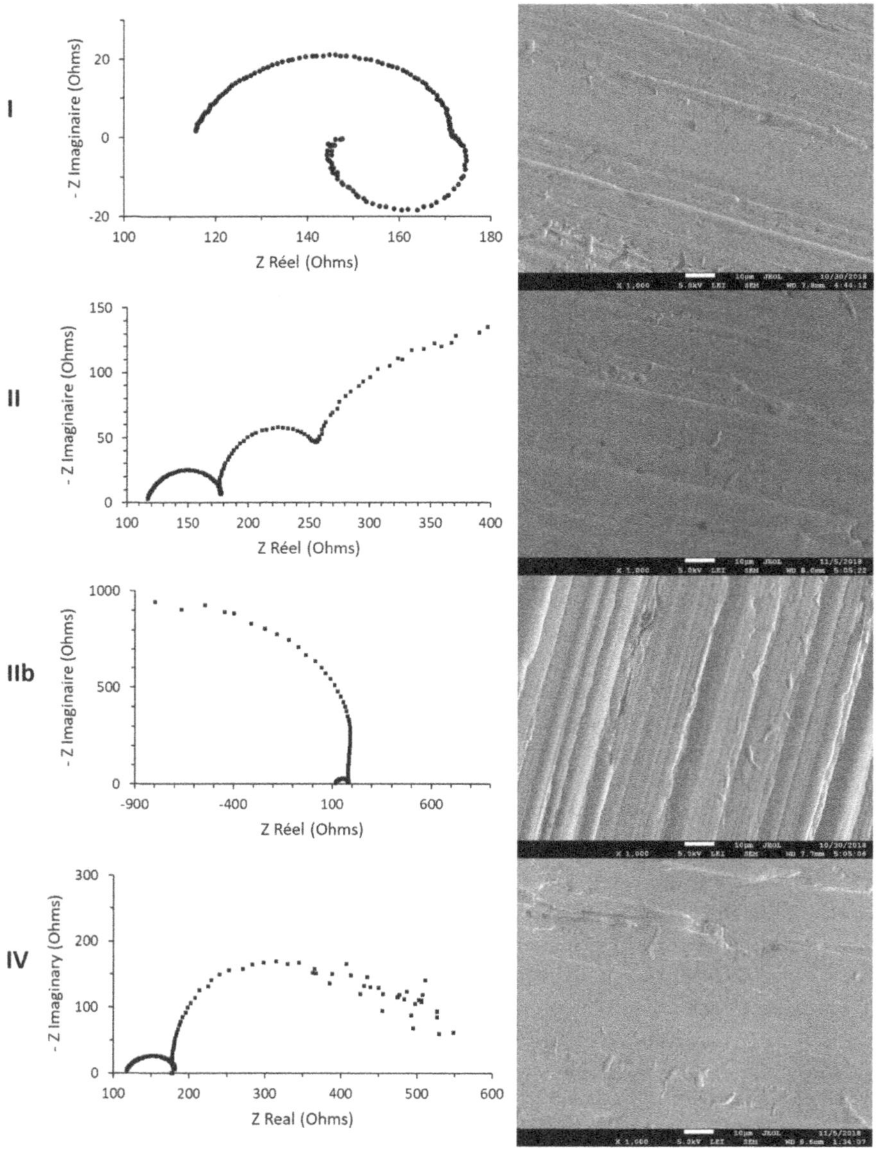

Fig. 3 SEM images corresponding to various EIS behavior. Scale bar is 10 μm

substrates, like Ni and Ni alloys, Bojinov et al. (1999) and Keddam et al. (2006) have explained the low-frequency capacitive loop by a negative resistance indicative of a second-order dynamic behavior resulting from the competition between a transpassive dissolution and a secondary passivation. On this basis, we assume that, in the case of AISI 304 stainless steel, this EIS signature in region IIb and the bad electropolishing result (Fig. 3) is the consequence of the nickel presence.

4 Conclusions

A comparative study of the electrochemical impedance spectroscopy signature between AISI 304 (Fe/Cr/Ni) and AISI 430 (Fe/Cr) supported by literature data has been performed. Additionally, a correlation has been established between the EIS data and the surface state obtained after polarization using scanning electron microscopy.

Various electrochemical behaviors corresponding to the transpassive dissolution and the secondary passivation of AISI 304 have been identified. The EIS signal characteristic of an efficient electropolishing process and consistent with the literature survey is reported. Moreover, for the first time, a new EIS signal is described for AISI 304. Using the literature evidences, this new signal is assigned to the presence of nickel in this alloy.

Future perspectives of this work will be the proposition of a mechanism of anodic dissolution able to explain the new EIS signal.

References

Betova I, Bojinov M, Laitinen T et al (2002) The transpassive dissolution mechanism of highly alloyed stainless steels I. Experimental results and modelling procedure. Corros Sci 44:2675–2697

Betova I, Bojinov M, Tzvetkoff T (2003) Role of surface reactions in the transpassive dissolution of ferrous alloys in concentrated H_3PO_4. Appl Surf Sci 220:273–287

Bojinov M, Tzvetkoff T (2003) The influence of solution anion on the mechanism of transpassive dissolution of ferrous- and nickel-based alloys. J Phys Chem B 107:5101–5112

Bojinov M, Betova I, Raicheff R (1997) Influence of molybdenum on the transpassivity of a Fe + 12%Cr alloy in H_2SO_4 solutions. J Electroanal Chem 430:169–178

Bojinov M, Betova I, Fabricius G et al (1999) The stability of the passive state of iron-chromium alloys in sulphuric acid solution. Corros Sci 41:1557–1584

Clerc C, Landolt D (1987) On the theory of anodic levelling: behaviour of macroprofiles

Datta M, Landolt D (1975) Surface brightening during high rate nickel dissolution in nitrate electrolytes. J Electrochem Soc 122:1466–1472

Keddam M, Takenouti H, Yu N (2006) Transpassive dissolution of Ni in acidic sulfate media: a kinetic model. J Electrochem Soc 132:2561

Landolt D (1987) Fundamental aspects of electropolishing. Electrochim Acta 32:1–11

Magaino S, Matlosz M, Landolt D (1993) An impedance study of stainless-steel electropolishing. J Electrochem Soc 140:1365–1373

Matlosz M, Magaino S, Landolt D (1994) Impedance analysis of a model mechanism for acceptor-limited electropolishing. J Electrochem Soc 141:410–418

Orazem ME, Tribollet B (2008) Electrochemical impedance spectroscopy. In: The electrochemical society series

Rotty C (2018) PhD thesis, 1–271

Chapter 17
Physicochemical, Electrochemical, and Theoretical Study of Three Synthesized Organic Inhibitors for 316L Stainless Steel in HCl

Messaoud Benamira, Yasmine Boughoues, Lyamine Messaadia, Nazih Ribouh, and Mohammed Himrane

Abstract This work deals with the synthesis and physicochemical, electrochemical characterization of three organic compounds, and then the study of their inhibitory efficiency against corrosion phenomenon of 316L stainless steel in hydrochloric acid solution medium. The structure of the as-prepared organic compounds was confirmed by FT-IR, ^{1}H, and ^{13}C NMR spectroscopy analyses. In parallel, electrochemical methods, impedance spectroscopy, and potentiodynamic polarization were of great interest to determine the inhibitive efficiency. In fact, the corrosion current density of 316L stainless steel without inhibitor is 76.60 μA/cm^2, and a significant decrease in the corrosion current densities was observed after the addition of the organic molecules, where an increase of 87% in terms of the inhibition efficiency was noted particularly with the Amine 3. Moreover, scanning electron microscopy images brought out the presence of the organic protective layer. In addition, following the concentration effect has shown that the phenomenon of stability and high inhibition efficiency could be reached even at small concentrations around of 10^{-5} M. The Langmuir isotherm was used to feature the adsorption on sample surface. Furthermore, in order to demonstrate the protective effect of organic compounds against 316L stainless steel corrosion, theoretical studies based on two different methods (density functional theory and molecular dynamic) were applied. Accordingly, the performed results through the different approaches are in good agreement with each other and provide a general insight into the phenomenon studied.

Keywords Corrosion · 316L stainless steel · Organic inhibitors · EIS · DFT

M. Benamira (✉) · Y. Boughoues · L. Messaadia · N. Ribouh · M. Himrane
Laboratory of Materials Interaction and Environment (LIME), University of Jijel, BP.98, Ouled Aissa, Jijel, Algeria
e-mail: benamira18@gmail.com; m_benamira@univ-jijel.dz

D. Trache et al. (eds.), *Materials Research and Applications*,
Materials Horizons: From Nature to Nanomaterials,
https://doi.org/10.1007/978-981-15-9223-2_17

1 Introduction

Corrosion is a universally known phenomenon that directly or indirectly causes each year significant material losses to industries and collectivities. More dangerously, corrosion is an increasing threat that causes irreversible damage to environment and intensifies health risks (pollution, contamination, etc.) that lead in many cases to death of human beings and animals.

In this context, corrosion is estimated to be responsible for the destruction of a quarter of the world's annual steel production (Akinlabi et al. 2014). Moreover, the phenomenon is not limited to steel, but rather it affects all types of polymers, ceramics, and metals. It is generated from chemical and/or physical interactions between the material and its environment. Clearly, it is the result of the action of a liquid or gaseous reagent on a metal or alloy surface.

In this regard, one of the most effective and practical alternatives against metal attack in corrosive environment is the use of corrosion inhibitors (Salarvand et al. 2017; Verma et al. 2019). Corrosion inhibitors can be defined as chemical compounds that if they are added to a corrosive solution, they will reduce metal losses.

It is well known that organic inhibitors are tied to the metal through an adsorption mechanism. Inhibitory efficiency is directly related to the structural properties of the used organic inhibitors (Argade et al. 2019; Goyal et al. 2018; Yıldız 2019). The inhibitors protect the metal surface against dissolution by the creation of a hydrophobic layer at the interface metal/solution. Several organic compounds have proven their usefulness as corrosion inhibitors during the industrial cleaning process with acid. The addition of organic compounds prevents the attack of the metal surface by H^+ ions as well as the adsorption of chloride ions and the substitution of adsorbed water molecules, thus blocking the active sites by forming a compact film enabling the decrease of the corrosion rate.

In this study, we focus on the corrosion phenomenon in an acid medium by using organic compounds. Three amine derivatives compounds with different structures and functional groups have been synthesized. The adsorption phenomenon is related to the electronic structure and in particular to the presence of functional groups, heteroatoms (N, O, S, P), aromaticity, multiple π orbital, and electron density at the donor–acceptor atoms.

The inhibition tests were performed on 316L stainless steel using the appropriate electrochemical methods and scanning electron microscopy (SEM) analysis to confirm the deposition of the organic layer on 316L stainless steel. Finally, theoretical studies by the density functional theory and molecular dynamic simulations were used in the hope of correlating between the electrochemical, scanning electron microscopy, and the molecular structures results.

2 Experimental

2.1 Materials Preparation

316L stainless steel was purchased from Good fellow SARL, France. The composition (wt.%) of the 316L stainless steel specimens as reported by the supplier and confirmed by energy-dispersive X-ray (EDX) spectroscopy analysis was Cr 17.49, Ni 8.94, Mo 2.05, Si 1.2, C 0.05, and Fe balance. The Cu-wires electrical contact was assured the samples by a silver lacquer. Then, the sample (1.5 cm^2) was placed in epoxy resin (CitoFix powder) purchased from Struers SARL, France. Before the experiments, the working electrodes surface (316L stainless steel) has been mechanically abraded with the SiC papers (500–1200 grits), then rinsed with a mixture of distilled water and acetone, and finally dried in air at ambient temperature. The aggressive solutions of hydrochloric acid (1 M) were prepared from analytical grade 37% HCl purchased from the Merck KGaA Corporation, Darmstadt, Germany.

2.2 Organic Inhibitors Preparation

The three amine compounds (Table 1) were synthesized according to a two-step synthesis. The first step was the synthesis of the corresponding imines from equimolar amounts of primary amine and 3-nitrobénzaldéhyde or 2-hydroxybenzaldehyde in methanol. 20 mmol of the two following primary amines: 4-aminophénol and aniline were mixed in a equimolar (1:1) molar ratio with 3-nitrobénzaldéhyde or 2-hydroxybenzaldehyde in ethanol to obtain the corresponding imines after 20 min at 60 °C. After that, the amine derivatives compounds as shown in Table 1 with different functional groups (NO_2, OH) were obtained from the chemical reduction of imines using $NaBH_4$ in MeOH at 0 °C for 60 min under magnetic stirring. The final products were recovered in the form of yellow-white crystals after filtration and then recrystallized from a water–methanol mixture. The purity and the structure of the three investigated organic compounds were confirmed by FT-IR, ^{13}C, and ^{1}H NMR spectral analyses. Hydrochloric acid (1 M) solutions were prepared with different concentrations of organic inhibitors. The name of the three organic compounds and their physical constants is shown in Table 1.

2.3 Electrochemical and Surface Analyses

The electrochemical tests (electrochemical impedance spectroscopy and potentiodynamic polarization) in hydrochloric acid solution (1 M) were carried out using a Solartron Analytical 1287A potentiostat and Frequency Response Analyzers (FRA)

Table 1 Chemical notations, chemical structures, molecular weight, and physical constants of the three organic compounds

	Amine 1	Amine 2	Amine 3
Name	N-(3-nitrobenzyl) aniline or (3-nitrobenzyl)-phenyl-amine	4-[(3-nitrophenyl) amino] phenol	2-{[(4-hydroxyphenyl) amino] methyl} phenol
Chemical structure	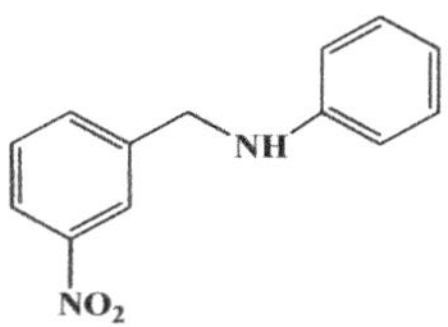	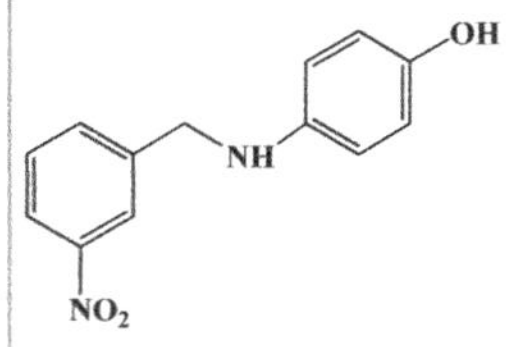	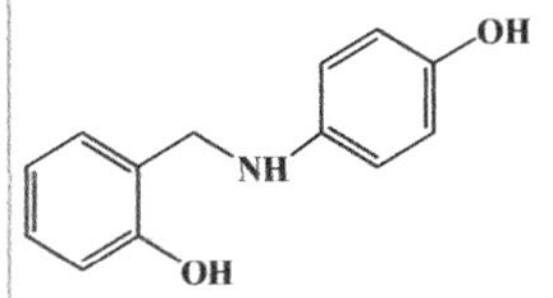
Molecular weight (g mol^{-1})	228.2466	244.2460	213.2319
IR (cm^{-1}) ν	1600 (N–H), 1450–1590 (C=C_{Ar}), 675–765 (C–H_{Ar}), 1400 (C–N), 1350 (N=O)	1450–1590 (C=C_{Ar}), 675–765 (C–H_{Ar}), 1400 (C–N), 3350–3500 (O–H), 1350 (N=O)	3200 (N–H), 3000–3100 (C–H_{Ar}), 1667–2000 (C–H str.)
^{1}H-NMR (CD_3OD) δ (ppm)	3.32 (s, 2H, CH_2), 4.42 (s, 1H, NH), 6.51–6.59 (m, 3H, Ar–H), 7.021–7.1 (m, 2H, Ar–H), 7.82 (t, 1H, Ar–H, J = 9.6 Hz), 8.094 (d, 1H, Ar–H, J = 25.7 Hz), 8.211 (s, 1H, Ar–H), 8.382 (d, 1H, Ar–H)	3.33 (s, 2H, CH_2), 4.33 (s, 1H, NH), 7.30 (d, 1H, Ar–H), 7.61 (t, 1H, Ar–H, J = 8 Hz), 8.08 (d, 1H, Ar–H), 8.36 (t, 2H, Ar–H, J = 23.5 Hz), 8.67–8.83 (t, 2H, Ar–H), 9.57 (s, 1H, OH)	3.39 (d, 2H, CH_2), 4.25 (s, 1H, NH, J = 106.5 Hz), 6.30 (d, 2H, Ar–H), 6.8 (d, 1H, Ar–H, J = 8 Hz), 6.85 (d, 1H, Ar–H), 7.04 (d, 2H, Ar–H), 9.48 (s, 1H, OH)
^{13}C-NMR (CD_3OD) δ (ppm)	40.058 (C_7), 113.206 (C_1), 116.862 (C_8), 122.261 (C_4), 123.61 (C_3), 128.564 (C_{10}, C_{11}, C_{12}), 130.428 (C_9, C_{13}), 134.319 (C_6), 143.626 (C_5)	40.249 (C_7), 114.499 (C_1), 116.459 (C_8), 122.536 (C_4), 125.708 (C_3), 130.265 (C_2), 141.769 (C_9, C_{13}), 144.38 (C_5), 149.507 (C_{10}, C_{12}), 157.819 (C_{11})	39.937 (C_7), 115.22 (C_1, C_2), 119.218 (C_8), 132.944 (C_6), 139.634 (C_9, C_{13}), 142.271 (C_{10}, C_{12}), 148.976 (C_5), 155.295 (C_3), 157.33 (C_{11}), 160.635 (C_4)

1260. 316L stainless steel specimen, saturated calomel electrode (SCE), and platinum were used as working, reference, and counter electrodes, respectively. The EIS measurements were recorded between 100 kHz and 10 MHz with a signal amplitude of 10 mV at open circuit conditions (OCP). The surface states of the 316L stainless steel with and without inhibitor were shown in order to confirm the presence of the thin layer. 316L stainless steel samples with organic inhibitors were immersed in HCl for 24 h. Field emission gun SEM (JEOL 7600) was used for the observation of the surface.

2.4 Theoretical Studies

The molecular structure of the three organic compounds was optimized by Gaussian 09 software (Frisch et al. 2009). Gaussian is released in 1970 by John Pople and used by scientists for theoretical study. Gaussian offers a variety of accurate methods used to predict energies, most molecular properties, and molecular structures of the organic compounds. The molecules were built using Gauss view. The study was conducted by DFT method at B3LYP functional (Becke three-term functional; Lee, Yang, Parr) with 6-31G (d, p) basis set (Becke 1993; Lee et al. 1988). The following parameters, energies of LUMO and HOMO (E_{LUMO}, E_{HOMO}), gap (ΔE), dipole moment (μ), and number of transferred electron (ΔN) of the organic compounds were calculated using the following relations:

$$\Delta N = \frac{\chi_{\text{Fe}} - \chi_{\text{inh}}}{2(\eta_{\text{Fe}} + \eta_{\text{inh}})} \tag{1}$$

where χ_{Fe} and χ_{inh} signify the electronegativity of Fe and inhibitors, respectively. η_{Fe} and η_{inh} represent the hardness of iron and inhibitors, respectively.

$$\chi = \frac{\text{IP} + \text{EA}}{2} = -\frac{(E_{\text{HOMO}} + E_{\text{LUMO}})}{2} \tag{2}$$

The parameters IP and EA are related to E_{HOMO} and E_{LUMO} estimated according to the following relations (Koopmans 1934):

$$\text{IP} = -E_{\text{HOMO}} \tag{3}$$

$$\text{EA} = -E_{\text{LUMO}} \tag{4}$$

$$\eta = \frac{\text{IP} - \text{EA}}{2} = -\frac{(E_{\text{HOMO}} - E_{\text{LUMO}})}{2} \tag{5}$$

The values of electronegativity $\chi_{Fe} = 7$ eV/mol and hardness $\eta_{Fe} = 0$ eV/mol are very often reported in the literature (Martinez 2003; Musa et al. 2011; Pearson 1988).

The Monte Carlo simulations were used to explain the adsorption processes of the amines, water (solvent), and the ions of HCl (H^+, Cl^-) on the iron surface. Material Studio 7.0 software used in this study is presented as one of the performed softwares to determine the relationships of a molecular structure with its properties and behavior (Dassault Systemes 2017). The Monte Carlo method within the Material Studio application that integrates adsorption locator module is used to find the low energy adsorption sites and the preferential adsorption between the substrate, 316L stainless steel surface, and the structure of the adsorbed molecules.

Fe (110) surface was cleaved with slab of 5 Å depth, and 30 Å vacuum slab. The Fe (110) plane that contains five layers was enlarged to a (10 × 10) supercell for strong iron–inhibitor interactions. The geometry-optimized structures were performed by COMPASS force field (Sun 1998). A supercell with 100 H_2O, 1 HCl (H_3O^-, Cl^-), and one molecule of inhibitor was created. The study was conducted in a simulation box (24.82 Å × 24.82 Å × 25.14 Å) with periodic boundary conditions at 298.15 K with a time step of 1 fs. The interaction energy of molecules with Fe (110) surface was obtained by using Eq. (6):

$$E_{\text{interaction}} = E_{\text{Total}} - \left(E_{\text{Surface+H}_2\text{O+H}_3\text{O}^+\text{+Cl}^-} + E_{\text{inhibitor}}\right) \tag{6}$$

where E_{Total} is the total energy of the simulation system, $E_{\text{Surface+H}_2\text{O+H}_3\text{O}^+\text{+Cl}^-}$ is denoted as the energy of the metal surface and solution without the inhibitor, and $E_{\text{inhibitor}}$ signifies the total energy of inhibitor molecule.

The binding energy is equal to $-E_{\text{interaction}}$ and is given as follows:

$$E_{\text{binding}} = -E_{\text{interaction}} \tag{7}$$

The high negative value of adsorption energy confirmed that the system presented a strong and stable adsorption (Cao et al. 2014).

The Fukui indices analyses were presented to predict the reactive sites of inhibitors. The Fukui functions f^+ and f^- were calculated using Dmol3 module through Hirshfeld population analysis (HPA) method and finite difference (FD) approximation (Hirshfeld 1977). BLYP (Becke, Lee, Yang, Parr) basis set was used for title molecules (Cao et al. 2014). Fukui functions are defined as:

$$f(r) = \left[\frac{\partial \rho(\vec{r})}{\partial N}\right]_{\vec{v}} \tag{8}$$

where $\rho(\vec{r})$ is the electronic density and N is the number of electrons in a constant external potential $\vec{v}$.

The use of the finite difference (FD) approximations approach allows us to determine the Fukui functions (Contreras et al. 1999; Yang and Mortier 1986):

$$f_k^+(r) = q_k(N+1) - q_k(N) \approx f_k^{\text{LUMO}}(r) \text{ (Nucleophile attack)} \tag{9}$$

$$f_k^-(r) = q_k(N) - q_k(N-1) \approx f_k^{\text{HOMO}}(r) \text{ (Electrophile attack)} \tag{10}$$

where $q_k(N)$, $q_k(N+1)$, and $q_k(N-1)$ are denoted as the charge of the neutral, anionic, and cationic species, respectively. q_k is the gross charge of atom k in a molecule.

3 Results and Discussion

3.1 Electrochemical Measurements

Figure 1a shows the potentiodynamic polarization curves with and without inhibitors after 2 h of immersion of working electrode (316L stainless steel) in 1 M HCl at 25 °C. The corrosion current densities (I_{cor}) and corrosion potentials (E_{cor}) listed in Table 2 were determined by fitting the plots using the CView® software. The electrochemical parameters such as inhibition efficiency (η %) and surface coverage (θ) were calculated from the related polarization curves using Eqs. (11) and (12), respectively:

$$\eta\% = \frac{I_{\text{corr}} - I_{\text{corr(inh)}}}{I_{\text{corr}}} * 100 \tag{11}$$

$$\theta = \frac{\eta\%}{100} \tag{12}$$

where I_{corr} and $I_{\text{corr(inh)}}$ are the corrosion current densities without and with inhibitors, respectively.

The corrosion current density of 316L stainless steel without inhibitor is 76.60 μA/cm^2. The addition of the inhibitors (10^{-4} M) to the working solution causes a significant decrease of corrosion current densities which results in reducing corrosion rate and an inhibition efficiency between 64 and 87%. The increase in the corrosion current densities is in the order of Amine 3 > Amine 1 > Amine 2. It is convenient to note that the cathodic Tafel slopes are almost the same. This attests the presence of the same electrochemical phenomenon, reduction of hydrogen. Conversely, anodic curves differ from one organic compound to another. Moreover, the decrease of current densities observed at the passivation plateau is due not only to the formation of the passivating chromium layer (Cr_2O_3) but also to the formation of an organic layer with high resistance to anodic polarization potentials. These results attest that the introduction of either aromatic rings or NO_2 and OH groups to amines is very favorable.

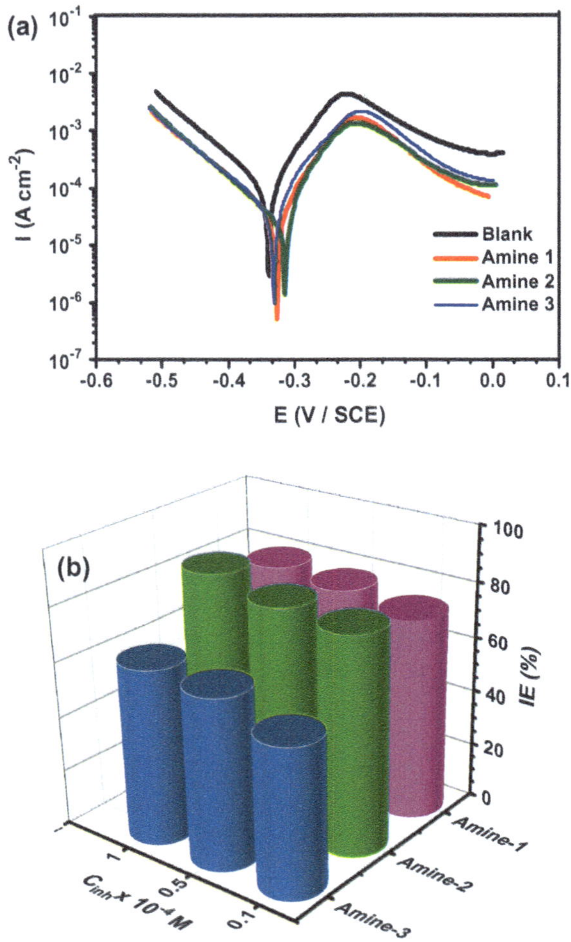

Fig. 1 Tafel polarization curves of 316L stainless steel in 1 M HCl solution with and without the organic compounds ($C = 10^{-4}$ M) at 25 °C (**a**) and the effect of concentration (**b**)

The concentration effect in inhibition efficiency is studied and the obtained results are shown in Fig. 1b and Table 2. Clearly, the obtained efficiency values increase by increasing the concentrations of the three organic compounds. The behavior of these organic compounds is manifested with the phenomenon of stability and high inhibition efficiency even in small concentrations for the Amine 1 and Amine 2.

The Nyquist plots and the equivalent circuits used to fit the experimental data of 316L stainless steel in HCl solution with and without inhibitors are shown in Fig. 2. It clearly indicates that the impedance spectra shape changed with the presence of organic compounds and the diameter of the semicircle increases. The impedance spectra of 316L stainless steel without inhibitors were fitted using the equivalent circuit of Fig. 2b. It consists of solution resistance (R_1 = Re) and constant phase element (CPE2) of double layer in parallel with R_2 resistance attributed to transfer

Table 2 Electrochemical parameters derived from polarization curves after immersion in 1 M HCl solution of mild steel without and with inhibitors at 25 °C

	C_{inh} ($\times 10^{-4}$ M)	E_{cor}/SCE (V)	I_{cor} ($\mu A/cm^2$)	IE (%)	θ	V_{cor} (μm/year)
Blank	0	−0.3392	76.600	–	–	396.942
Amine 1	0.1	−0.3324	20.612	73.09	0.73	122.639
	0.5	−0.3214	17.573	77.05	0.77	91.081
	1	−0.3295	15.762	79.42	0.79	81.695
Amine 2	0.1	−0.3204	16.055	79.04	0.79	83.213
	0.5	−0.3250	14.330	81.29	0.81	69.631
	1	−0.3132	10.168	86.72	0.87	52.701
Amine 3	0.1	−0.3333	3.5048	54.24	0.54	194.630
	0.5	−0.3328	2.9152	61.94	0.62	151.095
	1	−0.3324	27.810	63.69	0.64	144.140

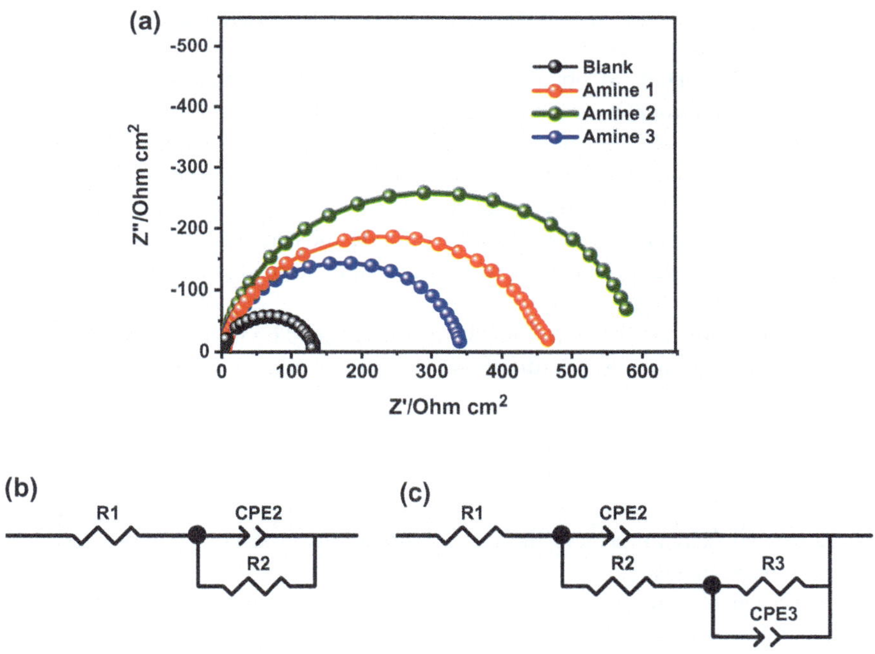

Fig. 2 Nyquist diagrams for mild steel in 1 M HCl with and without inhibitors (10^{-4} M) at 25 °C (**a**) and equivalent circuits used to fit the experimental data without inhibitor (**b**) and in the presence of inhibitors (**c**)

Table 3 EIS parameters for mild steel in 1 M HCl without and with inhibitors at 25 °C

	C_{inh} (× 10^{-4} M)	R_e (Ω cm^2)	R_{tc} (Ω cm^2)	R_f (Ω cm^2)	R_p (Ω cm^2)	IE (%)
Blank	0	0.380	133.9	–	133.9	–
Amine 1	0.1	0.1	1.575	200.2	119.0	58.05
	0.5	0.5	1.315	253.0	150.0	66.77
	1	1	1.202	300.3	161.3	70.99
Amine 2	0.1	1.994	232.0	120.5	352.5	62.01
	0.5	1.748	295.1	146.2	441.3	69.66
	1	1.930	410.4	171.2	581.6	76.98
Amine 3	0.1	1.812	179.4	110.0	289.4	53.73
	0.5	1.229	190.0	118.3	308.3	56.57
	1	0.133	217.3	120.1	337.4	60.31

resistance charge (R_{tc}). Constant phase element (CPE2) is used to describe non-ideal behavior (Popova et al. 2003; Singh et al. 2018).

On the other hand, to obtain a better fitting of the impedance spectra with inhibitors, an additional parallel R_3//CPE3 was introduced as shown in Fig. 2c. The solution resistance (R_1) was placed in series with CPE2//R_2 attributed to the pseudo-capacitance of the adsorbed organic layer onto the surface and to its resistance film (R_f), respectively. The resistance was placed in series with CPE3//R_3 attributed to the double layer response at the metal/electrolyte interface and transfer resistance charge (R_{tc}). The total polarization resistance values were calculated as the sum of R_2 and R_3 ($R_p = R_f + R_{tc}$).

The EIS parameters listed in Table 3 were determined by fitting the Nyquist plots using the ZView® software. Furthermore, $\eta\%$ was calculated according to Eq. (13):

$$\eta\% = \frac{R_{p(inh)} - R_p}{R_{p(inh)}} * 100 \tag{13}$$

where R_p and $R_{p(inh)}$ represent the total polarization resistances without and with inhibitor, respectively.

The presence of inhibitors increases the polarization resistance that is related to the substitution of the HCl ions and H_2O molecules by the organic molecules at the surface of the 316L stainless steel. Similarly, the increase of inhibitor concentrations lead to the same result and attest the growth of the thickness of the adsorbed organic film at the surface of 316L stainless steel.

As illustrated in Table 3, the presence of Amine 2 molecules increases significantly the charge transfer resistance and the film resistance in comparison with other inhibitors. This increase confirms the presence of an efficient thin layer formed after adsorption of amine molecules on the surface. Potentiodynamic polarization and EIS results correlate well with each other and confirm that Amine 2 presents the best inhibition efficiency values.

The adsorption of the organic compounds on the 316L stainless steel surface in HCl solution was fitted by Langmuir isotherm (Eq. 14):

$$\frac{C_{\text{inh}}}{\theta} = \frac{1}{K_{\text{ads}}} + C_{\text{inh}} \tag{14}$$

where K_{ads} and C_{inh} are the equilibrium constant and the inhibitor concentration, respectively.

K_{ads} is obtained from the plot of C_{inh}/θ versus C_{inh} and represents the intercept of the straight line on the C_{inh}/θ axis.

The equilibrium constant, the correlation coefficient (R^2), and the free energy of adsorption ($\Delta G^0{}_{\text{ads}}$) were calculated by Eq. (15) and presented in Table 4.

$$K_{\text{ads}} = \frac{1}{55.5}\exp\left(\frac{-\Delta G^0_{\text{ads}}}{RT}\right) \tag{15}$$

where $R = 8.314\ \text{J K}^{-1}\ \text{mol}^{-1}$ (universal gas constant), T (K) is the temperature, and 55.5 is the molar concentration of water.

Langmuir adsorption isotherm model was used to fit the experimental data as shown in Fig. 3. Correlation coefficient is closed to 1 ($R^2 > 0.997$) for all inhibitors.

Table 4 Thermodynamic parameters for the adsorption of inhibitor on mild steel in 1 M HCl at 25 °C

Inhibitors	K_{ads} (× 10^4 l/mol)	R^2	$\Delta G^0{}_{\text{ads}}$ (kJ/mol)
Amine 1	22.17	0.999	−40.45
Amine 2	18.52	0.997	−40.00
Amine 3	25.66	0.999	−40.81

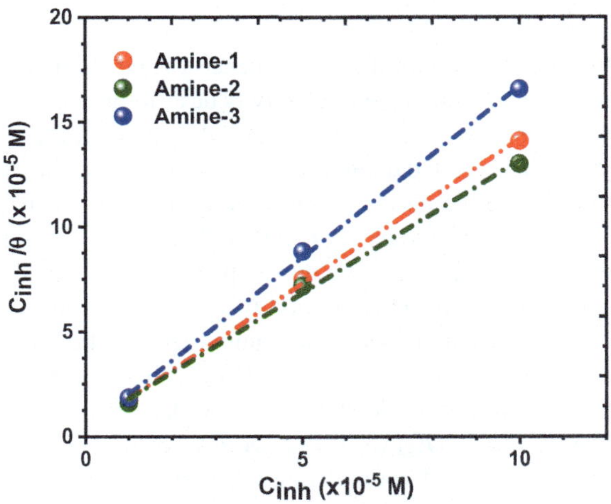

Fig. 3 Langmuir adsorption isotherm of inhibitors in 1 M HCl for mild steel at 25 °C

The values of $\Delta G^0{}_{ads}$ attest the spontaneity and the strong adsorption process of organic molecules on 316L stainless steel surface. Moreover, the $\Delta G^0{}_{ads}$ values are lower than -40 kJ mol^{-1} and can be assigned to the chemisorption adsorption related to the charge transfer between 316L stainless steel surface and organic molecules (Bentiss et al. 2005; Pavithra et al. 2012; Yıldız 2019).

3.2 316L Stainless Steel Surface Analysis

In order to confirm the electrochemical results, the 316L stainless steel surface is observed by scanning electron microscopy. Samples with and without the three organic inhibitors were immersed in the 1 M HCl solution for 24 h. Figure 4b shows the SEM micrograph of the 316L stainless steel after 24 h of immersion in the HCl solution without inhibitors. Meanwhile, it is shown the damage caused by the aggressive solution of HCl compared to the surface morphology before immersion (Fig. 4a). On the contrary, the damage was drastically limited in the corrosive solution with the presence of the three inhibitors. A protective film formed by the adsorbed organic molecules covers the surface of the 316L stainless steel. The total surface of the samples with Amine 1 and 2 is protected as shown in Fig. 4c, d. The formed layer limits the penetration of the water molecules and HCl ions to the surface and confirms again the high inhibition efficiency values obtained by the two inhibitors. The surface morphology of the sample with the Amine 3 is less covered with the presence of some porosity (Fig. 3e) and can explain the lower $\eta\%$ values compared to the other amine compounds.

3.3 Quantum Chemical Calculations and Molecular Dynamics (MD) Simulations

In order to explain the performance of the studied inhibitors, we investigated the molecular reactivity and selectivity descriptors for these molecules. Figure 5 presents the geometry optimization using DFT at B3LYP/6-31G (p, d) basis set in the gas phase. The related quantum chemical parameters are perceived as an important tool to examine the corrosion inhibition efficiency of the organic molecules (El-Raouf et al. 2018; Madkour et al. 2018). The E_{HOMO} is linked to the molecule ability to donate an electron to the unoccupied d-orbitals of the iron atoms, while E_{LUMO} is associated with molecule ability to accept electrons from the metal surface. The obtained results of DFT are summarized in Table 5. We find that Amine 2 and 3 have higher E_{HOMO} and lower E_{LUMO} than Amine 1. This suggests a tendency of those inhibitors to donate or accept electrons from the metal surface. The energy gap $\Delta E_{Gap} = (E_{LUMO} - E_{HOMO})$ is also another parameter to describe the chemical reactivity of inhibitor molecules. In general, small values of gap are related to organic

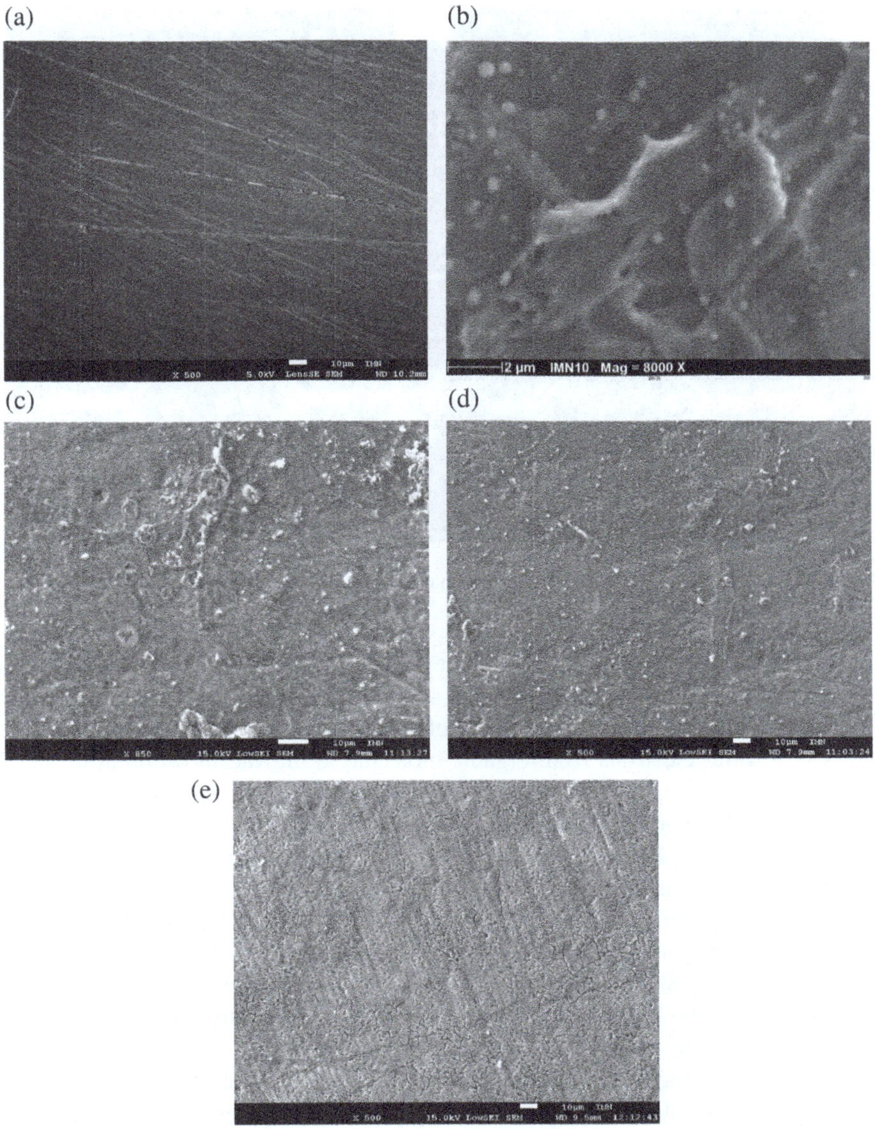

Fig. 4 SEM images of mild steel samples **a** before and **b** after 24 h immersion in 1.0 M HCl. After 24 h immersion in 1.0 M HCl. with 10^{-4} M of **c**: Amine 1, **d**: Amine 2 and **e**: Amine 3 inhibitors

molecules with high polarization, low kinetic stability, and high chemical reactivity (Fleming 1976). Furthermore, small value of ΔE_{Gap} increases and facilitates the adsorption on metal steel surface accompanied by an enhancement of inhibition efficiency values (Madkour et al. 2018; Olasunkanmi et al. 2015). Moreover, as shown in Table 5, the Amine 2 has a small energy gap (2.726 eV) than Amine 1 and

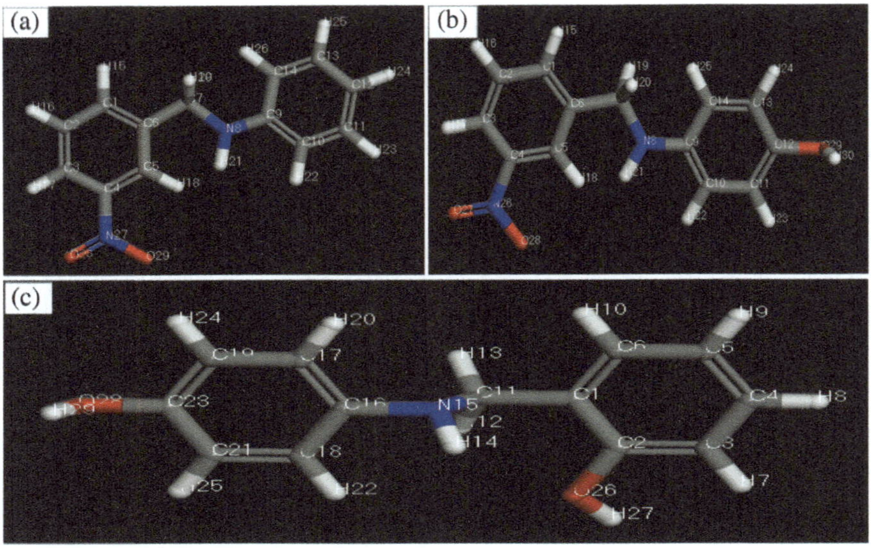

Fig. 5 Optimized molecular structures in gas phase of **a** Amine 1, **b** Amine 2, and **c** Amine 3

3 (3.168, 4.526 eV), which indicates that Amine 2 possesses the highest ability to accept electrons from d-orbital of metal surface. The obtained results corroborate the electrochemical results. The number of electrons transferred (ΔN) indicates the capacity to offer electrons from organic molecule to the vacant d-orbital of metal surface if $\Delta N > 0$. All values of ΔN are positive and confirm the electron-donating ability of the three organic inhibitors to the metallic surface.

The 3D plots of HOMO and LUMO frontier molecular orbitals and molecular electrostatic potential surface (MEPS) of the title amines are shown in Fig. 6. The three organic molecules have a similar HOMO in which the electron density is located on the phenol ring as well as over the (NH) atoms and the OH substituted group. This electronic distribution gives an idea about the preferred areas for an electrophilic attack. Whereas the LUMO orbitals are also similar, the electron density is located on the nitrobenzene group (NO_2) and the aromatic ring.

In this regard, the MEPS is a useful tool for understanding the chemical reactivity site of molecules. The electron density isosurface is shown by the same color. The red color shows the most negative electrostatic potential regions, electrophilic active region, the blue represents the most positive electrostatic potential regions, nucleophilic active region, and the green illustrates the zero potential regions. Electrostatic potential decreases in the following order: blue > green > yellow > orange > red (Asadi et al. 2019; Murray and Politzer 2011). It is clear from Fig. 6 that the electron-rich region is mainly localized around heteroatoms and conjugated double bonds. The nitrogen atom, OH, and NO_2 groups are indicators of the negative regions in the form of an orange color that favors electrophilic attacks. Hydrogen atoms have a positive sign (blue color) which favors nucleophilic attacks.

Table 5 Quantum chemical parameters calculated at DFT level using the 6-31G (d. p) basis set for inhibitor molecules in gas phase (eV)

Inhibitors	Total energy (eV)	E_{HOMO} (eV)	E_{LUMO} (eV)	ΔE (eV)	μ (D)	χ (eV)	η (eV)	ΔN
Amine 1	−20,747.166	−8.240	−5.0716	3.168	3.574	6.655	1.584	0.108
Amine 2	−22,793.826	−7.797	−5.0711	2.726	4.424	6.434	1.362	0.207
Amine 3	−19,276.106	−4.721	−0.195	4.526	3.718	2.458	2.263	1.003

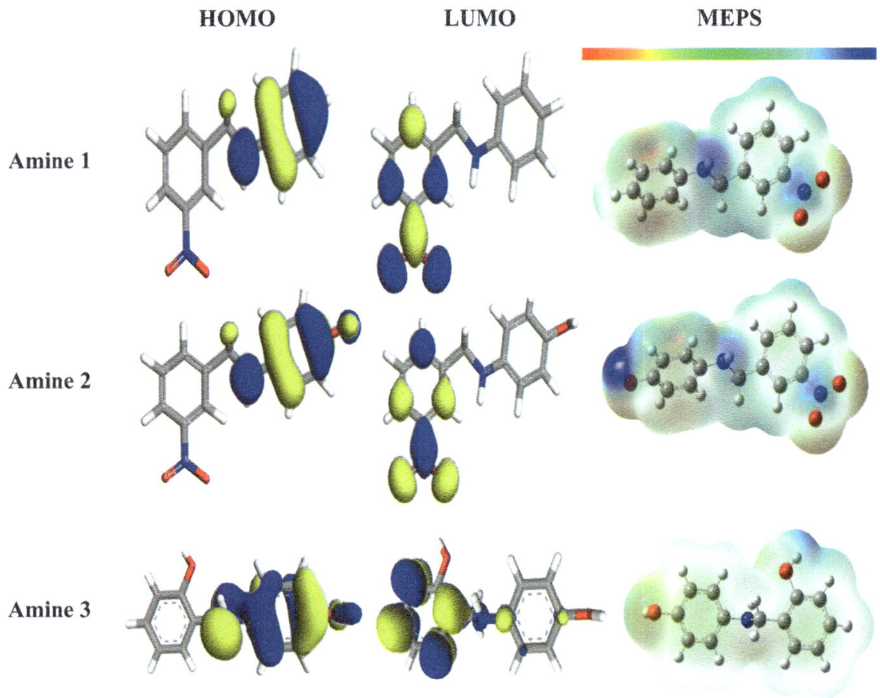

Fig. 6 HOMO and LUMO frontier molecular orbitals and molecular electrostatic potential surface (MEPS) of Amine 1, 2, and 3 using DFT at B3LYP/6-31G (d, p)

The condensed Fukui functions provide information about donor–acceptor adsorption type interactions between inhibitors and metal surface. Thus, it is necessary to determine the corresponding active sites in the organic molecules. The highly reactive atom possess the highest Fukui functions related to the other atoms (Jalaja et al. 2017). Higher is the negatively charge on the heteroatoms (N, O, etc.), more easily contribute to the donor–acceptor adsorption on the metal surfaces (Obi-Egbedi et al. 2011). Nucleophilic and electrophilic attack sites are calculated according to the Fukui functions f_i^+ and f_i^- (Contreras et al. 1999; Yang and Mortier 1986) and presented in Table 6 and Fig. 7. It is well illustrated that for Amine 1, the most positive part of f_i^+ function is localized in C (1), N (27), O (28), and O (29). Those atoms can be considered as the most reactive sites for nucleophilic attack. However, the most negative part of f_i^- function is localized in N (8), C (10), C (12), and C (14), suggesting that they are the most favorable sites for electrophilic attack. In Amine 2, N (26), O (27), and O (28) atoms contribute to the nucleophilic attack, where O (29) atom contribute to the electrophilic attack. In the case of Amine 3, C (1), C (3), C (4), and C (6) are contributed to the nucleophilic attack. On the other hand, N (15), C (16), C (21), and C (23) atoms are contributed to the electrophilic attack. According to these results, the three molecules present many active sites for donor–acceptor-type interaction.

Table 6 Calculated Fukui functions for Amine 1, Amine 2, and Amine 3

Atoms	Amine 1		Amine 2		Atoms	Amine 3	
	f_k^-	f_k^+	f_k^-	f_k^+		f_k^-	f_k^+
C (1)	0.007	0.076	0.007	0.075	C (1)	−0.018	0.069
C (2)	0.021	0.038	0.021	0.038	C (2)	0.003	0.033
C (3)	0.021	0.044	0.021	0.043	C (3)	0.019	0.097
C (4)	0.012	0.016	0.012	0.016	C (4)	0.028	0.095
C (5)	−0.005	0.038	−0.004	0.037	C (5)	0.021	0.046
C (6)	−0.010	0.032	−0.010	0.031	C (6)	0.002	0.098
C (7)	0.020	0.005	0.019	0.005	C (11)	0.028	0.017
N (8)	0.123	−0.001	0.117	−0.001	N (15)	0.175	0.008
C (9)	0.056	−0.004	0.055	−0.004	C (16)	0.052	0.008
C (10)	0.075	0.004	0.071	0.004	C (17)	0.049	0.032
C (11)	0.049	0.012	0.046	0.012	C (18)	0.004	0.015
C (12)	0.117	0.021	0.095	0.017	C (19)	0.047	0.026
C (13)	0.053	0.013	0.052	0.012	C (21)	0.056	0.039
C (14)	0.069	0.005	0.066	0.005	C (23)	0.081	0.028
N (26)	–	–	0.005	0.134	O (26)	−0.001	0.034
N (27)	0.005	0.134	–	–	O (28)	0.056	0.021
O (27)	–	–	0.032	0.195			
O (28)	0.032	0.196	0.011	0.194			
O (29)	0.011	0.195	0.062	0.014			

In practice, the molecular dynamic simulation (MD) is presented as a modern tool to understand the adsorption of organic compounds on the surface of the metal in the real conditions (H_2O, HCl). It can be observed from Fig. 8 that the three amines are adsorbed on the Fe (110) surface in parallel plane position, favorable configuration, with maximum interaction between the different groups of organic inhibitors and the metal surface. This position is due to the strong interactions between the molecules and the metal atoms. The closest molecules to the metal surface are the organic molecules of inhibitors. This confirms that the presence of the inhibitors in the corrosive solution of HCl allows the replacement of the solvent molecules (water) and the H^+ and Cl^- ions by the inhibitor molecules at the interface electrolyte/metal. The adsorbed molecules formed a protective film on the entire surface of the metal. Moreover, if we take into account the distance that separates the amines molecules from the surface of the metal, it appears that the Amine 2 is the closest, and the Amine 1 is closer than the Amine 3. This observation confirms again the high performance against corrosion of Amine 2. The calculated values of the interactions energy (E_{int}) and bending energy (E_{bind}) were calculated at 298 K and the obtained results are listed in Table 7. The $E_{\text{interaction}}$ values between the three inhibitors and Fe (110) surface are −449.61, −503.43, and −395.48 kJ/mol, respectively. The high negative values

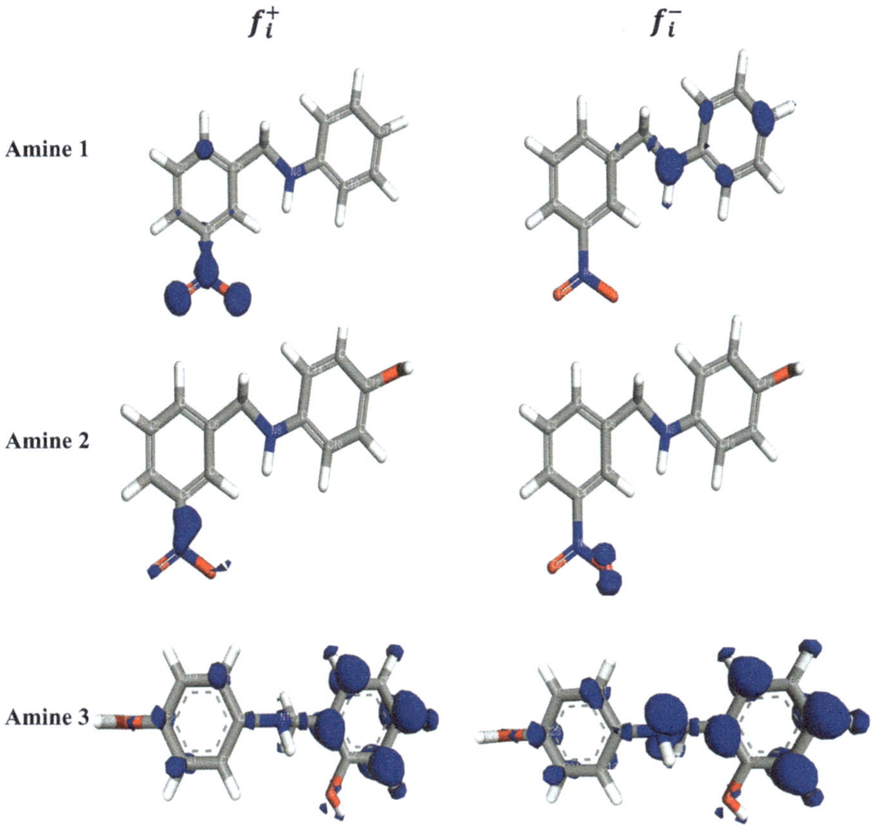

Fig. 7 Calculated Fukui functions for title molecules using the Dmol3 module of material studio

attest the strong and spontaneous adsorption process. In addition, the adsorption ability of the inhibitor molecules can be estimated by binding energy. A higher value of E_{bind} is a synonym of a stable adsorption (Kokalj 2015; Saha and Banerjee 2015). Accordingly, the interaction energy values confirm that Amine 2 adsorbs more spontaneously on the Fe (110) surface than the other amines. As a result, the high inhibition efficiency is expected.

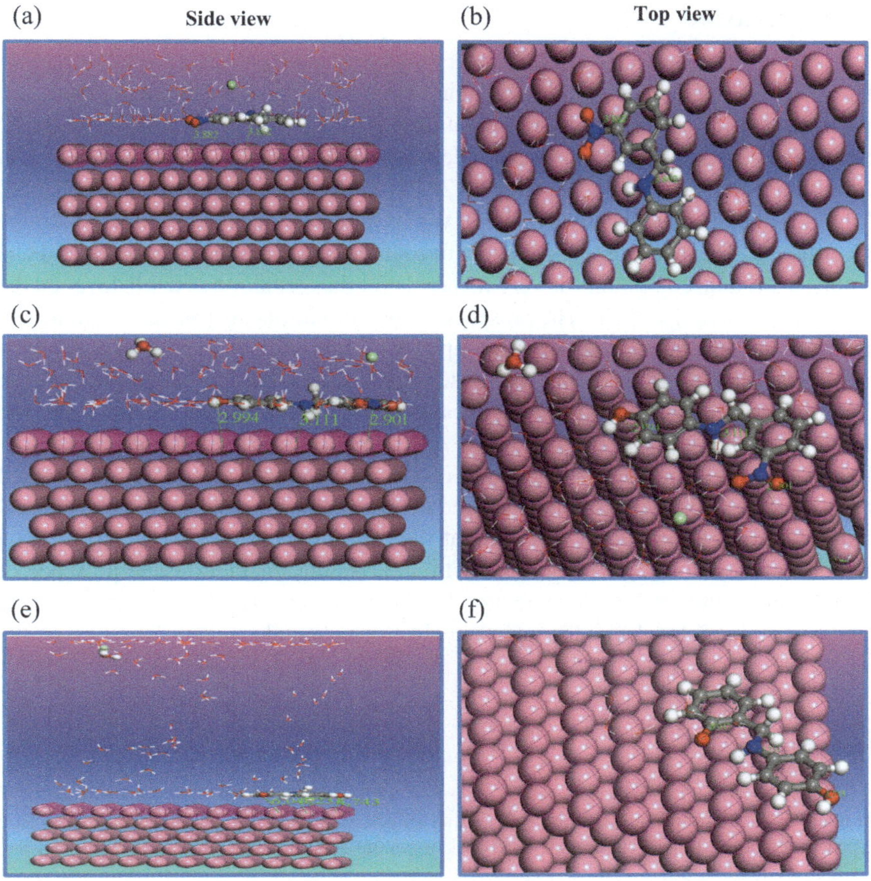

Fig. 8 Side and top views of equilibrium configurations for the adsorption of **a**, **b** Amine 1, **c**, **d** Amine 2, and **e**, **f** Amine 3 obtained by MD simulations on Fe (110)/100 H_2O/HCl surface

Table 7 Outputs and descriptors calculated by the MD simulation for the adsorption of inhibitors on Fe (110) surface

Inhibitors	$E_{\text{interaction}}$ (kJ/mol)	E_{Binding} (kJ/mol)
Fe (110)/Amine 1/100 H_2O/HCl	−449.61	449.61
Fe (110)/Amine 2/100 H_2O/HCl	−503.43	503.43
Fe (110)/Amine 3/100 H_2O/HCl	−395.48	395.48

4 Conclusion

The inhibition efficiency of three organic compounds was investigated using electrochemical tests, DFT, and MD simulations analyses. The addition of the organic compounds even at low concentrations decreases the corrosion current densities. An inhibition efficiency (IE) of 79.04, 86.72, and 63.69% for Amine 1, Amine 2, and Amine 3 is obtained, respectively. The IE values followed the order: Amine 3 < Amine 1 < Amine 2. These results attest that the introduction of aromatic rings, NO_2 and OH groups to organic compounds is very favorable for the formation of a thin layer on the 316L stainless steel surface. The presence of this film limits the iron dissolution. The appropriate equivalent circuit of the samples with inhibitors is used to fit the experimental data in order to consider the presence of the adsorbed organic molecules on the stainless steel/solution interface. The adsorption of the organic inhibitors obeyed the Langmuir adsorption isotherm. The ΔG^0_{ads} values of the inhibitors between −40 and −40.81 kJ/mol are the signature of the chemical adsorption. The theoretical studies by DFT and MD simulations predict and confirm the obtained experimental results. The MD simulation showed that the organic molecules were adsorbed on parallel position to the metal surface with a maximum interaction; the presence of the inhibitors allows the replacement of the water molecules and the H^+ and Cl^- ions by the organic molecules at the interface electrolyte/metal.

References

Akinlabi E-T, Andrews A, Akinlabi S-A (2014) Effects of processing parameters on corrosion properties of dissimilar friction stir welds of aluminium and copper. Trans Nonferrous Met Soc China 24:1323–1330

Argade GR, Sanders S, Mohandass G, Alsaleh A, D'Souza F, Golden TD, Mishra RS (2019) Corrosion inhibition study of Mg-Nd-Y high strength magnesium alloy using organic inhibitor. J Mater Eng Perform 28:852–862

Asadi N, Ramezanzadeh M, Bahlakeh G, Ramezanzadeh B (2019) Utilizing lemon balm extract as an effective green corrosion inhibitor for mild steel in 1M HCl solution: a detailed experimental, molecular dynamics, Monte Carlo and quantum mechanics study. J Taiwan Inst Chem Eng 95:252–272

Becke AD (1993) Becke's three parameter hybrid method using the LYP correlation functional. J Chem Phys 98:5648–5652

Bentiss F, Lebrini M, Lagrenée M (2005) Thermodynamic characterization of metal dissolution and inhibitor adsorption processes in mild steel/2, 5-bis (n-thienyl)-1, 3, 4-thiadiazoles/hydrochloric acid system. Corros Sci 47:2915–2931

Cao Z, Tang Y, Cang H, Xu J, Lu G, Jing W (2014) Novel benzimidazole derivatives as corrosion inhibitors of mild steel in the acidic media. Part II: theoretical studies. Corros Sci 83:292–298

Contreras RR, Fuentealba P, Galván M, Pérez PA (1999) Direct evaluation of regional Fukui functions in molecules. Chem Phys Lett 304:405–413

Dassault Systemes (2017) Materials studio tutorials. BIOVIA support. San Diego

El-Raouf MA, Khamis EA, Kana MT-A, Negm NA (2018) Electrochemical and quantum chemical evaluation of new bis (coumarins) derivatives as corrosion inhibitors for carbon steel corrosion in 0.5 M H_2SO_4. J Mol Liq 255:341–353

Fleming I (1976) Frontier orbitals and organic chemical reactions, vol 5. Wiley, New York, pp 05–27
Frisch M, Trucks G, Schlegel H, Scuseria G, Robb M, Cheeseman J, Scalmani G, Barone V, Mennucci B, Petersson G (2009) Gaussian 09, Revision A.1. Gaussian Inc., Wallingford
Goyal M, Kumar S, Bahadur I, Verma C, Ebenso EE (2018) Organic corrosion inhibitors for industrial cleaning of ferrous and non-ferrous metals in acidic solutions: a review. J Mol Liq 256:565–573
Hirshfeld FL (1977) Bonded-atom fragments for describing molecular charge densities. Theoret Chim Acta 44:129–138
Jalaja K, Mary YS, Panicker CY, Armaković S, Armaković SJ, Sagar BK, Van Alsenoy C (2017) Spectroscopic characterization of 4-[2-(5-Ethylpyridin-2-yl) ethoxy] benzaldehyde oxime and investigation of its reactive properties by DFT calculations and molecular dynamics simulations. J Mol Struct 1128:245–256
Kokalj A (2015) Ab initio modeling of the bonding of benzotriazole corrosion inhibitor to reduced and oxidized copper surfaces. Faraday Discuss 180:415–438
Koopmans T (1934) Über die Zuordnung von Wellenfunktionen und Eigenwerten zu den einzelnen Elektronen eines Atoms. Physica 1:104–113
Lee C, Yang W, Parr RG (1988) Development of the Colle-Salvetti correlation-energy formula into a functional of the electron density. Phys Rev B 37:785
Madkour LH, Kaya S, Guo L, Kaya C (2018) Quantum chemical calculations, molecular dynamic (MD) simulations and experimental studies of using some azo dyes as corrosion inhibitors for iron. Part 2: bis–azo dye derivatives. J Mol Struct 1163:397–417
Martinez S (2003) Inhibitory mechanism of mimosa tannin using molecular modeling and substitutional adsorption isotherms. Mater Chem Phys 77:97–102
Murray JS, Politzer P (2011) The electrostatic potential: an overview. Wiley Interdiscip Rev Comput Mol Sci 1:153–163
Musa AY, Kadhum AA, Mohamad AB, Takriff MS (2011) Molecular dynamics and quantum chemical calculation studies on 4, 4-dimethyl-3-thiosemicarbazide as corrosion inhibitor in 2.5 M H_2SO_4. Mater Chem Phys 129:660–665
Obi-Egbedi NO, Obot IB, El-Khaiary MI (2011) Quantum chemical investigation and statistical analysis of the relationship between corrosion inhibition efficiency and molecular structure of xanthene and its derivatives on mild steel in sulphuric acid. J Mol Struct 1002:86–96
Olasunkanmi LO, Obot IB, Kabanda MM, Ebenso EE (2015) Some quinoxalin-6-yl derivatives as corrosion inhibitors for mild steel in hydrochloric acid: experimental and theoretical studies. J Phys Chem C 119:16004–16019
Pavithra MK, Venkatesha TV, Kumar MP, Tondan HC (2012) Inhibition of mild steel corrosion by Rabeprazole sulfide. Corros Sci 60:104–111
Pearson RG (1988) Absolute electronegativity and hardness: application to inorganic chemistry. Inorg Chem 27:734–740
Popova A, Sokolova E, Raicheva S, Christov M (2003) AC and DC study of the temperature effect on mild steel corrosion in acid media in the presence of benzimidazole derivatives. Corros Sci 45:33–58
Saha SK, Banerjee PA (2015) A theoretical approach to understand the inhibition mechanism of steel corrosion with two aminobenzonitrile inhibitors. RSC Adv 5:71120–71130
Salarvand Z, Amirnasr M, Talebian M, Raeissi K, Meghdadi S (2017) Enhanced corrosion resistance of mild steel in 1 M HCl solution by trace amount of 2-phenyl-benzothiazole derivatives: experimental, quantum chemical calculations and molecular dynamics (MD) simulation studies. Corros Sci 114:133–145
Singh A, Ansari KR, Quraishi MA, Lgaz H, Lin Y (2018) Synthesis and investigation of pyran derivatives as acidizing corrosion inhibitors for N80 steel in hydrochloric acid: theoretical and experimental approaches. J Alloy Compd 762:347–362
Sun H (1998) COMPASS: an ab initio force-field optimized for condensed-phase applications overview with details on alkane and benzene compounds. J Phys Chem B 102:7338–7364

Verma C, Haque J, Quraishi MA, Ebenso EE (2019) Aqueous phase environmental friendly organic corrosion inhibitors derived from one step multicomponent reactions: a review. J Mol Liq 275:18–40

Yang W, Mortier WJ (1986) The use of global and local molecular parameters for the analysis of the gas-phase basicity of amines. J Am Chem Soc 108:5708–5711

Yıldız R (2019) Adsorption and inhibition effect of 2, 4-diamino-6-hydroxypyrimidine for mild steel corrosion in HCl medium: experimental and theoretical investigation. Ionics 25:859–870

The manufacturer's authorised representative in the EU is Springer Nature Customer Service Centre GmbH, Europaplatz 3, 69115 Heidelberg, Germany. If you have any concerns regarding our products, please contact ProductSafety@springernature.com

Printed and bound by CPI Group (UK) Ltd, Croydon, CR0 4YY
30/09/2026
02225288-0005